METHODS IN MOLECULAR BIOLOGY™

Series Editor
John M. Walker
School of Life Sciences
University of Hertfordshire
Hatfield, Hertfordshire, AL10 9AB, UK

For further volumes:
http://www.springer.com/series/7651

Plant Cell Morphogenesis

Methods and Protocols

Edited by

Viktor Žárský

Faculty of Science, Charles University, Prague, Czech Republic;
Institute of Experimental Botany, Academy of Sciences of the Czech Republic, Prague, Czech Republic

Fatima Cvrčková

Faculty of Science, Charles University, Prague, Czech Republic

Humana Press

Editors
Viktor Žárský
Faculty of Science, Charles University
Prague, Czech Republic

Institute of Experimental Botany
Academy of Sciences of the Czech Republic
Prague, Czech Republic

Fatima Cvrčková
Faculty of Science, Charles University
Prague, Czech Republic

ISSN 1064-3745 ISSN 1940-6029 (electronic)
ISBN 978-1-4939-6032-3 ISBN 978-1-62703-643-6 (eBook)
DOI 10.1007/978-1-62703-643-6
Springer New York Heidelberg Dordrecht London

Printed on acid-free paper

Humana Press is a brand of Springer
Springer is part of Springer Science+Business Media (www.springer.com)

Preface

The focus of plant biology in the post-genomic era is gradually leaving the confines of one major model, or even the growing number of sequenced model species. This opens new spaces for defining and solving major questions of basic plant biology. It should be appropriate in this situation to collect wisdom and skills accumulated mostly from work on *Arabidopsis thaliana*, the founding molecular biology model that helped in building a platform for applications in other species, including also crops. As editors of this volume on the methods in plant cell morphogenesis research, we are attempting to reflect this development in order to inspire future research in cellular aspects of morphogenesis in land plants in general.

Studying dynamics of plant shapes starting from the cellular level and advancing via tissues to organs and the whole plant is a truly fascinating perspective, which we share with the founders of our field in nineteenth and twentieth century. Here in Prague we acknowledge continuous inspiration by Jan Evangelista Purkyně, and in our field especially of his disciple Julius Sachs, who started his career at the German-speaking part of Charles University in Prague and became a father of modern plant physiology (including the pioneering studies of processes of plant morphogenesis). We are working on this volume in a building that was built in 1898 for the German plant physiology department, directed in those years by professor Hans Molisch—author of "Mikrochemie der Pflanzen" (published after his transition to Vienna in 1909). Many of the Czech contributors to this volume consider themselves "grandchildren" of professor Bohumil Němec—one of the fathers of experimental plant cell biology. When Němec discovered the decisive role of statocysts in root columella for root gravitropism in 1900, he immediately understood that to function in graviperception, columellar cells need to be not only internally dynamically polarized but also connected in a communicative (i.e. signalling) network with other root cells. This indicates an intricate internal cellular structure and intercellular communication, beyond the imagination of scientists of those times. Němec taught us, via his students (our teachers) and impressive published volumes on plant biology, to understand tissues and cells as products, not mere bricks, of living plant as a whole.

The first part of this volume (Chapters 1–7) presents a contemporary take on a classical approach that has been instrumental in establishing the plant cell biology field—namely light microscopy. First two chapters are directly linked to classic plant histochemistry methods book published by Bohumil Němec—"Botanical microtechnique" ("Botanická mikrotechnika" in Czech, Prague 1962). The following chapters present advanced methods of immunocytochemical analysis of cell walls, automated microscopy application in forward genetics screen, quantitative image analysis (including cytoskeletal structure and dynamics) and use of fluorescent markers to identify endomembrane compartments.

The following Chapters 8–11 are devoted to exciting new possibilities to monitor and quantify in detail structural dynamics of apical meristem on the cellular level, including details of cell shapes, gene expression, and mechanical features probed by Cellular Force Microscopy.

While light microscopy remains a central visualization method in plant cell biology, electron microscopy provides exciting insights into cellular ultrastructure. Chapters 12–15 describe a collection of useful electron microscopy techniques, including application of high-pressure freezing and low-temperature processing of samples, electron microscopy tomography, use of field emission scanning electron microscopy in the analysis of membrane structures, and immunogold localization procedures.

The choice of the experimental model is, as a rule, tightly interlinked with the choice of questions that can be studied, and it is hard to find a field where this would be more obvious than in case of cell morphogenesis. In Chapters 16–20, we present both essential and special techniques used to study model cell types as *Arabidopsis* root hairs, the moss *Physcomitrella patens*, cell lines and pollen tubes. The final three chapters introduce the use of advanced optical tools (optical tweezers and laser microdissection) and the application of heterologous expression in yeast in plant cell morphogenesis research.

We have approached our task in editing this collection of protocols in the hope that this volume may become a source of inspiration for further research quests into the plant cell, tissues, and organs morphogenesis research. We are especially grateful to many colleagues—best experts in their fields—from all over the world who accepted our invitation and contributed chapters to this volume, making it more likely that our hope may be fulfilled.

Prague, Czech Republic *Viktor Žárský*
Prague, Czech Republic *Fatima Cvrčková*

Contents

Preface . . . . *v*
Contributors . . . . *ix*

1 Essential Methods of Plant Sample Preparation for Light Microscopy. 1
Aleš Soukup and Edita Tylová

2 Selected Simple Methods of Plant Cell Wall Histochemistry and Staining for Light Microscopy. 25
Aleš Soukup

3 Resin Embedding, Sectioning, and Immunocytochemical Analyses of Plant Cell Walls in Hard Tissues. 41
Kieran J.D. Lee and J. Paul Knox

4 Automated Microscopy in Forward Genetic Screening of Arabidopsis. 53
Tereza Dobisová and Jan Hejátko

5 Image Analysis: Basic Procedures for Description of Plant Structures 67
Jana Albrechtová, Zuzana Kubínová, Aleš Soukup, and Jiří Janáček

6 Identifying Subcellular Protein Localization with Fluorescent Protein Fusions After Transient Expression in Onion Epidermal Cells. 77
Andreas Nebenführ

7 Visualizing and Quantifying the In Vivo Structure and Dynamics of the *Arabidopsis* Cortical Cytoskeleton Using CLSM and VAEM. 87
Amparo Rosero, Viktor Žárský, and Fatima Cvrčková

8 Sequential Replicas for In Vivo Imaging of Growing Organ Surfaces 99
Dorota Kwiatkowska and Agata Burian

9 Time-Lapse Imaging of Developing Meristems Using Confocal Laser Scanning Microscope. 111
Olivier Hamant, Pradeep Das, and Agata Burian

10 Quantifying Cell Shape and Gene Expression in the Shoot Apical Meristem Using MorphoGraphX. 121
Pierre Barbier de Reuille, Sarah Robinson, and Richard S. Smith

11 Mechanical Measurements on Living Plant Cells by Micro-indentation with Cellular Force Microscopy 135
Anne-Lise Routier-Kierzkowska and Richard S. Smith

12 High-Pressure Freezing and Low-Temperature Processing of Plant Tissue Samples for Electron Microscopy 147
Ichirou Karahara and Byung-Ho Kang

13 Reconstructing Plant Cells in 3D by Serial Section Electron Tomography. 159
Kiminori Toyooka and Byung-Ho Kang

14 Imaging Plant Nuclei and Membrane-Associated Cytoskeleton by Field Emission Scanning Electron Microscopy. 171
Jindřiška Fišerová and Martin W. Goldberg

15 Immunogold Labeling of Resin-Embedded Electron Microscopical Sections. 183
Ilse Foissner and Margit Hoeftberger

16 Live Cell Imaging of Arabidopsis Root Hairs. 195
Tijs Ketelaar

17 Morphological Analysis of Cell Growth Mutants in *Physcomitrella* 201
Jeffrey P. Bibeau and Luis Vidali

18 Plant Cell Lines in Cell Morphogenesis Research. 215
Daniela Seifertová, Petr Klíma, Markéta Pařezová, Jan Petrášek, Eva Zažímalová, and Zdeněk Opatrný

19 Antisense Oligodeoxynucleotide-Mediated Gene Knockdown in Pollen Tubes. 231
Radek Bezvoda, Roman Pleskot, Viktor Žárský, and Martin Potocký

20 Lab-on-a-Chip for Studying Growing Pollen Tubes. 237
Carlos G. Agudelo, Muthukumaran Packirisamy, and Anja Geitmann

21 Laser Microdissection of Plant Cells 249
Yvonne Ludwig and Frank Hochholdinger

22 Optical Trapping in Plant Cells 259
Tijs Ketelaar, Norbert de Ruijter, and Stefan Niehren

23 Heterologous Expression in Budding Yeast as a Tool for Studying the Plant Cell Morphogenesis Machinery. 267
Fatima Cvrčková and Michal Hála

Index. *283*

Contributors

CARLOS G. AGUDELO • *Optical Bio-Microsystem Lab, Mechanical Engineering Department, Concordia University, Montreal, QC, Canada*
JANA ALBRECHTOVÁ • *Department of Experimental Plant Biology, Faculty of Science, Charles University, Prague, Czech Republic*
RADEK BEZVODA • *Department of Experimental Plant Biology, Faculty of Science, Charles University and Institute of Experimental Botany, Academy of Sciences of the Czech Republic, Prague, Czech Republic*
JEFFREY P. BIBEAU • *Department of Biology and Biotechnology, Worcester Polytechnic institute, Worcester, MA, USA*
AGATA BURIAN • *Department of Biophysics and Morphogenesis of Plants, University of Silesia, Katowice, Poland*
FATIMA CVRČKOVÁ • *Department of Experimental Plant Biology, Faculty of Science, Charles University, Prague, Czech Republic*
PRADEEP DAS • *Laboratoire de Reproduction et Développement des Plantes, INRA, CNRS, ENS, UCB Lyon 1, and Laboratoire Joliot Curie, CNRS, ENS, Université de Lyon, Lyon, France*
PIERRE BARBIER DE REUILLE • *Max Planck Institute for Plant Breeding Research, Department of Comparative Development and Genetics, Köln, Germany*
NORBERT DE RUIJTER • *Laboratory of Cell Biology, Wageningen University, Wageningen, The Netherlands*
TEREZA DOBISOVÁ • *Laboratory of Molecular Plant Physiology and Functional Genomics and Proteomics of Plants, CEITEC, Masaryk University, Brno, Czech Republic*
JINDŘIŠKA FIŠEROVÁ • *Department of Experimental Plant Biology, Faculty of Science, Charles University and Institute of Molecular Genetics of the ASCR, Prague, Czech Republic*
ILSE FOISSNER • *Division of Plant Physiology, Department of Cell Biology, University of Salzburg, Salzburg, Austria*
ANJA GEITMANN • *Département de sciences biologiques, Institut de recherche en biologie végétale, Université de Montréal, Montreal, QC, Canada*
MARTIN W. GOLDBERG • *School of Biological and Biomedical Sciences, Durham University, Durham, UK*
MICHAL HÁLA • *Department of Experimental Plant Biology, Faculty of Science, Charles University and Institute of Experimental Botany, Academy of Sciences of the Czech Republic, Prague, Czech Republic*
OLIVIER HAMANT • *Laboratoire de Reproduction et Développement des Plantes, INRA, CNRS, ENS, UCB Lyon 1, and Laboratoire Joliot Curie, CNRS, ENS, Université de Lyon, Lyon, France*
JAN HEJÁTKO • *Laboratory of Molecular Plant Physiology and Functional Genomics and Proteomics of Plants, CEITEC, Masaryk University, Brno, Czech Republic*

Frank Hochholdinger • *Crop Functional Genomics, Institute of Crop Science and Resource Conservation (INRES), University of Bonn, Bonn, Germany*
Margit Hoeftberger • *Division of Plant Physiology, Department of Cell Biology, University of Salzburg, Salzburg, Austria*
Jiří Janáček • *Institute of Physiology, Academy of Sciences of the Czech Republic, Prague, Czech Republic*
Byung-Ho Kang • *Department of Microbiology and Cell Science and Interdisciplinary Center for Biotechnology Research, University of Florida, Gainesville, FL, USA*
Ichirou Karahara • *Department of Biology, Graduate School of Science and Engineering, University of Toyama, Toyama, Japan*
Tijs Ketelaar • *Laboratory of Cell Biology, Wageningen University, Wageningen, The Netherlands*
Petr Klíma • *Institute of Experimental Botany, Academy of Sciences of the Czech Republic, Prague, Czech Republic*
J. Paul Knox • *Centre for Plant Sciences, Faculty of Biological Sciences, University of Leeds, Leeds, UK*
Zuzana Kubínová • *Department of Experimental Plant Biology, Faculty of Science, Charles University, Prague, Czech Republic*
Dorota Kwiatkowska • *Department of Biophysics and Morphogenesis of Plants, University of Silesia, Katowice, Poland*
Kieran J.D. Lee • *Centre for Plant Sciences, Faculty of Biological Sciences, University of Leeds, Leeds, UK*
Yvonne Ludwig • *Crop Functional Genomics, Institute of Crop Science and Resource Conservation (INRES), University of Bonn, Bonn, Germany*
Andreas Nebenführ • *Department of Biochemistry and Cellular and Molecular Biology, University of Tennessee, Knoxville, TN, USA*
Stefan Niehren • *Molecular Machines and Industries GmbH, Eching, Germany*
Zdeněk Opatrný • *Department of Experimental Plant Biology, Faculty of Science, Charles University, Prague, Czech Republic*
Muthukumaran Packirisamy • *Optical Bio-Microsystem Lab, Mechanical Engineering Department, Concordia University, Montreal, QC, Canada*
Markéta Pařezová • *Institute of Experimental Botany, Academy of Sciences of the Czech Republic, Prague, Czech Republic*
Jan Petrášek • *Department of Experimental Plant Biology, Faculty of Science, Charles University and Institute of Experimental Botany, Academy of Sciences of the Czech Republic, Prague, Czech Republic*
Roman Pleskot • *Institute of Experimental Botany, Academy of Sciences of the Czech Republic, Prague, Czech Republic*
Martin Potocký • *Institute of Experimental Botany, Academy of Sciences of the Czech Republic, Prague, Czech Republic*
Sarah Robinson • *Max Planck Institute for Plant Breeding Research, Department of Comparative Development and Genetics, Köln, Switzerland*
Amparo Rosero • *Department of Experimental Plant Biology, Faculty of Science, Charles University, Prague, Czech Republic*
Anne-Lise Routier-Kierzkowska • *Max Planck Institute for Plant Breeding Research, Department of Comparative Development and Genetics, Köln, Germany*
Daniela Seifertová • *Institute of Experimental Botany, Academy of Sciences of the Czech Republic, Prague, Czech Republic*

RICHARD S. SMITH • *Max Planck Institute for Plant Breeding Research, Department of Comparative Development and Genetics, Köln, Germany*
ALEŠ SOUKUP • *Department of Experimental Plant Biology, Faculty of Science, Charles University, Prague, Czech Republic*
KIMINORI TOYOOKA • *RIKEN Plant Science Center, Yokohama, Japan*
EDITA TYLOVÁ • *Department of Experimental Plant Biology, Faculty of Science, Charles University, Prague, Czech Republic*
LUIS VIDALI • *Department of Biology and Biotechnology, Worcester Polytechnic institute, Worcester, MA, USA*
VIKTOR ŽÁRSKÝ • *Department of Experimental Plant Biology, Faculty of Science, Charles University and Institute of Experimental Botany, Academy of Sciences of the Czech Republic, Prague, Czech Republic*
EVA ZAŽÍMALOVÁ • *Institute of Experimental Botany, Academy of Sciences of the Czech Republic, Prague, Czech Republic*

Chapter 1

Essential Methods of Plant Sample Preparation for Light Microscopy

Aleš Soukup and Edita Tylová

Abstract

There are various preparatory techniques for light microscopy permitting access to the inner structure of plant body and its development. Minute objects might be processed as whole-mount preparations, while voluminous ones should be separated into smaller pieces. Hereby we summarize some of the "classical" techniques to cut more voluminous objects into slices and access their inner structure either for simple anatomical analysis or for further processing (e.g., histochemistry, immunohistochemistry, in situ hybridization, enzyme histochemistry).

Key words Paraffin, Sections, Freehand sectioning, Fixation, Whole-mount, Serial sections, Cryotome, Hand microtome

1 Introduction

There are various ways of preparation of plant objects to be investigated with light microscopy. Correct selection of appropriate technique largely depends on equipment available, but nature, optical character, complexity, and size of the object and purpose of the preparation take major part. Hereby we present a set of simple techniques which might provide vast, however not exhaustive, information on structural and cytological features of cells, tissues, and organs. Tissues, organs, or explants, which are not voluminous and optically dense, might be processed as a cleared whole-mount preparations. Such a way of preparation became very popular with advent of confocal microscopy and Arabidopsis as a model plant. However, there are many objects where cuttings or macerations are necessary to gain adequate information on internal structure. Available sectioning techniques allow preparation of sections with variable thickness according to intended application. Tissue preservation (fixation) and embedding into supporting matrix are common initial steps involved in most of sectioning methods

Viktor Žárský and Fatima Cvrčková (eds.), *Plant Cell Morphogenesis: Methods and Protocols*, Methods in Molecular Biology, vol. 1080, DOI 10.1007/978-1-62703-643-6_1, © Springer Science+Business Media New York 2014

determining quality and application of final microscopic sections. Sectioning of fresh (not fixed) and/or not embedded samples are valuable alternatives to consider.

1.1 Fixation

Fixation is commonly the initial step of the sequence. Choice of proper fixation is of great consequence for purpose of the preparation and its subsequent processing. We include here only the two very common basic procedures using FAA and buffered formaldehyde. FAA (formalin-acetic acid-alcohol) penetrates rapidly and is suitable for general anatomical or morphological work. However, preservation of cytological details is far less satisfactory comparing to formaldehyde. Fixation of samples with Clark's and Carnoy's fluids, alcohols, glutaraldehyde, acrolein, carbodiimides, chilled methanol or acetone, and others should be considered as alternatives according to goal of the preparation [1–4].

Process of fixation includes both penetration of the fixative into the tissue and its action within the tissue. While alcohol-elicited coagulation is a rapid process, saturation of the chemical linkages within the tissue by formaldehyde takes 1–2 days [5]. Diffusibility of fixative (distance that the fixative diffuse per 1 h within the object) varies strongly among tissues and fixatives, being about 25× higher for ethanol then for formaldehyde solutions [6]. This fact should be considered during sampling as size and character of the object strongly influence penetration of the fixative. Most of the experimental tests of fixative penetration use animal tissues (*see*, e.g., refs. [5, 7]), and only little data is available from plant tissues [8, 9]. The reasonable expectation of formaldehyde penetration rate does not exceed more than few mm per hour in plant tissues. It is difficult to get a coherent rule for estimation of fixation time regarding the size and character of the object. Low-pressure ("vacuum") infiltration of the tissue might be required to facilitate penetration of aqueous fixative with considerable surface tension into air-filled intercellular spaces. On the other hand, once filled with fixative, such intercellular spaces might provide important entrance pathway into more voluminous samples.

1.2 Cleared Whole-Mount Preparations

Cleared whole-mount preparations allow for focusing through the minute objects (usually not more than few hundreds of µm deep) and gain information on their inner arrangement. In fact there are several attitudes to clear the object. Removal of pigments, inclusions, and most cellular content decreases optical density of the object and improves transparency of the tissue and thus enables access to its inner structure [3, 10]. Treatment using sodium chloride [11], hydrogen peroxide [12], strong alkali or acids [13, 14], phenol [15], lactic acid [16–18], chloral hydrate [19–21], and their combinations are commonly used. Alternative saturation of the object with compounds of high refractive index decreases light dispersion and increases transparency of the tissue [22–24]. Various procedures combine both attitudes. Hereby we present simple protocol of gentle tissue clearing with high refractive index solution, which preserves most of the cellular content. We have introduced into usage sodium iodide solution [25] as a high refractive index nontoxic alternative to chloral hydrate, which is a regulated narcotic in most countries. The procedure is not self-reliant for highly pigmented and highly optically dense (e.g., secondary xylem) tissues and should be combined with pigmentation removal in such a case.

1.3 Hand Sectioning

Hand sectioning is fast and easy method of fresh/fixed specimen sectioning. While it might seem old fashioned in an equipment-loaded laboratory, if done with skill, it gains quickly substantial information on structure and in combination with various detection techniques also on composition and other parameters of tissue. Freehand sectioning with a razor blade is the simplest option and should be considered as a basic level laboratory skill. Hand microtome and straight razor blade (Fig. 1a) can push the sectioning further to achieve series of sections of standardized thickness (≥50 μm is realistic for most tissues). Hand sectioning has no necessity for infiltration and embedding. For smaller objects, additional reinforcement might be necessary to facilitate manipulation in hand or fixation in clamp of hand microtome. We commonly use elder pith (dead parenchymatous tissue), but other material (carrot, styrofoam, potato, roll of parafilm, paraffin encasing, etc.) or encasing into agarose block surrounding the object during sectioning [26, 27] might be used. In fact the hand sectioning can provide sections rather similar to vibratome. Quality of the cutting edge is the most limiting factor, and high-quality disposable razor blades (not the single-sided technical ones) or well-maintained straight razor (requires proper honing and stropping) is crucial for the sectioning.

The other procedures presented in this selection will involve specimen infiltration and embedding with supporting matrix to form blocks suitable for sectioning. Such embedding allows attaining thinner sections (less than 10 μm) and routine serial sections.

1.4 Paraffin Embedding

Paraffin is the very classical embedding medium introduced into microtechnique by Klebs [28]. Paraffin melts at rather high temperatures (54–60 °C), is strongly hydrophobic, and does not allow for routine sectioning below approx 3 μm. In spite of these disadvantages, it is still the most common embedding medium. Easy cutting and junction of sections into ribbons allow for straightforward routine of serial sections. Its high hydrophobicity requires strict dehydration of the object and use of intermedium (intermediate anhydrous paraffin solvent) to completely saturate tissue with paraffin before embedding. Butanol is the most commonly used intermedium, which substitutes the originally more common and

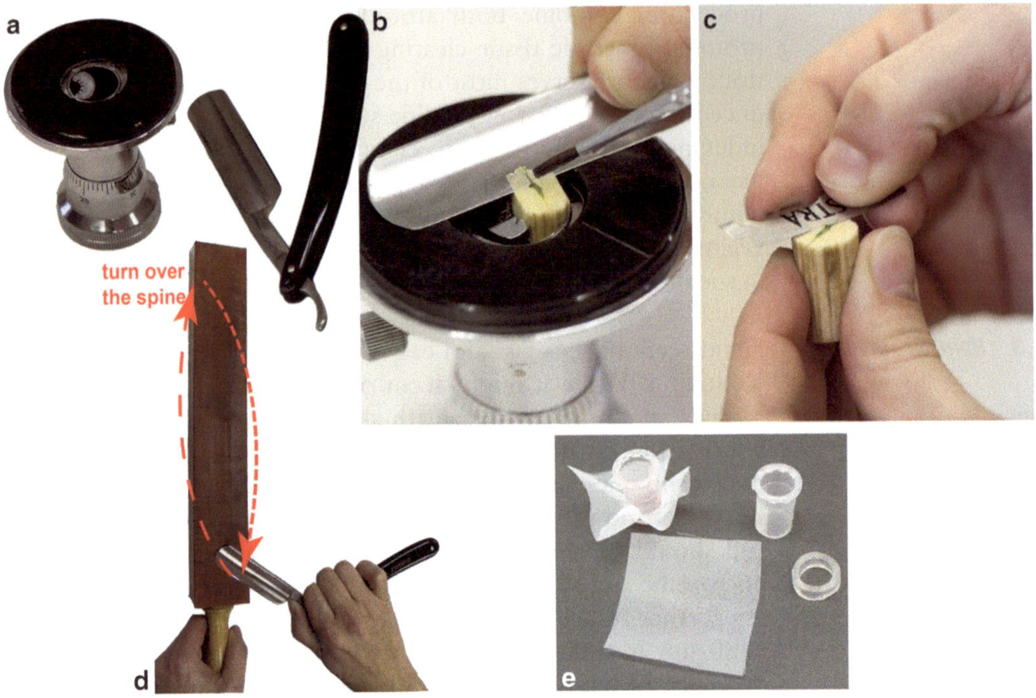

Fig. 1 Hand sectioning can be done either with bare hand or with hand microtome (**a**), a simple device with a micrometric screw, which allows stepwise adjustment of section thickness. Scale on the microtome is normally graduated in 10 μm increments. Object fixed into the central clamp is cut with a straight razor or other suitable blade. To slide the blade along the glass plate smoothly, the blade should have a flat grind on the glass touching side. Press the straight blade slightly on the glass plate of hand microtome with thumb and smoothly slide along to cut the sections (**b**). Notice the position of hands during freehand sectioning (**c**). Section should be kept permanently moistened with drop of water or buffer. To strop the straight blade (**d**), place the blade flat on the strop and draw it spine first along the strop so that the whole length of edge is treated. Rotate the blade over its spine, so the edge moves away from the strop and draw the blade back. Repeat as long as necessary. (**e**) Simple sections staining holders made of Eppendorf vials, tubing ring, and fine mesh

toxic xylene. In our laboratory we generally use n-butanol. t-Butanol is more efficient solvent of paraffin and more potent to be used for infiltration. However, high melting point (Tm 25 °C, frequently solid at lab temperature) and higher price of t-butanol make n-butanol the easier option.

There are various protocols for the paraffin infiltration and embedding, which might differ in tissue damage and time consumption. Dehydration and paraffin infiltration are steps which usually induce most of the tissue shrinkage. To minimize volume changes caused by intense solvent exchange, gradual series of solutions with decreasing water content are commonly employed. Various solvents were proposed for use in dehydration (isopropanol, acetone, methyl cellosolve, etc.) and paraffin infiltration (chloroform, xylene, n-propanol, n- or t-butanol; reviewed in ref. [3]). Ethanol-butanol dehydration series [29, 30] became a method of choice in our lab. Combination of gradual dehydration and concurrent introduction of intermedia minimizes distortion and saves time. Damage and/or hardening attributed to longer action of higher temperature during infiltration with paraffin reported in literature was not observed to be that significant in our hands, but damage caused by overly fast infiltration progress or improper elimination of intermedium was recorded rather frequently. Protocol of paraffin oil-regulated rate of infiltration should be mentioned [29] in this context. This protocol use mixture of butanol and paraffin oil (1:1) instead of pure butanol to saturate the objects before paraffin infiltration. Paraffin oil has higher viscosity and is therefore slowly replaced with melted paraffin in latter steps, reducing thus shrinkage of tissue related to paraffin infiltration. The whole procedure of sample processing towards paraffin embedding is a sequence of events. Each of them might introduce artifacts into the preparation, which cannot be corrected latter on and accumulation of errors therefore commonly takes place.

Alternatives to paraffin sections might be found in low temperature melting Steedman's wax (*see* ref. [31]), which is suitable for higher temperature sensitive objects (e.g., sections for immunodetection) and infiltration protocol is significantly shorter. On the other hand, sectioning, flattening of sections on slides, and storage of samples are slightly more complicated with hygroscopic nature of the polyethylene glycol distearate-based wax. There are various types of resins used for sample embedding and sectioning (Technovit, LR White, Lowicryl, GMA, Spurr, etc.) that differ in hydrophobicity, hardness, and sectioning properties. Resin-embedded objects can be sectioned to thinner slices (less than 1 μm) to achieve higher degree of cytological details. Because of specific requirement for sectioning of the resin, this topic is out of scope of this chapter.

1.5 Sectioning of Frozen Material

Sectioning of frozen material does not require extensive sample dehydration and embedding medium infiltration. Cryosections are suitable for a wide range of light-microscopy applications (e.g., immunohistochemistry, in situ hybridization, enzyme histochemistry), but it should be mentioned that it might not be straightforward to gain good quality sections for plant tissues. Fixed or fresh (unfixed) samples might be processed according to intended application. Standard thickness of sections is 8–20 μm, but thickness down to 3 μm is attainable for some samples using standard cryomicrotome. Objects are encased into cryoembedding medium, which acts as an object-surrounding matrix for sectioning.

Freezing is critical step of the procedure, which strongly determines quality of sections. Freezing procedure should prevent formation of large ice crystals inside the sample, structural damage, and related sectioning problems. Highly vacuolated plant tissues are therefore rather complicated objects from this point of view. There are two principal approaches to minimize freezing distortion—flash freezing or cryoprotection pretreatment. Flash freezing approach prevents formation of large hexagonal ice crystals due to high freezing rate and small cubic crystals or even vitreous ice should form. Efficiency of the procedure can be further increased under high-pressure conditions [32]. Isopentane supercooled with liquid nitrogen or with solid carbon dioxide is frequently used to ensure proper heat transfer from object. Supercooling is suitable only for small specimens and even in such a case low thermal conductivity of biological samples presents the limitation for freezing rate [33]. The other approach restricts formation of large ice crystal due to presence of rather high concentrations of cryoprotective solutes. Sucrose is common cryoprotectant used in wide range of concentrations from 10 to over 75 % [34–36]. Infiltration with 8–15 % glycerol [37], 10 % dimethyl sulfoxide [38] or polyvinyl alcohol, and polyethylene glycol mixtures [39] can be used. Freezing rate of the cryoprotected specimen is far less critical, and freezing directly in the cryostat chamber (freezing shelf) is thus possible. The process of antifreeze treatment takes several hours and therefore requires foregoing fixation of samples to minimize processing-related artifacts. Besides freezing of the object, proper setup of cryotome (temperature, anti-roll plate, blade settings) is crucial for successful sectioning.

2 Materials

2.1 Fixation

1. FAA (formalin-acetic acid-alcohol): Mix together 50 % (v/v) of ethanol, 5 % (v/v) of acetic acid, 5 % (v/v) of formalin, and 40 % (v/v) of distilled water (should be adjusted according to stock ethanol and acetic acid concentration; for variations *see* **Note 1**).
2. 4 % formaldehyde in 50 mM phosphate buffer (pH 7.2): Dissolve 8 % (w/v) of paraformaldehyde (PFA) in distilled water; to facilitate dissolution, add minimal volume of 1 M KOH solution (approx. 200 µl per 100 ml) and warm the solution up to ~60 °C in a fume hood. When PFA is dissolved (the solution comes clear), add equal volume of 100 mM phosphate buffer of proper pH (selected according to purpose). Check pH and titrate to required pH with 1 M HCl if necessary (*see* **Note 2**).
3. Phosphate buffer: Mix together x ml of 0.2 M acid sodium phosphate (27.8 g NaH_2PO_4 in 1,000 ml) + y ml of 0.2 M middle sodium phosphate (53.65 g $Na_2HP0_4.7H_20$ in 1,000 ml) fill up to 200 ml with distilled water to gain 100 mM buffer. The values of x and y are specified in Table 1.

Table 1
Phosphate buffer composition (mixing ratios) to gain required pH

x	y	pH	x	y	pH
93.5	6.5	5.7	45.0	55.0	6.9
92.0	8.0	5.8	39.0	61.0	7.0
90.0	10.0	5.9	33.0	67.0	7.1
87.7	12.3	6.0	28.0	72.0	7.2
85.0	15.0	6.1	23.0	77.0	7.3
81.5	18.5	6.2	19.0	81.0	7.4
77.5	22.5	6.3	16.0	84.0	7.5
73.5	26.5	6.4	13.0	87.0	7.6
68.5	31.5	6.5	10.5	89.5	7.7
62.5	37.5	6.6	8.5	91.5	7.8
56.5	43.5	6.7	7.0	93.0	7.9
51.0	49.0	6.8	5.3	94.7	8.0

Table 2
Composition of individual steps of ethanol-butanol dehydration series

Step no.	Distilled water (%)	Ethanol (%)	Butanol (%)
1	70	20	10
2	60	25	15
3	45	30	25
4	30	30	40
5	20	25	55
6	10	20	70
7	–	15	85
8	–	–	100

2.2 Whole-Mount Preparation

NaI-based clearing solution for whole mounts: Dissolve 0.04 g of $Na_2S_2O_3$ in 20 ml of 65 % (aq. v/v) glycerol. Add and dissolve 17 g of NaI, 2 % (v/v) of DMSO to final solution. The final solution should be clear and colorless with refractive index close to 1.5.

2.3 Dehydration, Paraffin Infiltration, and Embedding

1. Ethanol-butanol dehydration series: Composition of individual steps is specified in Table 2.
2. Anhydrous ethanol and butanol: To efficiently remove water from the standard stock butanol or 96 % ethanol (*see* ref. [40]),

pour the solvent into a flask and introduce enough of desiccant (approx. 1/5 of volume). Desiccant can be either solvent-drying molecular sieve (3 Å for both butanol and ethanol) or anhydrous salt (e.g., K_2CO_3, $CaSO_4$, or $CuSO_4$), which binds the water but does not dissolve in alcohol. Let the capped flask stand overnight. Filtrate or decant water-free solvent and keep it in tightly closed flask to prevent air humidity entrance. To regenerate the molecular sieves as well as hydrated salt, place them into drying oven at 250 °C in thin layer for about 2 h.

3. Paraffin: paraffin is a mixture of long-chain alkanes. There are various types of paraffin suitable for embedding and sectioning. The classical method is based on recycling of suitable paraffin and alchemy of preparation of such paraffin [29]. Most laboratories use commercially available and easily accessible paraffin these days. Various brands are on the market (e.g., Paraffin, Tissue-Tek, Paramat, Paraplast, Histoplast, Sasol; *see* **Note 3**).
4. Gelatine-subbing coated microscopic slides (alum gelatin adhesive, chrome alum; [41]): Place 0.5 g of pure gelatin in 100 ml of distilled water and heat to approx. 45 °C to dissolve it completely. Add 0.05 g of $KCr(SO_4)_2.12H_2O$ (the usage of other alums is also possible) and dissolve and filter the solution. Immerse set of clean slides in staining rack into the solution for 10 s, blot excess of solution, and let the slides dry (48 h at room temperature or 12 h at 50 °C); protect slides from dust. Slides can be submerged several times (2–5 times) to heighten coating layer if necessary (*see* **Note 4**).
5. Poly-l-lysine-coated slides: Dilute 10× Poly-l-lysine stock solution (0.1 % w/v) to prepare working solution. Immerse clean slides into the solution for 10 min to 1 h. Dry and store coated slides in dust-free dry place; 4 °C is recommended for longer storage (*see* **Note 4**).
6. Glycerol albumen: Mix carefully egg white with equal volume of pure glycerol. Filter the mixture over glass wool or few layers of gauze. Add 1 % of sodium salicylate or thymol as a preservative (causes background autofluorescence!). Alternatively use 0.5–1 g NaN_3 (be careful, toxic). Smear a tiny amount (pinhead volume) evenly over a clean grease-free slide with your finger to make very fine (not wet) coating. Protein precipitate forms on the slides if high amount of albumen adhesive was used (*see* **Note 4**).

2.4 Cryosections

1. High-viscosity cryoembedding medium ([39]; *see* **Note 5**): Dissolve 65–75 g of polyvinyl alcohol (PVA) 56–98 in 1 l of distilled water or phosphate buffer (pH 7.4; 50 mM). Warm up to 100 °C to completely dissolve it. Add 10 ml of Tween 20, 0.5–1 g NaN_3 (preservative for long-term storage) and 40 ml polyethylene glycol 400. Optionally supplement

carboxymethylcellulose (CMC) to increase medium thickness up to semisolid gel. Add 7–10 g CMC powder on the surface of the medium, leave to rehydrate overnight, and mix well. Centrifuge to eliminate air bubbles.

2. Sucrose solutions: 3, 10, and 20 % w/v solutions of sucrose in 0.1 M phosphate buffer (pH 7.4).

3 Methods

3.1 Fixation of the Samples

1. Cut samples of adequate size to allow rapid access of fixative to inner tissues. In general the smaller is the better. On the other hand, size of structure of interest and/or cell size and investigator desire should be considered during sampling. Use sharp razor blade to minimize damage in vicinity of cut edge.
2. Submerge samples into adequate volume (*see* **Note 6**) of fixative solution immediately after excision or cut under suitable buffer, water, or cultivation solution to avoid drying. We found 20 ml scintillation vials to be the convenient vessels for fixation.
3. Alcoholic solutions easily fill in intercellular spaces due to low surface tension. If aqueous solutions are used, application of lower pressure ("vacuum infiltration") might be necessary to substitute air with fixative solution. Vacuum pump connected to plastic desiccator via regulator allows for controlled gradual decrease of pressure within the chamber. The rate of pressure drop depends on the nature of object and fixative used and should be adjusted accordingly. In general decrease should not cause boiling of the solution, but only slowly escaping stream of bubbles should be stimulated. Bring the samples slowly down to minimum pressure of the pump (approx. 5 mBar), turn off the vacuum line, and let the samples to equilibrate within the chamber for 10–20 min. Then let air slowly in to fill up the chamber again. The reintroduction of pressure should be gradual and as gentle as possible to fill "vacuum" within the sample intercellular spaces with solution during this period. Quick release of pressure difference might cause collapse of the intercellular spaces.
4. Let the samples to be fixed for selected period of time (*see* **Notes 7** and **8**).

3.2 Simple Protocol of Whole-Mount Samples or Thick Sections Clearing

Procedure is optimized for Arabidopsis seedlings and might need minor readjustment for other samples. Multiwell culture plates are convenient to process larger sets of samples (*see* **Note 9**).

1. Fix samples in 4 % formaldehyde buffered to pH 7.2 (25 mM) overnight.

2. Wash out fixative with 15 % (aq. v/v) glycerol containing 2 % (v/v) of dimethyl sulfoxide and leave for 30 min.
3. Replace the solution with 30 % glycerol containing 2 % of DMSO and leave for 30 min.
4. Transfer into 50 % glycerol and leave for 30 min.
5. Replace solution with 65 % glycerol and leave for 30 min.
6. Mount the objects into NaI clearing solution and apply cover slip. Let the objects to clear up. In most cases 24 h is sufficient, for more voluminous objects time should be prolonged.
7. Preparations can be saved for weeks at 4 °C.

3.3 Freehand Sectioning

Good quality double-sided razor blade is indispensable to successfully cut objects in bare hands. Quality of the blade makes strong limitation to the quality and attainable thickness of the sections. Longitudinal splitting of the blade is rather convenient practice. Besides better handling it is easier to control which side is still fresh and having good edge.

1. Grip the sample as indicated in Fig. 1c. For larger axially symmetric objects, it does not make much sense to intent to cut complete sections. More convenient is to get partial but thinner sections. If the object is too thin to be griped, it should be supported with some moist material and cut within such a material. We prefer soft elder pith soaked with appropriate buffer of water.
2. Wet the blade and cut the section holding the object more/less vertically (Fig. 1c). It is convenient to use fine brush to collect and manipulate sections. Always keep the sections in solution as drying of the tissue is destructive and rapid at lab temperature.
3. Holders made of Eppendorf vials (with conical part cut off), a ring of tubing (inside diameter 10 mm), and fine mesh (Fig. 1e) might be used for convenient handling of sections (*see* also ref. [42]).

3.4 Hand-Microtome Sectioning

Razor blade should be kept very sharp during the sectioning (for maintenance *see* **Note 10** and Fig. 1d).

1. Preparation of the sample is identical to freehand sectioning (see above).
2. Clamp the specimen in the central position of microtome so that it extends over the flat glass plate. If necessary, use supporting material (e.g., water-soaked elder pith or carrot sticks) to fix small specimens in appropriate position similarly to freehand sectioning (Fig. 1b).
3. Carefully place the flat side of straight blade on the glass plate and cut the object to align it with the plate (sectioning plane).

The straight razor blade should be laid down completely and slide smoothly when drawn along the glass plate. Be careful not to touch the glass plate with blade edge as it can be easily damaged this way.

4. Keep the specimen moist all the time. Wet it with small drops of water from brush to prevent drying and allow sections to float effortlessly up onto the razor blade (Fig. 1b).
5. Add drop of water on blade and collect floating sections with fine brush (or dropper in case of very small specimens) for further processing.

3.5 Paraffin Embedding and Sectioning

1. Fix samples as indicated above. Label samples with pencil on slip of cardboard, as graphite lead is stable in any solvent. The cardboard will pass together with samples through the dehydration and infiltration series and will be finally embedded into paraffin.
2. Wash the fixative out of your samples for 2 × 15 min. For wash use the same water content as used in the fixative. In case of formaldehyde, use the buffer included in the fixative, for FAA use ethanol of approximately the same concentration as in the fixative.
3. Gradually dehydrate objects and exchange dehydrating solvent for paraffin-dissolving intermedium. Thorough dehydration is indispensable for later successful paraffin infiltration. Starting point of the dehydration series should be selected according to fixative as in previous step (e.g., third step of ethanol-butanol series for 50 % FAA). Pass samples through higher steps of ethanol-butanol dehydration series with adequate time in each dehydration step (*see* **Note 11**). The sample MUST NOT dry out during any dehydration step. Use adequate volume of dehydration solution comparative to sample volume to keep its dehydration capacity (*see* **Note 12**). At least 100× sample volume might be a good thumb rule.
4. Repeat the anhydrous butanol bath (2× in total) to completely remove remaining ethanol from samples before starting paraffin infiltration.
5. Gradually introduce paraffin to fully infiltrate the objects and exhaustively eliminate butanol (or any other intermedium) from the samples in the end. Too rash infiltration is the most common reason of object shrinkage. That is because butanol escapes faster from the object than paraffin is able to replace it and compensate for volume changes. Timing of individual steps presented below is informative and should be adjusted according to the object. Place samples in 100 % (waterless) butanol in suitable vessels (we use 50 ml vessels with cap for infiltration). Add chips of paraffin (approx. 1/5 of butanol volume) and let them stand for 1 day at laboratory temperature.

6. Place caped dishes at 40 °C oven and let paraffin dissolve. Add enough paraffin to keep a few undissolved chips on the bottom and let over day.
7. Open the dishes in the end of the day and let overnight. If the paraffin is completely dissolved add more. Part of the butanol will evaporate slowly—be careful not to let the samples dry.
8. Increase temperature to 58 °C, add paraffin, and let stand for 1 day (samples can stand even over weekend in this or latter steps).
9. Pour off approx. one third of the paraffin-butanol mixture and bring to original volume with melted paraffin. Let samples infiltrate for 3 h to half a day.
10. Replace half of the mixture with melted paraffin (3–12 h). Repeat this step once more.
11. Let open vessels stand overnight in oven to completely evaporate residuum of butanol. It is convenient to apply low pressure in vacuum oven (paraffin must not solidify) to facilitate complete butanol removal. Butanol should not be smelled from the samples at the end of this step.
12. Replace melted paraffin for a pure one and let stand for half a day (paraffin can be recycled for this step) to eliminate rest of butanol from samples.
13. Replace melted paraffin for a pure one and proceed to embedding.
14. Pour the last paraffin change with objects into the paper origami dish (Fig. 2a) or suitable mold (*see* **Note 13**). Top up with pure melted paraffin (do not exceed 65 °C) and arrange the samples on the bottom so that it is easy to separate them latter (Fig. 2b).
15. Individual samples or their groups (e.g., sets of root segments) might be organized into rows and columns. Arrangement can be achieved with hot needle or tweezers, which are also convenient to remove potential air bubbles from hot paraffin surface. Arrangement can be done on hot plate to extend period of time for manipulation, but fast work in ceramic dishes is usually sufficient. Suitable orientation of objects for sectioning has to be considered during embedding. Cardboard tag should be placed into the block together with samples.
16. Cool paraffin blocks. Rapid cooling rate is recommended (e.g., cool water bath). Paraffin blocks can be stored for long time (many years at room temperature).
17. Cut individual samples from the paraffin block and fix them to wooden, metal, or plastic chucks suitable for your microtome clamp. Heat the block and chuck at the site of contact to melt surface layer of paraffin and press them together.

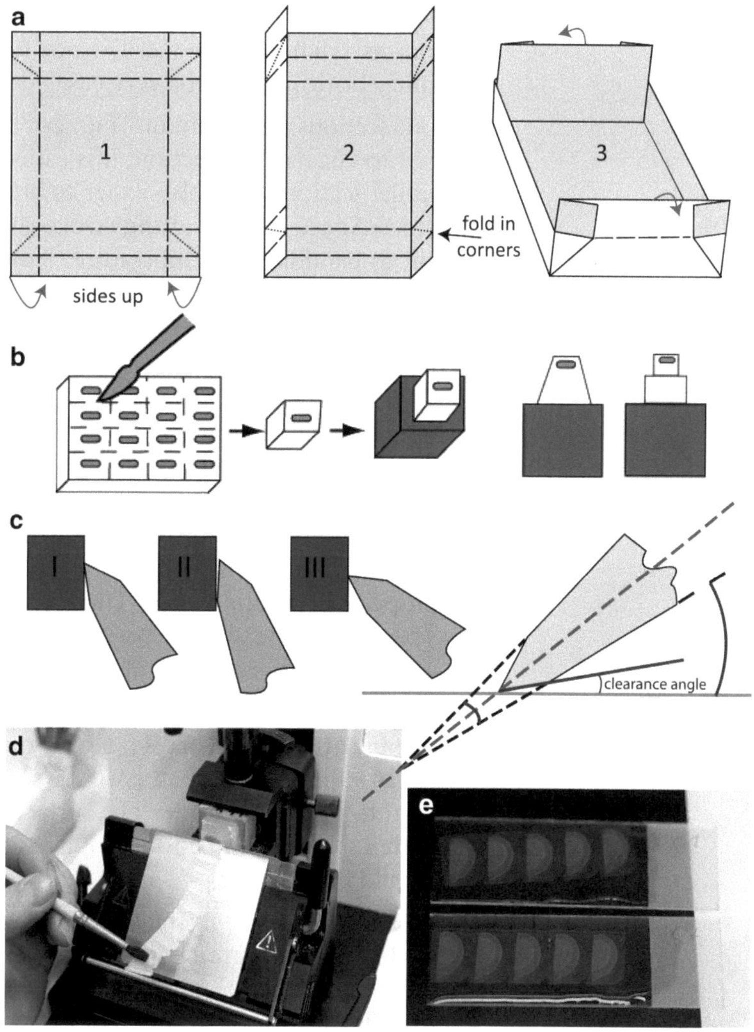

Fig. 2 Cardboard origami for paraffin embedding (**a**)—bend along the dashed lines to get a boat suitable for paraffin embedding. Grey indicates upper sided of the cardboard. (**b**) Arrange the objects into paraffin block so that they might be easily separated after embedding. (**c**) To adjust correct blade clearance angle, the shape of the blade should be respected. Incorrect angle of the blade can either crush (*II*) or scrape (*III*) the block instead of cutting (*I*). (**e**) Sectioning of paraffin-embedded object on the microtome and ribbon formation, (**d**) subsequent flattening of sections on slides

18. Trim paraffin blocks into desirable size and shape for easy sectioning. Proper trimming allows easier achievement of ribbons during sectioning (Fig. 2b).
19. Fix the chuck into the microtome clamp. Use microtome clamp to adjust object into optimal position according to cutting plane. Reassure that the clamping mechanism is tightened securely before trimming and sectioning.
20. Adjust proper microtome knife angle (*see* **Note 14**; Fig. 2c).

21. To clean the knife/blade, use petrol; this is not as dry and noxious as xylene or toluene. Do not touch the fine edge of the knife/blade as it can be very easily damaged.
22. Cut sections into a ribbon (Fig. 2d). As it is always more difficult to cut the first section, it is easier to cut ribbon than individual sections. It is also easier to arrange pieces of ribbon on slides. Speed of the cutting stroke should be adjusted according to paraffin and temperature. Slow and steady strokes usually result in best sections with least compression. Use moistened brush to manipulate ribbons, as it is easier then forceps and less probable to cause blade damage. For troubleshooting of the most common problems (*see* Table 3).
23. Transfer ribbon on black cardboard and cut it into equal pieces to be placed on the glass slides. Their length should be less then length of available cover slips. If series of sections is required, take care to maintain their proper order. Nick in the paraffin block, which can be seen in the ribbon, can make proper orientation easier. There are two sides of the ribbon. The glossy one should be placed towards the slide (down), while the matt site is the upper one.
24. Use pre-coated slides to ensure adhesion of sections for further manipulation (*see* **Note 4**). Cover the glass slide with distilled water so that only small part stays without water and can be used to handle the slide. The surface tension of water helps to flatten sections, and enough of free space surrounding ribbons should be available.
25. Float the ribbons on the water surface, arrange it, and heat it on hot plate to stretch and flatten the sections (Fig. 2e). The temperature of the plate should be approx. 5 °C lower than paraffin melting temperature. Let the slides on the plate for 5–10 min, as stretching the ribbon should be slow and gradual to be efficient. Temperature can be adjusted also experimentally so that it is gradually increased till the paraffin of sections melts, then the temperature is adequately lowered. If the temperature is too high, the ribbons will melt (objects are lost); if too low, flattening does not get complete (lines and wrinkles are still discernible on ribbons). Stretching of sections in water bath is more convenient for large individual sections. If small bubbles form under the ribbon, use boiled distilled water to eliminate dissolved gasses.
26. Remove the slide from hot plate, let it cool down, and rearrange the ribbons if necessary.
27. Gently remove most of the water and let the slides dry to attach sections to the slides on warm plate (40 °C overnight). Protect slides from dust.
28. When dry, sections can proceed to staining or store the slides in box before further processing.

Table 3
Troubleshooting the most common problems with paraffin sectioning

Problem	Cause	Remedy
Separate sections curl up, cracks parallel to blade edge may appear	The block is too cold Sections are too thick for used temperature Wrong clearance angle of the knife resulting in irregular section thickness	Straighten the first section using a soft brush; subsequent sections within the ribbon usually do not roll Warm up the block by breathing on it, touching it with your finger or placing incandescent bulb into its vicinity Modify section thickness Change the angle of the blade
Individual sections do not ribbon	Incorrectly prepared block (opposite sites are not parallel, side of the block is not parallel to the blade edge) Cold block or knife Cutting is too slow (sections are glued together with heat as blade hits the block)	Realign block edges and position according to the knife Warm up the block and knife Cut faster
Individual sections are strongly compressed, folded, and may stick on the knife	Temperature is too high Dull blade Blade is dirty with paraffin Too thin sections for the type of paraffin Too low clearing angle	Cool down the block Resharpen knife or change blade Clean the knife Increase section thickness Increase knife angle Decrease speed of sectioning
Objects are separating from the section	Improperly embedded object (improper dehydration, incomplete infiltration, incomplete elimination of ethanol or intermedium) Object is too hard for used paraffin	Re-embed the object (if possible) Use harder paraffin (higher melting temperature Cool down the block to make it harder Soften the object
Sections catch on the block when travelling back	Improper knife angle Dirty or dull knife	Modify knife angle Carefully clean the knife from both sides
Ribbon is not straight but turns	Sides of the block are not parallel (mutually or to the knife edge) Object is heterogeneously hard	Realign and trim the block
Longitudinal lines on the ribbon	Nicks on the blade edge Dirty edge Hard particles in object (sclerenchyma) Dust in paraffin	Use other part of the edge, change blade, resharpen knife Clean the edge Decrease the knife angle Soften the object Re-embed into clean/harder paraffin

3.6 Cryosectioning

1. Fix the specimen in an appropriate fixative (e.g., 4 % formaldehyde in phosphate buffer). Use lowered pressure ("vacuum infiltration") to substitute fixative for the air within the tissue if necessary.
2. Wash out fixative for 15 min with phosphate buffer used to prepare fixative.
3. Infiltrate samples gradually with 3, 10, and 20 % sucrose solutions. Each step takes at least 30 min at room temperature. Agitate gently and apply 0.1 % of surfactant (Triton or Tween 20) with 3 % sucrose solution to facilitate infiltration. Individual steps should be prolonged to infiltrate properly compact and more voluminous samples (*see* **Note 15**).
4. Freeze pretreated samples directly on the specimen chuck. Let the cryostat cool down to working temperature and turn on cryobar (freezing shelf) boost to minimize the bar temperature first. Add small amount of semisolid cryoembedding medium (e.g., OCT or high-viscosity cryoembedding medium) on the specimen chuck and use heat extractor to make flat base (the extractor frequently stick to the medium if not properly frozen; apply Teflon coating spray to the extractor to minimize this problem). Add more medium on the top of frozen platform and transfer the sample into this medium. Quickly arrange sample into desired position and freeze the block on cryobar. Sample should be covered with tiny layer of cryoembedding medium (*see* **Note 16**). Frozen samples can be stored at −80 °C in closed container if necessary. Do not store them in the cryotome chamber as the samples dry out rather quickly.
5. Trim frozen medium encasing object to adjust the specimen block size for easier cutting. Leave enough medium in surrounding of the sample. However, it is more difficult to cut thin sections from larger block.
6. Prepare all needed equipment (brushes, forceps, etc.) into cryotome chamber to get to right working temperature before use.
7. Mount the chuck with object on the microtome head and let its temperature to equilibrate. Working temperature should be selected according to desired thickness, character of the sample, and composition of embedding medium beside others. Independent setup of chamber (knife) and sample temperature is of advantage; it might be convenient to use 2–3 °C lower temperature of knife than sample (e.g., [43]). Commonly we use specimen temperature between −8 and −20 °C for standard section of 8–20 μm. Thinner sections might require lower temperature. It is reasonable to start with −15 °C and adjust the temperature according to appearance of sections. If the sections wrinkle and smear on knife, the working temperature is too high. If sections crumble, temperature is too low. For troubleshooting of the most common problems (*see* Table 4).

Table 4
Troubleshooting cryotome sectioning

Problem	Cause	Remedy
Sections smear or crumple on the blade edge	Temperature of specimen is too high The space under anti-roll plate is too low	Select a lower temperature Wait to equilibrate the object temperature and try sectioning again
Sections shatter at the tip of the blade	Specimen is too cold The anti-roll plate is not correctly adjusted The blade might be dull or its clearance angle is too steep The specimen surface is large	Select a higher temperature and let the object equilibrate; if from the cryobar, let it to adopt temperature of the clamp Knife clearance angle of 2–5° is recommended for disposable blades Adjust the anti-roll plate correctly Use another area of the blade or a new blade Trim the specimen parallel and increase the section thickness
Sections curl up when the anti-roll plate is raised up	The anti-roll plate is too warm	Lay down the anti-roll plate on knife and let its temperature stabilize Minimize air exchange within the chamber
Sections do not run flat under the anti-roll plate	Dirty anti-roll plate and/or knife Dull blade	Clean with dry cloth or brush; it is convenient to have some frozen in the chamber If necessary, use ethanol to clean blade and anti-roll plate Change blade or use another area of the edge
Sections curl in front of the anti-roll plate and do not go underneath	The anti-roll plate is too far below the edge of blade	Readjust the anti-roll plate
Section smear on the top of the anti-roll plate	Anti-roll plate goes beyond the blade edge and crushes the object	Readjust the anti-roll plate
Chatter on sections	The chuck or blade is not secured correctly Specimen is too hard, too cold or too big The clearance angle is incorrect Cutting speed is too high	Check and fix the stabilization of block holder and blade Modify temperature; let the object equilibrate with specimen head Trim the object to decrease its size Reset clearance angle of the blade Decrease the speed of cutting
Variable thickness of sections	The chuck or blade is not secured correctly The clearance angle is incorrect	Recheck and fasten the microtome head and blade holder Reset the blade angle
Sections are torn perpendicularly to the blade edge	Dust or nick on the blade Leading edge of the anti-roll plate is dirty or damaged	Clean front and back side of the blade Replace the blade or move to another part of the blade Clean or replace anti-roll plate

Fig. 3 Detail of cryotome head with chuck and object. In the lower part of the picture is the anti-roll plate with section underneath (**a**). To pick up the section, rise up the anti-roll plate and place the slide very close to the section (resting the corner of slide on the blade holder helps). The section "jumps" on the slide and melts on its surface (**b–d**). Do not press the slide on section, as it would freeze onto the blade holder. (**e**) Slide with two collected sections

8. Fix the chuck into holder and adjust its axial position with microtome head according to the blade. Approach slowly with objects towards the blade. Do not trim too quickly as damage may be caused to blade and object can break off.
9. Set the position of anti-roll plate parallel with blade edge. The edge of the anti-roll plate should be close enough to the edge to allow emerging sections to slide underneath (up to 0.5 mm) but not too close to crush the block. It is not easy to adjust position visually, so while very gently touching the top of the plate, one should feel the cutting edge. Be careful not to hurt your fingers! Cut few sections (it is reasonable to start with section of 15 μm) and further correct the plate position. If the section rolls in front of the plate, move it slightly up. If the plate is too high, object touches the plate lightly in passing (can be sensed with finger touching lightly the plate holder)—lower the plate.
10. Cut several sections to make smooth block surface.
11. Cut the sections; they run individually or in row of few under the plate (Fig. 3a).
12. Pick up sections with brush chilled in cryotome chamber or collect them with subbed (adhesive coated) and marked slide of room temperature. The slide should be approached very closely to sections but not direct touching them. Sections will

"jump" and melt onto the slide within few seconds (Fig. 3b–e). If you press the slide on section, the section melts on the blade holder and it is not easy to clean it off.

13. Brush away the condensed ice from blade holder before further sectioning.

4 Notes

1. There are two commonly used options of FAA regarding to final ethanol concentration –70 % and more delicate 50 % (v/v). Content of acetic acid can be also modified between 2 and 6 %. Material can be stored in solution for considerable period of time. Use formalin (commercial ~40 % formaldehyde solution) for preparation.
2. Buffers used with aldehyde fixatives must not react with them (e.g., TRIS, EDTA amino groups will react with aldehydes). Phosphate, HEPES, PIPES-based buffers, or others of Good's buffers are recommended. Phosphate might precipitate some divalent cations (Mg^{2+}, Ca^{2+}). Osmolarity of the buffer should be selected according to particular object. For most of plant samples, we use 25–100 mM buffers. Be aware that 4 % formaldehyde itself is a 1.33 M solution.
3. Melting temperature is closely related to hardiness of the paraffin. It normally stays between 56 and 58 °C, but 54 and 60 °C mixtures are available too. Composition of the embedding paraffin blends differs mainly in content and composition of plasticizers (plastic polymers) improving sectioning properties and "hardness." We prefer the use of paraffin with minimum or no additives and stabilizers as we have experienced easier infiltration and no separation/precipitation of plasticizers at higher temperatures. Some of the high polymer content mixtures seem to be rather sensitive to higher temperatures, and it is better not to exceed 65 °C. Paraffin without additives can be cut down to approximately 5 μm. It is claimed that with additives sections down to 2–3 μm are accessible. This might be valid only for soft plant tissues, and we prefer resin embedding for semi-thin sections. We do not recycle paraffin with additives.
4. Adhesives are compounds used to glue sections on the glass slides. Unlike animal tissues, plant tissues have relatively lower protein content, and presence of cell wall and vacuole makes them less adhesive. That is why they float away from the slides easily during staining or other processing. Selection of the right adhesive depends on intended use. Glycerol albumen is the easiest to use, alum gelatin is standard subbing that holds

well the sections, and Poly l-lysine is good solution for immuno and other more sensitive applications. To prepare adhesive (subbing)-coated slides, cleaning and degreasing of slides is of high importance. Even new slides should be washed with detergent followed with 96 % ethanol and distilled water. Alternatively the slides can be washed in dishwasher and carefully rinsed with distilled water before use.

5. Cryoembedding media are commercially available or can be prepared in the laboratory. We have positive experience with both options. OTC (optimum cutting temperature compound) is a commercially available cryoembedding medium (e.g., Tissue-Tek OCT) based on polyvinyl alcohol (PVA) and polyethylene glycol (PEG). Cryo-gel™, Cryomatrix™, and PolyFreeze™ are further commercial options differing in viscosity.
6. It might be accepted as a common rule of thumb that volume of fixative should not be less than 50× volume of the fixed tissue. Otherwise, the buffering capacity of the solution (pH, concentration of fixative, molarity) might not be sufficient.
7. There are very few experimental data to estimate the time needed for aldehyde penetration and fixation. In animal tissues (e.g., liver), the penetration rate normally does not exceed 1 mm per h. In the work of Mersey and McCully [8], the acrolein fixation passed about 140 μm per min along the root hair. The formation of linkages (incorporation of formaldehyde) within the tissue might be also rather slow process, taking hours to be saturated (e.g., [5]).
8. Rotary vane vacuum pump with pressure regulator and plastic desiccator allow for controlled gradual drop of pressure within the desiccator chamber. Evaporating fixative (or any other solution) accumulates within the oil and can cause corrosion (damage) of the pump chamber. It is necessary to let the pump run long enough to warm up the oil and evaporate the condensate from pump. It is condensate induced corrosion induced corrosion that most frequently damages the rotary vane vacuum pump. During vacuum infiltration, make sure to vent pump exhaust into the fume hood and not into laboratory. Fixatives are toxic. If the fixative fixes your samples, your own tissues might be fixed as well!
9. Tissues of high density or pigmentation might require removing the cytoplasm content with 2 % NaOH in 30 % EtOH or dimethyl sulfoxide. The latter is more efficient and can remove complete protoplasts. Extraction of chlorophyll and lipidic compounds might be done in methanol/chloroform (1:1) mixture. To clear colored phenolic depositions, alkalized hydrogen peroxide or sodium hypochloride-based protocols mentioned in introduction should be used.

10. Sharpening with fine-grained stone (e.g., Japanese whetstone, grit 8,000) is part of the regular maintenance but does not replace stropping. The finest cutting edge is affected during sectioning, and frequently it is enough to straighten and polish the blade by stropping (Fig. 1d) to recover its sharpness. While sharpening requires edge-forward movement of the blade on the stone, it should be carefully drawn spine first to avoid cutting of the strop during stropping. The hard steel of blade is highly sensitive to corrosion and should be therefore kept away from acids and stored clean and dry.
11. Timing of individual dehydration steps depends on the size of the objects and their texture [3]. For easy objects (e.g., roots samples with diameter up to 2 mm, leaves 5×5 mm segments, etc.), 3 h in the step should be safe. Small objects up to 1 mm might require only 15–30 min, while larger objects (minimal dimension 10–15 mm) might require days per step. Objects of high tissue density or covered with low permeability cuticle have significantly impeded exchange of solutions. Timing of changes should be prolonged accordingly.
12. The dehydration series solutions lose their properties with use, and contamination with compounds extracted from the samples takes place. It is recommended to keep record on the number and type of processed samples and replace solutions regularly. Precaution is more important for anhydrous (latter) steps of the series.
13. The embedding can be done using commercially available molds, cassettes, or origami made of smoothen cardboard which does not soak up much of the paraffin (Fig. 2a). Ceramic dishes with flat bottom are also good option. Spray/smear the ceramic dishes with 50 % glycerol or commercial detergent before embedding to facilitate latter separation of paraffin block from the ceramic surface.
14. Because the knife (as well as disposable blades) is not a simple wedge but has facets on the edge (Fig. 2c), the setup of knife clearance angle should respect the shape. Too low angle will crush sample; too steep adjustment might cause rolling of sections and chatter over hard objects. Clearance angle is usually between 3° and 5° or more acute approx. 10° for thinner sections and harder objects. Scale for adjustment is normally marked on the blade holder.
15. Transfer samples into mixture of 20 % sucrose solution and cryoembedding medium (1:1) and let them infiltrate overnight within refrigerator. This step is optional, but introduction of cryoembedding medium (e.g., OCT) into the object might further improve quality of sections [44].

16. Alternatively, place samples into a small aluminum foil mold filled with cryoembedding medium. To prepare the mold, fold a rectangular sheet of thicker Al foil around a small cover slips box. It is easier to handle small specimens and arrange them into desirable position using forceps or needle. Freeze the mold on cryobar inside the cryostat chamber or immerse the mold into isopentane supercooled with liquid nitrogen. Peel away the Al foil and fix frozen block on the chuck with drop of cryoembedding medium.

Acknowledgment

This work has been supported by the COST-LD11017 project.

References

1. Pearse AG (1980) Histochemistry (theoretical and applied): preparative and optical technology. Churchill Livingstone, Edinburg
2. Pearse AG (1985) Histochemistry (theoretical and applied): analytical technology. Churchill Livingstone, Edinburg
3. O'Brien TP, Mccully ME (1981) The study of plant structure: principles and selected methods. Termarcarphi Pty LTD, Melbourne
4. Ruzin SE (1999) Plant microtechnique and microscopy. Oxford University Press, Oxford
5. Fox CH, Johnson FB, Whiting J et al (1985) Formaldehyde fixation. J Histochem Cytochem 33:845–853
6. Medawar PB (1941) The rate of penetration of fixatives. J Royal Micro Soc 61:46–57
7. Bancroft JD, Gamble M (2008) Theory and practice of histological techniques. Churchill Livingstone, London
8. Mersey B, Mccully ME (1978) Monitoring of the course of fixation of plant cells. J Micro 114:49–76
9. Coetzee J, van der Merwe CF (1985) Penetration rate of glutaraldehyde in various buffers into plant tissue and gelatin gels. J Micro 137:129–136
10. Gardner RO (1975) An overview of botanical clearing technique. Biotech Histochem 50: 99–105
11. Bybd DW Jr, Kirkpatrick T, Barker KR (1983) An improved technique for clearing and staining plant tissues for detection of nematodes. J Nematol 15:142–143
12. Stebbins GL Jr (1938) A bleaching and clearing method for plant tissues. Science 87:21–22
13. Malamy JE, Benfey PN (1997) Organization and cell differentiation in lateral roots of *Arabidopsis thaliana*. Development 124: 33–44
14. Shobe WR, Lersten NR (1967) A technique for clearing and staining gymnosperm leaves. Bot Gaz 128:150–152
15. Sporne KR (1948) A note on a rapid clearing technique of wide application. New Phytol 47:290–291
16. Simpson JLS (1929) A short method of clearing plant tissues for anatomical studies. Biotech Histochem 4:131–132
17. Lux A, Morita S, Abe J et al (2005) An improved method for clearing and staining free-hand sections and whole-mount samples. Ann Bot 96:989–996
18. Peterson CA, Fletcher RA (1973) Lactic acid clearing and fluorescent staining for demonstration of sieve tubes. Biotech Histochem 48:23–27
19. Lersten NR (1986) Modified clearing method to show sieve tubes in minor veins of leaves. Biotech Histochem 61:231–234
20. Herr JM Jr (1971) A new clearing-squash technique for the study of ovule development in angiosperms. Am J Bot 58:785–790
21. Beeckman T, Engler G (1994) An easy technique for the clearing of histochemically stained plant tissue. Plant Mol Biol Rep 12: 37–42
22. Bougourd S, Marrison J, Haseloff J (2000) An aniline blue staining procedure for confocal microscopy and 3D imaging of normal and perturbed cellular phenotypes in mature Arabidopsis embryos. Plant J 24:543–550
23. Cunningham JL (1972) A miracle mounting fluid for permanent whole-mounts of microfungi. Mycologia 64:906–911

24. Truernit E, Bauby H, Dubreucq B et al (2008) High-resolution whole-mount imaging of three-dimensional tissue organization and gene expression enables the study of phloem development and structure in Arabidopsis. Plant Cell 20:1494–1503
25. Dubrovsky JG, Soukup A, Napsucialy-Mendivil S et al (2009) The lateral root initiation index: an integrative measure of primordium formation. Ann Bot 103:807–817
26. Zelko I, Lux A, Sterckeman T et al (2012) An easy method for cutting and fluorescent staining of thin roots. Ann Bot 110:475–478
27. de Almeida Engler J, Van Montagu M, Engler G (1994) Hybridization in situ of whole-mount messenger RNA in plants. Plant Mol Biol Rep 12:321–331
28. Klebs E (1869) Die Einschmelzungs-Methode, ein Beitrag zur mikroskopischen Technik. Arch micro Anat Entw 5:164–166
29. Johansen DA (1940) Plant microtechnique. McGraw-Hill Book Co. Inc., New York
30. Sass JE (1940) Elements of Botanical microtechnique. McGraw-Hill Book Co. Inc., New York, London
31. Vitha S, Baluska F, Jasik J et al (2000) *Steedm*an's wax for F-actin visualization. Dev plant soil sci 89:619–636
32. Sartori N, Richter K, Dubochet J (1993) Vitrification depth can be increased more than 10-fold by high-pressure freezing. J Micro 172:55–61
33. Quintana C (1994) Cryofixation, cryosubstitution, cryoembedding for ultrastructural, immunocytochemical and microanalytical studies. Micron 25:63–99
34. Beneš K (1973) On the media improving freeze-sectioning of plant material. Biol Plant 15:50–56
35. Tirichine L, Andrey P, Biot E et al (2009) 3D fluorescent in situ hybridization using Arabidopsis leaf cryosections and isolated nuclei. Plant Methods 5:11–18
36. Knapp E, Flores R, Scheiblin D et al (2012) A cryohistological protocol for preparation of large plant tissue sections for screening intracellular fluorescent protein expression. Biotechniques 52:31–37
37. Zhang Z, Niu L, Chen X et al (2012) Improvement of plant cryosection. Front Biol 7:374–377
38. Knox RB (1970) Freeze-sectioning of plant tissues. Biotech Histochem 45:265–272
39. Cocco C, Melis GV, Ferri GL (2003) Embedding media for cryomicrotomy: an applicative reappraisal. Appl Immunohistochem Mol Morphol 11:274–280
40. Williams DBG, Lawton M (2010) Drying of organic solvents: quantitative evaluation of the efficiency of several desiccants. J Org Chem 75:8351–8354
41. Pappas PW (1971) The use of a chrome alum-gelatin (Subbing) solution as a general adhesive for paraffin sections. Biotech Histochem 46:121–124
42. Brundrett MC, Enstone DE, Peterson CA (1988) A berberine–aniline blue fluorescent staining procedure for suberin, lignin, and callose in plant tissue. Protoplasma 146:133–142
43. Ferri GL, Cocco C, Melis GV et al (2002) Equipment testing and tuning: the cold-knife cryomicrotome microm HM-560. Appl Immunohistochem Mol Morphol 10: 381–386
44. Barthel LK, Raymond PA (1990) Improved method for obtaining 3-microns cryosections for immunocytochemistry. J Histochem Cytochem 38:1383–1388

Chapter 2

Selected Simple Methods of Plant Cell Wall Histochemistry and Staining for Light Microscopy

Aleš Soukup

Abstract

Histochemical methods allow for identification and localization of various components within the tissue. Such information on the spatial heterogeneity is not available with biochemical methods. However, there is limitation of the specificity of such detection in context of complex tissue, which is important to consider, and interpretations of the results should regard suitable control treatments if possible. Hereby we present set of selected simple staining and histochemical methods with comments based on our laboratory experience.

Key words Cell wall, Histochemistry, Lignin, Suberin, Pectin, Cellulose, Callose, Antibody, Staining

1 Introduction

Rigid plant cell wall is a prominent structure tightly related to cell shape, function, and interactions in the context of a multicellular body and in communication with surrounding environment. In fact, plant cell walls are structures most frequently followed studying tissue and organ anatomical organization. Combination of simple methods of cell wall staining and histochemistry might provide substantial and easily accessible information on cell wall composition, modifications, and changes related to development and tissue differentiation. However, unlike the biochemical detection, it does not allow for specific separation of cross-linked, complex mixture of components, which significantly increase probability of nonspecific results and interactions during detection. Therefore, higher probability of incorrect interpretation should be compensated with use of proper controls and independent parallel reactions if possible.

Histochemical detection should not be confused with procedures of "anatomical staining" because the affinity of pigment to target structure (e.g., safranin staining of lignified cell walls) depends highly on particular conditions (pH, polarity of solvent, temperature, time of dyeing, etc.) and is far less specific than colored product gained during specific reaction with substrate.

Viktor Žárský and Fatima Cvrčková (eds.), *Plant Cell Morphogenesis: Methods and Protocols*, Methods in Molecular Biology, vol. 1080, DOI 10.1007/978-1-62703-643-6_2,

Even such "anatomical staining" can gain useful results, but interpretation should be very cautious.

1.1 General Cell Wall Staining Methods

Toluidine blue O polychromatic staining is a simple and very useful oversight staining procedure disseminated in plant microtechnique by O'Brien et al. [1]. Besides good overall contrast of most structures, it also renders information on properties of the stained material. That is because the pigment interacts with stained material and shifts its absorption spectra towards longer wavelengths according to density of surface polyanions and subsequent dye aggregation [2]. Such coloration is referred as a metachromatic [3]. Therefore cell walls with low pectin content will stain blue (orthochromatic color), while pectin rich material will stain purple to pink (metachromatic color). Because lignified/phenolics containing cell walls present lower concentration of acidic groups, their staining is usually greenish. Cell wall material can be stained metachromaticaly above pH 3. Besides, cell wall tannin-containing vacuoles might stain green to bright blue, DNA-containing nucleus green.

PAS (periodic acid—Schiff's reagents) reaction is a nonselective polysaccharide detection procedure [4]. Periodic acid is a strong oxidizing agent cleaving vicinal diol linkages of polysaccharides and producing dialdehydes. These are subsequently detected with Schiff's reagent or its fluorescence alternatives [5, 6]. However, there is often background signal of some aldehydes present in the tissue (e.g., lignin monomers), and some others might be introduced during treatment with aldehyde fixative, which should be therefore used with consideration. That is why control sections without previous periodic acid treatment should be always included to ascertain about the origin of aldehydes. Optionally the autochthonous aldehydes might be eliminated before periodic acid treatment with borohydride reduction [6]. The PAS reaction scheme has been used recently also for staining of whole mount objects in combination of fluorescent leucobases of propidium iodide [7].

Calcofluor staining might be considered another general procedure of fluorescent cell wall accentuation. Calcofluor (synonyms are Tinopal, Fluorescent brightener) is nonspecific UV excited fluorochrome with high affinity to plant and fungi cell walls [8]. Its selectivity is considered to be related to $(1\rightarrow3)$, $(1\rightarrow4)$, -β-D-glucan chains of polysaccharides similarly to Congo red [9].

Besides procedures of general cell wall detection, there are methods aimed for specific components of cell wall.

1.2 Staining of Pectins, Callose, and Hemicelluloses

Alcian Blue is a basic dye, which can be used to rather specifically stain dissociated acidic carboxyl groups of pectins [10]. Acidic environment used for staining further narrows spectrum of potentially dissociated (stainable) acidic groups. In fact, there are not many other compounds that might react with the dye in plant cell walls under such conditions. Staining mechanism of Alcian Blue is rather similar to ruthenium red [11]. Ruthenium red is a hexavalent cation, which binds to variety of polyanions; that is why this classical reaction with pectin should be considered typical rather than specific. Ruthenium red also has its traditional use in electron microscopy (e.g., [12]).

Specificity of the staining can be further verified after pointed carboxyl blockage via methylation [4]. Blockage of acidic carboxyl should also abolish most of toluidine blue metachromasy discussed above. Pectins can be often (depending on the linkages within the cell wall context) extracted with hot aqueous solutions, Ca^{2+} chelating agents, and weak alkali solutions. Therefore such treatment should be avoided prior to pectin staining with any of the methods described. On the other hand extracting agents and their sequences might be used in connection with detection methods to further specify or confirm composition of extracellular material according to specific extractability (e.g., [13, 14]).

Callose is highly dynamic polymer (e.g., [15]). Its presence in tissues might be easily induced, for example, with chemical fixation of samples. Callose deposition is one of rather fast responses to stress or plant cell injury. Also aldehyde fixation (in fact it is a kind of chemical injury) induces deposition of callose into the plasmodesmata containing pit fields in order of minutes. That is why usage of cold methanol fixation or callose synthase inhibitors proved to be convenient to approach in vivo presence of callose. There are two most common ways of callose detection in tissues.

The most frequently used is staining with aniline blue [16, 17], respective its common impurity—the UV excited fluorochrome Sirofluor. Because the content of Sirofluor in the raw dye is variable according to brand and batch, it is reasonable to test your dye stock with known material first or use purified (and far more expensive) fluorochrome, which form highly fluorescent complexes with (1–3) β-D-glucans [18]. Advantage of purified fluorochrome might be seen also in the extended staining pH range from 3 to 10, while aniline blue staining should be at higher pH [18]. There are several reports on compromised specificity of the reaction and possible interaction with other polymers [19]. Control unstained section are convenient to ensure about nature of the fluorescence emission. Besides the fluorescent staining, bright field visualization of callose with reasonable specificity might be gained with Resorcin blue [17]. Since the antibodies are commercially available (e.g., Biosupplies, Australia), callose immunolocalization provides easily accessible highly consistent, sensitive, and specific detection alternative.

As far as we are aware there is no reliable and specific histochemical test for identification of hemicelluloses. This gap can be efficiently filled with the use of specific antibodies, which allow for precise distinction of various cell wall components (including pectins, hemicelluloses, and proteins). There are several sources of the antibodies currently available (e.g., http://www.plantprobes.

net, http://cell.ccrc.uga.edu/~mao/wallmab/Antibodies/antib.htm). Some of them were used in our lab in protocol similar to protocol presented bellow for callose.

1.3 Staining of Cell Wall Lipids and Lignin

Presence of lipidic compounds in cell wall is frequently connected with formation of cell layers with modified permeability of apoplast. Two principal insoluble cell wall lipids were historically distinguished by their position. Suberin is located in internal and secondary dermal tissues, while cutin constitutes cuticular part of epidermis on the surface of plant organs [20]. Considerable variation in monomeric composition of suberin and proportion of aromatic and lipidic domain were reported in between species and during development [21, 22]. There are several staining procedures used for detection of lipidic compounds in cell walls. Lipidic Sudan dyes (Sudan III, Sudan IV, Sudan Black B) are traditionally utilized in alcoholic solutions. Polyethylene glycol/glycerol-based staining solution of Sudan red 7B introduced by Brundrett [23] proved to be far more efficient and is the method of choice. Lipidic dyes partition from the slightly polar dyeing solution into the lipidic compartments of the tissue. It should be emphasized that intensity of staining depends highly on lipidic nature of cell wall material (quantity as well as molecular context of derivatives of fatty acid). Therefore sensitivity of the detection should be considered during interpretation, and nonspecific precipitation should be avoided. Improvement of sensitivity was reported due to use of lipidic fluorochrome Fluorol yellow [23]. However, background staining and autofluorescence can be sometimes difficult to distinguish from specific Fluorol yellow signal. That is why for some objects (e.g., maize roots) Sudan red 7B is preferred in our hands. Commonly we use fresh sections after aldehyde or no fixation. Several pitfalls are known (see method description below). Modified method of Fluorol yellow staining combined with lactic acid clearing of object was published by Lux [24]. Finally a very old technique of concentrated sulphuric acid digestion of cell wall material might also provide valuable information as only suberin and cutin impregnated material should resist it [25, 26].

Berberine–toluidine blue staining procedure was introduced by Brundrett et al. [27] to detect material of Casparian bands, suberin lamellae, and lignified tissue. It is a very frequently used staining based on acidophilic nature of berberine, which stains aromatic domains of lignified and suberised cell walls. The staining of berberine is combined with counter stain of toluidine blue (alternatively aniline blue, Evans blue, or Crystal violet) to quench background fluorescence. Selectivity of such a quenching is most likely related to physical properties of the cell wall (decreased accessibility of the material, e.g., due to suberinization). In fact counterstaining itself provides valuable information and combination with other acidophilic fluorochromes (e.g., acridine orange) or observation of autofluorescence of suberised cell walls is possible.

It is still far from clear to which extend the aromatic domain of suberin is identical or similar to lignin and how to distinguish those. Abovementioned staining can be considered as indication but not as a proof of lignifications. There are several others historically established methods of detection of lignification.

The most frequently used is Wiesner's reaction [28] using phloroglucinol condensation with cinnamic aldehydes (coniferyl aldehyde) in acidic environment and formation of cherry red product [29, 30]. There is potential cross-reactivity with other aliphatic and aromatic aldehydes [31], but in standard conditions the specificity can be considered rather high. Alternatively aniline sulphate [28, 32] is proposed. The output of the reaction and localization seems to be very similar to phloroglucinol reaction, but with lower contrast of resulting yellow coloration.

Another often-used lignin test is Mäule's reaction [33]. Syringyl moieties of lignin are considered to be the reaction target [29, 30]. The lignin composition related difference in detection, comparing to phloroglucinol can be strongly pronounced during development [30, 34] as well as in between taxonomic groups [35]. Schiff's reagent staining might be also used for detection of aldehydes of lignin [25].

There is wide spectrum of acidophilic dyes that have some affinity to lignified cell walls (PI, DAPI, Hoechst, basic Fuchsine, etc.). However, because of dependence on staining conditions and low specificity of such staining, it should be considered as informative and further confirmation of lignification is recommended. Autofluorescence of aromatic compounds is another very useful approach to follow phenolic compounds within the cell walls [36, 37].

1.4 Detection of Enzyme Activities

Apoplastic plant peroxidases play a key role in various metabolic processes—e.g., lignin and suberin formation, cross-linking of cell wall components, auxin metabolism, and metabolism of reactive oxygen species [38]. Peroxidase enzymatic activity might be probed with various co-substrates in presence of H_2O_2. The most common is diaminobenzidine (DAB), which yields upon oxidation brownish polymer [39, 40] and tetramethylbenzidine (TMB)—chromogen which yields a blue reaction product upon oxidation [41, 42]. Substrate does not have specific selectivity for particular heme protein, and therefore distinction of catalase and peroxidase is based on their different pH optimum. Peroxidase has its optimum at neutral range (pH ~6.5) while for catalase it is above pH 10 (*see* ref. [43]). To optimize the reaction progress, higher temperature (37 °C) is recommended which increases the enzyme activity, and adequately reduced exposure decreases spontaneous precipitation of DAB in presence of H_2O_2. Precipitation is further decreased if the reaction proceeds in dark as light induces spontaneous decomposition of H_2O_2. It is very important to include suitable controls (e.g., reaction mixture without H_2O_2 and sections where peroxidase activity was inhibited) [7].

2 Materials

All solutions should be prepared in distilled water unless stated otherwise.

2.1 Various Single-Component Dyes

1. Toluidine blue O: 0.01–0.025 % (w/v) toluidine blue O in water. Store at 4 °C. It can last for long rather time (months). Check periodically for mold.
2. Calcofluor white stock solution: 1 % solution of Calcofluor in distilled water. Gently heat the solution and add minimum of 1 M sodium hydroxide (final pH 10–11) to dissolve the dye completely. Aliquots of stock solution can be stored at -20 °C for a long period of time.
3. Alcian Blue: 0.1 % (w/v) Alcian Blue in 3 % acetic acid (alternatively citrate buffer of pH 3.5, 100 mM can be used, but is less selective).
4. Ruthenium red: 0.05 % (w/v) aqueous solution. Do not use phosphate and some other anionic buffers as those might precipitate the dye.
5. Aniline blue fluorochrome: 0.005–0.01 % solution of water-soluble aniline blue buffered to pH above 8.5 (e.g., 100 mM K_2HPO_4 with pH 9). Stock solution of purified aniline blue fluorochrome Sirofluor (1 mg/ml) in distilled water can be stored in aliquots at –20 °C.
6. Resorcin blue: Dissolve 3 g of resorcinol (p.a.) in 200 ml of distilled water. Add 3 ml of concentrated ammonia. Heat up in steam bath for 10 min (do not boil!). Let the red–brown solution cool down to lab temperature. The solution will gradually gain blue color (after approx 6 h). Heat again in the steam bath for about 30 min till no more ammonia escapes (test with wet pH indicator paper). Dilute prepared solution 1:50 with distilled water for staining.
7. Sudan red 7B and Fluorol yellow: Dissolve Sudan red 7B (0.1–0.2 % w/v) or Fluorol yellow 088 (0.01 % w/v) in PEG-400 heating the solution up to 90 °C. Do not allow to go over 100 °C as overheating change staining properties. Add equal volume of 90 % aq. glycerol. Filter solution through coarse filter paper or let stand overnight, and decant supernatant or centrifuge to sediment crystals of undissolved dye (if present it contaminates surface of sections).
8. Aniline sulphate solution: Dissolve 1 g of aniline sulphate (toxic and dangerous to environment) in 10 ml of 0.05 M H_2SO_4 and 90 ml of 70 % EtOH.

2.2 PARS Reaction

1. Periodic acid solution: 1 % (w/v) H_5IO_6. Significantly lower concentration (0.2 %) proved to be also efficient.
2. Schiff's reagent according to de Tomasi: Dissolve 1 g of basic fuchsine in 200 ml of boiling distilled water. Stir the solution for 5 min and let it cool down to 50 °C, filtrate with paper. Add 20 ml of 1 M HCl. Cool down to 25 °C. Add and dissolve 1 g of $Na_2S_2O_5$ (potassium metabisulfite). Leave in dark for 14–24 h to gain a pale yellowish-orange clear solution. Add enough active coal (approx. 2 g) and shake for few minutes, filter on paper to gain clear colorless solution. Store in a dark tightly stoppered bottle at 4 °C. The solution deteriorates with time. Discard when it turns colored.
3. SO_2 water: Mix 5 ml of 1 M HCl with 5 ml of 10 % $K_2S_2O_5$ and 100 ml H_2O before use. Solution remains efficient for a few days in a closed bottle.
4. Reducing solution: Dissolve 1 g of KI and 1 g of $Na_2S_2O_3.5H_2O$ in 50 ml of H_2O, add 0.5 ml of 2 M HCl. Prepare fresh before use.

2.3 Callose Immunodetection

1. Primary antibody solution: Dilute monoclonal antibody towards (1–3)-β-glucan (Biosupplies Australia PTY Ltd) 1:100 in 1× PBS with addition of 10 μl of BSA stock per 1 ml of final solution.
2. Secondary antibody solution: Select anti-mouse or anti-rabbit IgG antibody of your choice (we use Invitrogen anti-mouse IgG Alexa Fluor 488; 1:1,000) and dilute accordingly in 1× PBS with addition of 10 μl of BSA stock per 1 ml of final solution.
3. 10× PBS (phosphate-buffered saline) stock solution: Weight 80.1 g NaCl, 2 g KCl, 14.7 g Na_2HPO_4. $2H_2O$, and 2.38 g KH_2PO_4 to prepare 1 l of solution. 10× PBS stock has pH 6.8, after dilution to 1× PBS should be pH 7.3. It is recommended to check with pH meter before use.
4. 10 % BSA (bovine serum albumin) stock solution: Dissolve 1 g of powdered BSA (Fraction V) in 10 ml of distilled H_2O. Store in 1 ml aliquots at −20 °C.
5. Casein 3× stock solution: Add 3.33 % (w/v) of casein into distilled water and titrate to pH 10 with minimal amount of 2 M KOH. Let casein to dissolve at 40 °C with constant stirring (approx. 2 h). When it is completely dissolved, titrate to pH 7 with minimum of 2 M HCl. Add 10× PBS stock (11 % of volume of solution prepared aforetime) to gain 3 % casein solution in PBS. Aliquots can be stored at −20 °C. Before usage dilute (1:2) with PBS.
6. Buffered glycerol with n-propyl gallate: Add 3 % (w/v) of n-propyl gallate (antifade reagent) into glycerol and stir overnight at room temperature (it is not readily soluble in aqueous solutions). Mix 8:2 with TRIS buffer (0.1 M, pH 9.0).

Centrifuge to remove undissolved propyl gallate. Solution can be stored in dark at 4 °C for about a year.

7. TRIS buffer (0.1 M, pH 9.0): Dissolve 12.1 g TRIS base in approx. 750 ml of distilled water. Titre with 1 M HCl to pH 9.0, and fill with distilled water to 1 l of final volume.
8. High-humidity chamber is used to prevent evaporation of low volumes of antibodies from slides. Simple chamber can be made of large Petri dish with soaked tissue or filter paper on the bottom. Glass rods are used to separate slides from the soaked tissue and prevent their contact.

2.4 Berberine: Toluidine Blue Staining

1. Berberine dye solution: 0.2 % Berberine hemisulphate in water. The solution is close to the saturation and crystals will form when stored at 4 °C, those should be redissolved before use. 0.1 % solution is used in most publications but higher concentration does not cause overstaining.
2. Toluidine blue O dye solution: 0.05 % w/v of toluidine blue O in water.
3. Crystal violet solution: 0.05 % w/v Crystal violet in water.

2.5 HCl: Phloroglucinol (Wiesner's Reagent)

1. Acidified phloroglucinol solution: Phloroglucinol (1 % w/v, saturated) solution in 18 % aq. HCl. The solution oxidizes with time, turns deep yellow–brown and the intensity of reaction decreases. That is time to change it for fresh one.
2. Acidic glycerol to mount sections: Mix 75 % (final volume) of glycerol with 15 % of H_2O and 10 % of concentrated H_2SO_4.

2.6 Mäule Reaction

1. $KMnO_4$ solution: Prepare fresh 1 % w/v aqueous solution of $KMnO_4$.
2. Alkalized glycerol: 15 % (w/v) solution of Na_2CO_3 in 50 % aqueous glycerol (alternatively 15 % ammonium hydroxide in 75 % aq. glycerol can be used).

2.7 Peroxidase Activity Detection

Two optional co-substrate mixtures are described.

1. DAB reaction mixture: Prepare fresh solution just before incubation, containing 500 μl DAB stock (1 mg in 1 ml of distilled water, *see* **Note 1**), 499 μl acetate buffer (pH 5; 0.1 M), 50 μl $NiCl_2$ (8 % w/v in distilled water). Add 1 μl of H_2O_2 (30 % in distilled water) just before usage.
2. TMB reaction mixture: Prepare fresh solution just before incubation, containing 10 μl TMB stock (10 mg in 1 ml 96 % ethanol) and 989 μl acetate buffer (pH 5; 0.1 M). Add 1 μl of H_2O_2 (30 % in distilled water) just before usage.

3. Acetate buffer (pH 5; 0.1 M): Mix 14.8 ml of 0.2 M acetic acid with 35.2 ml 0.2 M sodium acetate and make up to 200 ml with distilled water.
4. Solutions for peroxidase inhibition: Use either (1) fresh solution of 3 % H_2O_2 in methanol, (2) acetate buffer containing 0.1 % sodium azide and 0.5 % H_2O_2, or (3) acetate buffer containing 0.1 % phenylhydrazine.

3 Methods

3.1 Toluidine Blue Staining

1. Stain fresh sections in toluidine blue O solution for 1–5 min.
2. Wash carefully in water.
3. Mount into water or low percentage glycerol (less than 25 % aqueous solution) to maintain metachromatic staining (*see* **Note 2**). If resin sections are used, let them air dry and mount them with nonaqueous media.
4. Observe in bright field optics.

3.2 PARS Reaction for Detection of Cell Wall Polysaccharides

We most commonly use fresh sections. Other types of sections should be fully hydrated before treatment.

1. Select parallel control section and skip H_5IO_6 oxidation step for those.
2. Oxidize sections in 1 % w/v H_5IO_6 for 1 min in laboratory temperature. Time should be adjusted properly if bigger objects (wholemounts) are treated.
3. Wash sections in distilled water 3×.
4. Optionally apply for 3 min the reducing solution and wash again with distilled water. Solution can be applied to clear away remainings of periodic acid, not necessary for sections but can be useful for bigger objects as wholemounts.
5. Stain in Schiff's reagent for 10 min.
6. Wash very carefully in SO_2 water 3 × 10 min to prevent oxidation of reduced colorless fuchsin and unspecific background staining.
7. Mount into 50 % v/v glycerol in SO_2 water.
8. Purple coloration of the tissue is tightly bound, so it is also possible to dehydrate objects and use permanent mounting. The background staining strongly depends upon efficient Schiff reagent wash out.
9. Observe in bright field. Presence of polysaccharides should be indicated with purple coloration, compare with control sections. If indigenous aldehydes are present before periodic acid treatment (in control sections), their reduction might be performed in the beginning of procedure (*see* **Note 3**).

3.3 Calcofluor Staining

1. Dilute the stock solution 1:100 with water and stain objects for 0.5–5 min. We use fresh sections but other types of sections should work if fully rehydrated.
2. Wash in water.
3. Mount in water, 50 % glycerol or other aqueous mounting media.
4. Observe under UV excitation. The cell wall material should yield pale blue signal.

3.4 Alcian Blue Staining

1. Rinse the sections in acetic acid (3 % aq. solution).
2. Control sections might be methylated to block free carboxyl groups in acidified methanol (1 M HCl in MetOH) for 4 h at 60 °C [4]. Methylation should mask free carboxyl and therefore inhibit polyanionic staining. Methylation of the carboxyl can be reverted with alkalized ethanol (1 % KOH in 70 % EtOH, 10 min in laboratory temperature).
3. Stain in Alcian Blue for 30 min at laboratory temperature.
4. Thoroughly wash in 3 % acetic acid (at least 10 min).
5. Mount into 75 % glycerol (sections could be also dehydrated and mount permanently).
6. Observe in bright field optics. Polyanionic compounds stain cyan color.

3.5 Ruthenium Red Staining

1. Wash fresh or fully rehydrated sections with water or suitable buffer.
2. Control sections might be methylated to block free carboxyl groups in acidified methanol (1 M HCl in MetOH) for 4 h at 60 °C [4]. Methylation should mask free carboxyl and therefore inhibit polyanionic staining. Methylation of the carboxyl can be reverted with alkalized ethanol (1 % KOH in 70 % EtOH, 10 min in laboratory temperature).
3. Stain the sections until the walls are red (normally within 5 min).
4. Rinse the sections with water.
5. Mount in water or 50 % glycerol.
6. Observe in bright field optics. Polyanionic compounds stain intensely red. Limited penetration of the dye was reported, which should be considered evaluating the results on thicker sections.

3.6 Aniline Blue Fluorochrome (Sirofluor) Staining

1. Stain the sections for 5–10 min in solution of aniline blue or flood sections with solution of Sirofluor (stock diluted 1:50 in distilled water or suitable buffer). Aqueous solutions low in ionic solutes decrease background staining of cellulose [43].
2. Rinse carefully with water or suitable buffer.

3. Mount into water or 50 % glycerol (alternatively it is possible to observe directly in staining solution due to low fluorescence of unbound fluorochrome in water solutions).
4. Optionally the background autofluorescence can be decreased counterstaining Sirofluor with toluidine blue or Crystal violet (as described for Berberine staining).
5. Observe in fluorescence setup. Fluorochrome yields yellow–green fluorescence with blue excitation and pale yellow fluorescence with UV excitation.

3.7 Resorcin Blue Staining

1. Dilute prepared solution 1:50 with distilled water and stain sections for 1–2 min.
2. Carefully wash 3× in water.
3. Mount into citrate buffer pH 3.2 (or 50 % buffered glycerol).
4. Observe in bright field optics. Callose is stained blue while lignified structures change the color to red.

3.8 Callose Immunodetection

1. Fix pieces of tissue in −20 °C methanol for 5–10 min.
2. Wash 2 × 5 min v PBS.
3. Prepare sections, preferentially fresh hand sections. Cryo-sections dried to slides should be rehydrated for 30 min in PBS. If paraffin sections are used, paraffin should be removed from sections in toluene, rehydrated via alcohol series down to water, and PBS.
4. Block the nonspecific protein binding with 1 % casein solution in PBS, 15 min.
5. Wash in PBS for 2 min and blot excess of solution from the edge of slide.
6. Apply primary antibody in high-humidity chamber at laboratory temperature for 2 h (time of application might be prolonged).
7. Wash 2 × 10 min in PBS, blot excess of solution.
8. Apply secondary antibody in humidity chamber at laboratory temperature for 2 h.
9. Wash 2 × 5 min in PBS.
10. Counter stain with toluidine blue for 10 min.
11. Mount into glycerol with propyl gallate.

3.9 Sudan Red 7B or Fluorol Yellow Staining Procedure

1. Fresh hand sections or cryotome sections are best suitable for the staining.
2. Blot sections to minimize transfer of water into staining solution, optionally 75 % glycerol wash might be included before staining to minimize precipitation of dye.

3. Stain in the solution for minimum of 1.5 h at laboratory temperature (the time can be extended significantly without a risk of overstaining; warming the staining solution up to 60 °C might accelerate the staining process).
4. Quickly rinse excess of dye solution with detergent solution (e.g., 0.5 % aq. solution of SDS).
5. Carefully wash with water.
6. Mount into 75 % aq. glycerol.
7. Sudan red gives intense red coloration of lipidic compounds, while Fluorol yellow yields green/yellowish fluorescence with UV excitation (for possible pitfalls *see* **Note 4**).

3.10 Berberine: Toluidine Blue Staining

1. Stain sections (we preferentially use fresh ones after aldehyde fixation) for at least 1 h in Berberine solution.
2. Wash twice with water.
3. Counter stain in 0.05 % toluidine blue O in water for 5–10 min. Alternatively Crystal violet (syn. Gentian violet) can be used to efficiently quench background fluorescence.
4. Wash carefully with water.
5. Mount into water or 25–50 % glycerol.
6. Observe under UV excitation as a yellow fluorescence or under blue excitation as a green emission.

3.11 Wiesner's (HCl: Phloroglucinol) Reaction

1. Stain the sections (we preferentially use fresh ones, but paraffin-embedded sections might be used after rehydration) at laboratory temperature with acid phloroglucinol solution till the sections turn red (within few minutes).
2. Mount the sections into glycerol acidified with sulphuric acid to maintain the reaction product, which last for several days or even weeks. Hydrochloric acid (escaping hydrogen chloride) is highly aggressive to metallic and optical parts of the microscope. That is why it is strongly recommended not to use it in close vicinity of the microscope.
3. Observe with bright field optics. Lignin modified cell walls stain cherry red.

3.12 Aniline Sulphate Procedure

1. Treat the section with aniline sulphate solution for 5 min in lab temperature.
2. Mount into acidified glycerol described for Wiesner's reaction.
3. Observe in bright field. Lignins are stained bright yellow.

3.13 Mäule Reaction for Lignin

1. Oxidize sections in solution of $KMnO_4$ for 10–20 min.
2. Wash 3× with distilled water.

3. Flood with 1 M HCl until the dark precipitate disappears (normally it takes about 30–60 s).
4. Wash gently with water.
5. Mount into alkalized glycerol.
6. Lignin is colorized red or brown red.

3.14 Peroxidase Activity Detection

1. Fix the object in 4 % formaldehyde in phosphate buffer (25 mM, pH 6.8) for 2–4 h at room temperature.
2. Carefully wash fixative out of sections with phosphate buffer (25 mM, pH 6.8) 2× for 15–20 min.
3. Prepare sections (we normally use hand sections) and select parallel sections for controls.
4. Recommended controls are as follows: (1) sections treated with reaction mixture without H_2O_2; (2) sections with peroxidase inhibited with H_2O_2 in methanol, 10 min at laboratory temperature; and (3) sections with peroxidase inhibited with phenylhydrazine, 10 min at laboratory temperature.
5. Wash sections carefully with acetate buffer 2 × 5 min.
6. Treat the section with the incubation medium at 37 °C for 1 h (or longer activity is weak).
7. Wash section carefully with acetate buffer 2 × 5 min.
8. Mount into 50 % glycerol.
9. Observe with bright field optics.

4 Notes

1. DAB is commonly used as hydrochloride, which is more soluble. If DAB is not in the form of hydrochloride it should be dissolved first in a drop of dimethylformamide and then add to buffer. Low concentration of DMF (up to ~0.5 %) should not affect peroxidase activity. DAB is carcinogenic. Waste should be oxidized (commercial bleach, $KMnO_4$) before being discarded. $NiCl_2$ catalyze precipitation of the product and decrease its run from reaction site improving accuracy of localization.
2. Toluidine blue staining is very convenient for fresh sections. Metachromasy is stable only in aqueous (highly polar) solutions and disappears in organic solvents [44]. It fades even if mount in stronger (we normally do not exceed 25 %) glycerol solutions. The intensity of the staining (concentration of dye solution) should be adjusted according to type and thickness of the section. As thicker freehand sections might be overstained with presented dye concentration, it is reasonable to dilute staining solution 5–10×. Other types of sections, e.g., paraffin [45] or hydrophilic resin sections [46], work well if

fully hydrated before staining and air-dried afterwards before mounting into nonaqueous mounts. Acetate buffer pH 4.4 can be used instead of water to prepare dye solution for more consistent results.

3. To reduce aldehydes on sections, dissolve 5 mg of $NaBH_4$ in 10 ml of borate buffer (pH 7.6) and treat section for 1 h in lab temperature [4].
4. There are several pitfalls of the Sudan red staining procedure. First, there might be problem with unspecific precipitation of Sudan red pigment on sections. The primary reason might be in water contact with the dyeing solution, which might produce crystals of Sudan. The dyeing solutions remain stable for considerable period of time (months), but is sensitive to water absorption and deteriorates if let open for a long time. We have also experienced staining problems dues to long-term storage (several years) of PEG 400 used for preparation of the solution. Be also careful with microscope setup to localize well cell wall response as plasma membrane staining may in some cases cause seeming coloration of cell walls due to refraction. The possibility of nonspecific staining of strongly acid structures (chromosomes) was indicated by Lillie [47].

Acknowledgment

This work has been supported by the MSM0021620858 project and COST- LD11017.

References

1. O'Brien TP, Feder N, McCully ME (1964) Polychromatic staining of plant cell walls by toluidine blue O. Protoplasma 59:368–373
2. Sylvén BENG (1954) Metachromatic dye-substrate interactions. Q J Microsc Sci 93–95:327–358
3. Bergeron JA, Singer M (1958) Metachromasy: an experimental and theoretical reevaluation. J Biophys Biochem Cytol 4:433–457
4. Pearse AG (1985) Histochemistry (theoretical and applied). Churchill Livingstone, Edinburgh
5. Rost FWD (1995) Fluorescence microscopy. Cambridge University Press, Cambridge
6. Kasten FH, Burton VIVI, Glover PEGG (1959) Fluorescent Schiff-type reagents for cytochemical detection of polyaldehyde moieties in sections and smears. Nature 184:1797–1798
7. Truernit E, Bauby H, Dubreucq B et al (2008) High-resolution whole-mount imaging of three-dimensional tissue organization and gene expression enables the study of phloem development and structure in Arabidopsis. Plant Cell 20:1494–1503
8. Herth W, Schnepf E (1980) The fluorochrome, calcofluor white, binds oriented to structural polysaccharide fibrils. Protoplasma 105:129–133
9. Wood PJ, Fulcher RG, Stone BA (1983) Studies on the specificity of interaction of cereal cell wall components with Congo Red and Calcofluor. Specific detection and histochemistry of (1–3), (1–4), -β-D-glucan. J Cereal Sci 1:95–110
10. Benes K (1968) On the stainability of plant cell walls with alcian blue. Biol Plant 10:334–346
11. Luft JH (1971) Ruthenium red and violet. I. Chemistry, purification, methods of use for electron microscopy and mechanism of action. Anat Rec 171:347–368
12. Muhlethaler K (1950) Electron microscopy of developing plant cell walls. Biochim Biophys Acta 5:1–9

13. Soukup A, Votrubová O (2005) Wound-induced vascular occlusions in tissues of the reed Phragmites australis: their development and chemical nature. New Phytol 167:415–424
14. Redgwell RJ, Selvendran RR (1986) Structural features of cell-wall polysaccharides of onion *Allium cepa*. Carbohydr Res 157:183–199
15. Evert RF, Derr WF (1964) Callose substance in sieve elements. Am J Bot 51:552–559
16. Currier HB, Strugger S (1956) Aniline blue and fluorescence microscopy of callose in bulb scales of *Allium cepa* L. Protoplasma 45:552–559
17. Eschrich W, Currier HB (1964) Identification of callose by its diachrome and fluorochrome reactions. Stain Technol 39:303–307
18. Evans NA, Hoyne PA, Stone BA (1984) Characteristics and specificity of the interaction of a fluorochrome from aniline blue (sirofluor) with polysaccharides. Carbohydr Polym 4:215–230
19. Smith MM, McCully ME (1978) A critical evaluation of the specificity of aniline blue induced fluorescence. Protoplasma 95:229–254
20. Priestley JH (1921) Suberin and cutin. New Phytol 20:17–29
21. Soukup A, Armstrong W, Schreiber L et al (2007) Apoplastic barriers to radial oxygen loss and solute penetration: a chemical and functional comparison of the exodermis of two wetland species, *Phragmites australis* and *Glyceria maxima*. New Phytol 173:264–278
22. Pollard M, Beisson F, Li Y, Ohlrogge JB (2008) Building lipid barriers: biosynthesis of cutin and suberin. Trends Plant Sci 13: 236–246
23. Brundrett MC, Kendrick B, Peterson CA (1988) Efficient lipid staining in plant material with sudan red 7B or fluorol yellow 088 in polyethylene glycol. Biotech Histochem 66:133–142
24. Lux A, Morita S, Abe J et al (2005) An improved method for clearing and staining free-hand sections and whole-mount samples. Ann Bot 96:989–996
25. Jensen WA (1962) Botanical histochemistry. Freeman, San Francisco
26. Zimmermann A (1892) Die botanische mikrotechnik: Ein handbuch der mikroskopischen präparations-reaktions- und tinktions-methoden. Verlag der H.Laupp'schen Buchhandlung, Tübingen
27. Brundrett MC, Enstone DE, Peterson CA (1988) A berberine–aniline blue fluorescent staining procedure for suberin, lignin, and callose in plant tissue. Protoplasma 146: 133–142
28. Wiesner J (1878) Note über das Verhalten des Phloroglucins und einiger verwandter Körper zur verholzten Zellmembrane. Sitzungsber Akad Wiss Math-naturw Kl 77:60–66
29. Akin DE (1989) Light microscopy and histology of lignocellulose related to biodegradation. In: Chesson A, Ørskov ER (eds) Physico-chemical characterization of plant residues for industrial and feed use. Elsevier applied science, London, New York, pp 58–64
30. Pomar F, Merino F, Barcelo AR (2002) O-4-Linked coniferyl and sinapyl aldehydes in lignifying cell walls are the main targets of the Wiesner (phloroglucinol-HCl) reaction. Protoplasma 220:17–28
31. Clifford MN (1974) Specificity of acidic phloroglucinol reagents. J Chromatogr 94: 321–324
32. Gahan PA (1984) Plant histochemistry and cytochemistry—an introduction. Academic, London
33. Johansen DA (1940) Plant microtechnique. McGraw-Hill Book Co. Inc., New York
34. Stafford HA (1962) Histochemical and biochemical differences between lignin-like materials in *Phleum pratense* L. Plant Physiol 37:643–649
35. Ibrahim RK, Towers GHN, Gibbs RD (1962) Syringic and sinapic acids as indicators of differences between major groups of vascular plants. J Linn Soc Lond Bot 58:223–230
36. Harris PJ, Hartley RD (1976) Detection of bound ferulic acid in cell walls of the Gramineae by ultraviolet fluorescence microscopy. Nature 259:508–510
37. Harris PJ, Hartley RD (1980) Phenolic constituents of the cell walls of monocotyledons. Biochem System Ecol 8:153–160
38. Almagro L, Gómez Ros LV, Belchi-Navarro S et al (2009) Class III peroxidases in plant defence reactions. J Exp Bot 60:377–390
39. Frederick SE (1987) DAB procedures. In: Vaughn KC (ed) CRC handbook of plant cytochemistry–cytochemical localization of enzymes. CRC, Boca Raton, FL, pp 3–23
40. Graham RC, Karnovsky MJ (1966) The fine structural localization of peroxidase activity. J Histochem Cytochem 14:291–302
41. Brand JA, Tsang VC, Zhou W et al (1990) Comparison of particulate 3, 3', 5, 5'-tetramethylbenzidine and 3, 3'-diaminobenzidine as chromogenic substrates for immunoblot. Biotechniques 8:58–60
42. Mesulam MM (1978) Tetramethyl benzidine for horseradish peroxidase neurohistochemistry: a non-carcinogenic blue reaction product with superior sensitivity for visualizing neural afferents and efferents. J Histochem Cytochem 26:106–117
43. Evans NA, Hoyne PA (1982) A fluorochrome from aniline blue: structure, synthesis and fluorescence properties. Aust J Chem 35: 2571–2575
44. Pal MK (1965) Effects of differently hydrophobic solvents on the aggregation of cationic

dyes as measured by quenching of fluorescence and/or metachromasia of the dyes. Histochem Cell Biol 5:24–31
45. Sakai WS (1973) Simple method for differential staining of paraffin embedded plant material using toluidine blue O. Biotech Histochem 48:247–249
46. O'Brien TP, McCully ME (1981) The study of plant structure: principles and selected methods. Termarcarphi Pty LTD, Melbourne
47. Lillie RD (1977) HJ Conn's Biological Stains: A handbook on the nature and uses of the dyes employed in the biological laboratory. Sigma Chemical Company, St. Louis

Chapter 3

Resin Embedding, Sectioning, and Immunocytochemical Analyses of Plant Cell Walls in Hard Tissues

Kieran J.D. Lee and J. Paul Knox

Abstract

Plant cell walls are structurally diverse macromolecular composites. One of our best methodologies to determine the temporal and spatial regulation of cell wall polysaccharides in relation to development are monoclonal antibody (MAB) and carbohydrate-binding module (CBM) probes and their detection by immunofluorescence microscopy. Here we describe resin embedding, sectioning, and in situ chemical and enzymatic cell wall disassembly and their use with immunocytochemical analyses as a means to unravel the complexity of cell wall molecular architecture in hard tissues and seeds.

Key words Monoclonal antibody, Carbohydrate-binding module, Acrylic resin embedding, Cell wall immunocytochemistry, Immunofluorescence microscopy, Plant cell walls

1 Introduction

Plant cell walls are complex macromolecular composites comprised of several interconnected networks. In primary cell walls cellulose microfibrils are tethered by noncellulosic polysaccharides such as xyloglucan, xylans, heteromannans, and mixed-linkage glucans, which are often referred to as hemicelluloses; this fibrous network is embedded in a gel-like matrix of pectic polysaccharides, glycoproteins, proteins, ions, and water. Pectic polysaccharides [1] are structurally diverse and the constituent polymers are currently classified as homogalacturonan (HG), rhamnogalacturonans I and II (RG-I and RG-II), and xylogalacturonan (XGA). Noncellulosic polysaccharides possess numerous structural elaborations that impact on polymer properties and functions [2]. A current challenge is to understand how these structural variants influence polymer-polymer interactions in individual cell wall architectures in the context of tissues and organs.

Approaches to map cell wall architecture in relation to development often make use of monoclonal antibody (MAB) and carbohydrate-binding module (CBM) probes that recognize

Viktor Žárský and Fatima Cvrčková (eds.), *Plant Cell Morphogenesis: Methods and Protocols*, Methods in Molecular Biology, vol. 1080, DOI 10.1007/978-1-62703-643-6_3, © Springer Science+Business Media New York 2014

defined polysaccharide features. These probes may be directly coupled to a fluorescent tag or used in conjunction with a tagged secondary antibody with specificity to the primary probe. Whole-mount labelling is a relatively quick and useful technique to assess cell wall composition at unadhered surfaces, whereas sectioning of embedded material enables mapping of cell wall polymer microdomains in internal cells and tissues without loss of context. By using immunocytochemical analyses in combination with chemical and enzymatic cell wall disassembly, one can investigate whether specific cell wall polymers have the capacity to block probe access to, or "mask," underlying cell wall architectures, thereby giving insight into possible polymer interactions in muro [3, 4]. With the increasing number of probes now available making analyses costly both in terms of time and money, strategies for choosing suitable probes and treatments are discussed. Here we focus on resin embedding and sectioning of hard tissues; wax embedding of soft tissues is discussed elsewhere [5].

2 Materials

Prepare all solutions using deionized water and analytical grade reagents. Prepare and store all reagents at 4 °C (unless indicated otherwise).

2.1 Molecular Probes

1. Large panels of MAB probes with specificities to plant cell wall polysaccharides and proteoglycans are available from the following suppliers: Biosupplies (http://www.biosupplies.com.au), Carbosource Services (http://www.carbosource.net) and PlantProbes (http://www.plantprobes.net). Biosupplies and Carbosource MABs are raised using mouse hybridoma technology, and thus the probes require anti-mouse secondary reagents; PlantProbes provides mostly rat MABs and thus require anti-rat secondary reagents. Secondary reagents with a range of tags depending on the particular application are available at Sigma-Aldrich (http://www.sigmaaldrich.com). The available panel of probes is now considerable (see supplier websites for full details), thus one must choose carefully depending on the system under study (*see* **Note 1**).
2. Recombinant CBMs are derived from microbial glycosyl hydrolases and engineered with a polyhistidine (His) tag to allow detection with anti-His secondary reagents. They may also be engineered with a directly coupled fluorescent protein, such as GFP, allowing direct visualization of probe binding by epifluorescence microscopy. Although not yet widely used, some are available commercially (*see* http://www.plantprobes.net).

2.2 A Short Guide to Probe Selection

The range of available MABs and CBMs is increasing rapidly and therefore it can be challenging to decide where to start and with which probes. It is important to consider probe selection when embarking upon an immunochemical survey of cell walls, especially if an overview is required rather than a focus on a particular subset of polymers. A good place to start would be with probes directed to HG, the major pectic polysaccharide, and also the major hemicellulose that is known for that system/taxon. There are now several probes that recognize HG with differing patterns and degrees of methyl esterification. It may be useful to start with at least a couple of probes with specificities requiring the presence of methyl esters on HG in addition to a probe for unesterified HG, e.g., JIM5, JIM7, and LM19 (see PlantProbes and Carbosource websites). The highly structurally heterogeneous RG-I can be detected to some extent with probes directed to the arabinan and galactan side chains of the polymer and with backbone-directed probes. The major non-pectic, noncellulosic polymers are now covered with a range of probes. These include probes for a range of epitopes present in heteroxylans, e.g., LM10 and LM11 (PlantProbes), and xyloglucan, e.g., LM15 (PlantProbes) and CCRC-M1 (Carbosource). There are also widely used probes for mixed-linkage glucan and heteromannans (Biosupplies). For glycoproteins such as extensins and arabinogalactan proteins, the glycan epitopes can be highly regulated and restricted in occurrence in relation to taxon. It should be noted that for the commonly used glycoprotein glycan MABs, e.g., JIM13, MAC207, and LM2 (PlantProbes and Carbosource), the epitope has not always been characterized in detail. Moreover, the extent to which glycan epitopes are associated with specific protein core sequences has not been determined. If details cannot be located in the literature, then trial and error to identify the appropriate probe may be required.

2.3 Fixation and Preparation of Plant Materials for Sectioning

For immunochemical analyses of plant cell walls, aldehyde fixatives are most commonly used:

1. 4 % (w/v) paraformaldehyde (PFA) in PEM buffer: 4 % (w/v) PFA, 50 mM piperazine-1,4-bis(2-ethanesulfonic acid) (PIPES), 5 mM ethylene glycol-bis(2-aminoethylether)-*N,N,N′,N′*-tetraacetic acid (EGTA), 5 mM $MgSO_4$; pH adjusted to 7.0 with KOH. Prepare first a 16 % (w/v) stock solution of PFA in water by heating to 60 °C and adding 1 M NaOH dropwise until the cloudy solution turns clear. Cool to RT and add a PEM buffer concentrate to the desired concentration. Alternatively a 16 % (w/v) formaldehyde solution is available (Agar Scientific, Stansted, UK). Aliquots of PFA fixative can be stored at −20 °C for up to 6 months, although it is preferable to use freshly prepared fixative.

2. 2.5 % (w/v) Glutaraldehyde (GA) in 0.1 M sodium cacodylate buffer pH 7.0. GA is commercially available as 8, 25, and 50 % (w/v) solutions. Alternatively a fixative containing a mixture of PFA and GA may be employed (*see* **Note 2**). GA fixatives should be freshly prepared, stored at 4 °C and used within 1 month.
3. Ethanol: aqueous solutions 30–100 % (v/v) for sample dehydration.
4. LR White resin, hard grade, containing 0.5 % (w/v) benzoin methyl ether catalyst (Agar Scientific).
5. Gelatin capsules (Agar Scientific).
6. Vectabond (Vector Laboratories, Peterborough, UK)-coated Multitest 8-well glass slides (MP Biomedicals, Solon, USA).
3. Nickel grids (Agar Scientific) for electron microscopy.
4. In-house prepared glass knives for trimming blocks.
5. A diamond knife: a Histo Diamond knife for semi-thin sections and/or an Ultra Diamond knife for ultrathin sections (Diatome, PA, USA). Alternatively the highest quality in-house prepared glass knives may be fitted with a boat made from waterproof tape and sealed with wax.

2.4 Immunomicroscopy

1. Super PAP hydrophobic pen (Agar Scientific) for outlining each well on the slide.
2. Phosphate-buffered saline (PBS): prepare a 10× stock with 1.37 M NaCl, 27 mM KCl, 100 mM Na_2HPO_4, 18 mM KH_2PO_4 (pH 7.4) and autoclave before storage. Alternatively, use prepared 10× PBS (Severn Biotech, Kidderminster, UK). Prepare a working solution by dilution of one part stock solution with nine parts water.
3. Blocking/antibody dilution buffers: PBS with 3 % (w/v) milk protein (e.g., Marvel, Premier Beverages, UK) (PBS/MP) or 3 % (w/v) bovine serum albumin (Sigma-Aldrich) in PBS (PBS/BSA).
4. Secondary antibodies: anti-rat immunoglobulin (IgG) (whole molecule) reagents coupled to FITC and gold; mouse anti-His; anti-mouse IgG coupled to FITC (Sigma-Aldrich), anti-His coupled to AlexaFluor 488 (Serotec, Kidlington, UK); anti-rat coupled to AlexaFluor 488 (Invitrogen).
5. Anti-fade reagent: Citifluor glycerol/PBS AF1 (Agar Scientific).
6. Microscope slide cover slips.
7. Calcofluor White: 0.25 % (w/v) in PBS (Fluorescent Brightener 28, Sigma-Aldrich).
8. Toluidine Blue O: 0.1 % (w/v) in 0.1 M sodium phosphate buffer pH 5.5.

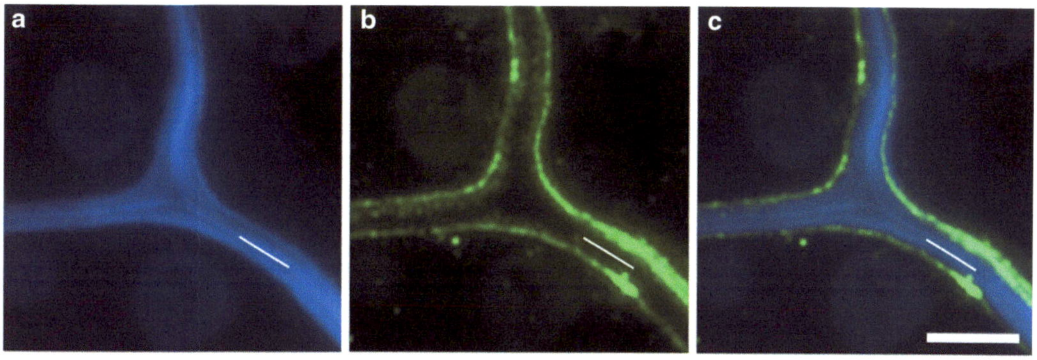

Fig. 1 Epifluorescence micrographs of tobacco seed endosperm. (**a**) Calcofluor White labelling indicates all cell walls. (**b**) A MAB specific to β-1-4-linked galactan labels inner cell walls. (**c**) shows merged image. *White lines* indicate positions of middle lamellae. Bar = 6 μm

2.5 Chemical and Enzymatic Pretreatments

1. 0.1 M sodium carbonate pH 11.4. For removal of methyl groups from polysaccharides.
2. 0.1 M KOH. For the removal of acetyl groups from polysaccharides [6].
3. Pectate lyase (from *Cellvibrio japonicus*) or polygalacturonase (from *Aspergillus niger*) (Megazyme, Bray, Ireland). For enzymatic removal of polysaccharides (*see* **Note 3**).
4. 3-(Cyclohexylamino)-1-propanesulfonic acid (CAPS) buffer: 50 mM CAPS, 2 mM $CaCl_2$, pH 10 for pectate lyase treatments.
5. Sodium acetate buffer: 50 mM sodium acetate, pH 4.0 for polygalacturonase treatments.

3 Methods

Immunofluorescence microscopy is a sensitive and rapid technique for the analysis of cell wall architectures in organs and tissues. Analysis with a 100× oil immersion lens provides an excellent level of detail and, in a good quality section, one can even resolve individual cell walls and middle lamellae using fluorescent probes and epifluorescence microscopy (Fig. 1). The methods presented here focus on immunofluorescence techniques; preparation of samples for electron microscopy is also discussed and should be employed when higher-resolution imaging is required to localize cell wall components in specific cell wall domains or organelles. Carry out all procedures at room temperature unless otherwise specified.

3.1 Plant Material Preparation and Fixation

1. The polymerization of LR White resin is inhibited by the presence of air. It is therefore crucial that samples are well fixed and of a small enough size to allow infiltration of fixative solution.

2. For hard materials such as plant seeds, it is necessary to create a hole in the sample using a needle to allow solutions to penetrate to the center of the sample (*see* **Note 4**).
3. For large, stiff materials, such as stems, the best results will be obtained by hand sectioning the material into blocks or slices no larger than 5 mm thick prior to fixation and embedding (*see* **Note 5**).
4. Use 4 % (w/v) PFA in PEM buffer for epifluorescence microscopy of semi-thin (0.5 μm) resin sections. Use 2.5 % (w/v) GA in 0.1 M sodium cacodylate buffer for electron microscopy analyses of ultrathin (~80 nm) resin sections.
5. Small samples are fixed under vacuum (to expel air) for at least 1 h at room temperature. For larger samples it may be necessary to increase the incubation time, but for no more than overnight. Incubation of samples in fixative for longer periods increases sample autofluorescence and can make the sample brittle. After fixation samples should be transferred to PEM or PBS buffer and stored at 4 °C until use.

3.2 Resin Embedding

1. Wash fixed material in PEM or PBS buffer three times, each for 10 min (or overnight at 4 °C).
2. Dehydrate by incubation in an ascending ethanol series—10, 20, 30, 50, 70, 90 %, and two times 100 % (v/v)—with 30 min incubation at 4 °C for each solution. Ensure a sufficiently large volume of liquid is used for the sample and place samples in tubes on a rotator.
3. Infiltrate with resin by incubation in an ascending resin series of 10, 20, 30, 50, 70, and 90 % (v/v) resin in ethanol with a 1 h incubation at 4 °C for each solution. Finally transfer samples to 100 % (v/v) resin and incubate overnight, then 8 h, then overnight.
4. Transfer samples to gelatin capsules containing fresh resin and ensure appropriate orientation of plant material (capsules may be examined under a dissecting microscope if samples are very small). Fill to the top with resin and seal to exclude air. It is useful to prepare several capsules containing the same sample as some may have poor orientation and so may not be suitable for sectioning.
5. Allow polymerization of resin either at 60 °C for 24–48 h, 37 °C for 5 days, or by the action of UV light at −20 °C.

3.3 Sectioning of Resin-Embedded Material

These instructions relate to the use of a Reichert-Jung Ultracut E Ultramicrotome (Reichert, Vienna, Austria).

1. Using an in-house prepared glass knife, trim the resin block until your sample is visible at the cut face of the block.

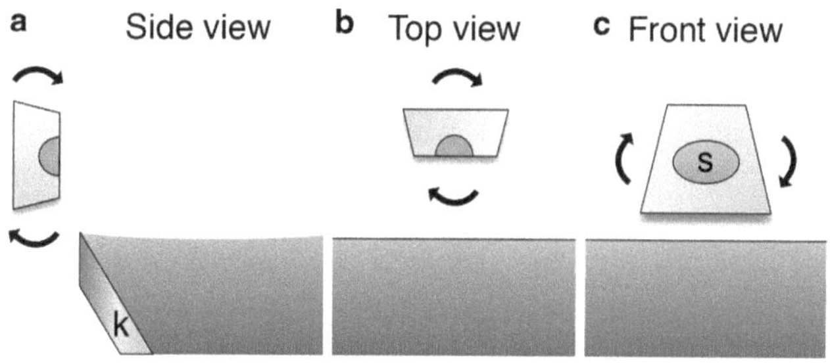

Fig. 2 Sample preparation and positioning for sectioning with an ultramicrotome. (**a**) Align block parallel to the cutting direction of the knife (*k*). For acrylic resin, such as LR White, the water level (*blue*) should form a concave meniscus in the boat. (**b**) Align block face parallel to the knife. (**c**) Ensure upper and lower sides of the block are parallel with the knife. (*s*) = sample in block (color figure online)

2. Using a fresh stainless steel razor blade, trim as much resin from the block as possible to produce a trapezium leaving your sample in the middle of the block face with the widest edge of the trapezium at the bottom of the block. The upper and lower edges of the block should be parallel with the knife edge. The procedure is outlined in Fig. 2.
3. For light microscopy, cut sections to a thickness of 0.5–2 μm onto water.
4. Transfer sections to a drop of water on Vectabond-coated Multitest slides and dry sections onto the slide using a slide drying hot plate. It may be useful to analyze some of the sections under the light microscope by staining with aqueous 1 % (w/v) Toluidine Blue O containing 2 % (w/v) Borax (filtered before use) to determine section quality and that the features of interest are present.
5. For electron microscopy, cut ultrathin sections to a thickness of ~80 nm when they appear silvery gold in color and collect sections on nickel grids.

For troubleshooting issues with sectioning *see* **Note 6**.

3.4 Immunolabelling of Sections Using MABs

These procedures are for the indirect immunofluorescence labelling of sections of plant material (*see* **Note** 7). Always include a negative control (omission of the primary MAB) to assess the extent of cell wall autofluorescence present in the sample. Here we focus on immunofluorescence procedures, but there are very effective alternatives such as immunogold with silver enhancement for light microscopy [7]:

1. Use the PAP pen to isolate individual wells on the Multitest slide.
2. Block nonspecific binding sites by incubation with PBS/MP for at least 30 min (a 30 μl volume is sufficient to ensure coverage of the sample).

3. Incubate with PBS for 5 min.
4. Incubate with primary MAB diluted in PBS/MP for at least 1 h at RT or overnight at 4 °C. A 5–10-fold dilution of a hybridoma cell culture supernatant is a good starting point for the primary MAB; however, a range of dilutions should be assessed (*see* **Note 8**). For overnight incubations it is useful to incubate slides in a sealed container on wetted filter paper to prevent drying out.
5. Wash with three changes of PBS with at least 5 min for each change.
6. Incubate with a secondary antibody diluted in the region of 100-fold in PBS/MP for at least 1 h at RT. For example, anti-rat IgG (whole molecule) linked to FITC is widely used for rat MABs. Alternatively anti-rat IgG linked to AlexaFluor 488 may be used, which is brighter and more photostable.
7. Wash with three changes of PBS with at least 5 min for each change.
8. Incubate with a 10-fold dilution of the Calcofluor White stock solution for 5 min (*see* **Note 9**).
9. Wash with three changes of PBS, each 5 min.
10. Mount samples using a small drop of anti-fade reagent (for Multitest slides 2 μl is a suitable volume), cover with a cover slip, and examine. To prevent slippage and slides drying out, the edges of the cover slip can be sealed.
11. Examine with a microscope fitted with epifluorescence optics and filters (e.g., UV, FITC, and TRITC). Sample autofluorescence can be assessed by examining the no-primary-antibody-control in the FITC and TRITC channels (*see* **Note 10**).

3.5 Immunolabelling of Plant Cell Walls Using Recombinant CBMs

1. Isolate individual wells on slides and block nonspecific binding sites (*see* Subheading 3.4, **steps 1** and **2**).
2. Incubate with the CBM diluted in PBS/MP for at least 1 h at RT. The most effective working concentration should be determined by trial studies for each CBM, but most CBMs can be used effectively in the range of 5–20 μg/ml.
3. Ensure that there is a no-CBM-control to assess cell wall autofluorescence in the section.
4. Wash with three changes of PBS, each 5 min.
5. In the case of a CBM fused with a fluorescent protein, proceed directly to **step** 7. In the case of a CBM with a His tag, incubate with anti-His coupled with FITC at 100-fold dilution in MP/PBS for at least 1 h. Alternatively anti-His linked to AlexaFluor 488 may be used.
6. Wash with three changes of PBS, each 5 min.

7. Incubate with Calcofluor White if required (*see* Subheading 3.4, **step 8**).
8. Mount slides using anti-fade reagent and examine.

3.6 Immunogold Labelling for Electron Microscopy

1. Block to prevent nonspecific binding by floating the EM grid section side down on a droplet (at least 20 μl) of PBS/BSA on Parafilm® for 30 min.
2. Transfer grid to a droplet of primary antibody diluted in PBS/BSA. MAB cell culture supernatants should be diluted between 5- and 200-fold.
3. Wash grids by incubation in a minimum of three changes of PBS, each 10 min.
4. Transfer grids to secondary antibody diluted 20-fold in PBS/BSA. We routinely use anti-rat IgG coupled to 10 nm gold.
5. Wash as in **step 3** and then extensively in distilled water.
6. Allow the grid to dry and then examine in an electron microscope.

3.7 Section Pretreatments Prior to Immunolabelling

Cell wall polysaccharide epitope masking has been demonstrated in a number of parenchyma systems [3, 4, 7, 8]. Pectic HG is often methyl esterified, and to effect its most efficient removal by pectate lyase or polygalacturonase enzymes, a pretreatment of the section with a high pH solution is required (*see* **Note 3**):

1. Incubate the section with a solution of 0.1 M sodium carbonate (pH 11.4) for 2 h.
2. Wash two times with deionized water, each 10 min.
3. Incubate with pectate lyase (10 μg/ml) in CAPS buffer for 2 h or polygalacturonase (20 μg/ml) in sodium acetate buffer for 2 h.
4. Wash two times with deionized water, each 10 min.
5. Sections are now ready for immunolabelling (*see* Subheadings 3.4, 3.5 or 3.6).
6. Do not let sections dry prior to labelling.

4 Notes

1. In the absence of a priori knowledge about the plant tissues of interest, one may wish to perform a nitrocellulose-based assay such as a dot blot [9] of extracted cell wall material to get an overview of the polysaccharides present and their relative abundance. Such an approach will avoid the issues associated with epitope masking and is a more rapid alternative than systematically pursuing enzymatic pretreatments of sections when an immunochemical survey of the tissue is desired.

2. The most important step in preparing material for sectioning and microscopic analyses is the killing and preservation of cells and tissues. Chemical fixation of tissues can produce structural artifacts (swelling and shrinkage) and so it is important to be aware of this prior to microscope analyses of sections. Fixatives work by cross-linking proteins, lipids, and nucleic acids. They do not cross-link polysaccharides and so some polysaccharides may be lost during sample preparation. If one suspects this to be the case, it may be useful to analyze a fresh hand-cut section of the sample or by performing a tissue print [9]. GA forms a dense, extensive cross-linked matrix and is therefore better than PFA at preserving the fine structure of cells. Furthermore, GA cross-linking reactions reach end point much quicker than PFA. However, GA has a slower infiltration rate than PFA and so more time should be allowed for sample fixation. The extensive cross-linked matrix of GA can impact on antibody reactions, for this reason it may be preferable to use GA/PFA in combination to take advantage of the best qualities from both aldehydes. Common compositions are 2 % (w/v) PFA + 1 % (w/v) GA and 2.5 % (w/v) PFA + 0.25 % (w/v) GA; the latter ensures tissue stability while retaining probe accessibility. Both GA and PFA fixation of tissue can generate fluorescent compounds, although the effect is more pronounced with GA. In general this is not a problem when analyzing resin-embedded samples. However, sample autofluorescence can be quenched by post-staining sections with 0.1 % (w/v) Toluidine Blue O in 0.1 M sodium phosphate buffer, pH 5.5 after immunolabelling.
3. Pectic HG has been shown to mask XG and heteromannan polysaccharides. Pectic HG may be removed from resin sections by treatment with either pectate lyase or polygalacturonase. Both enzymes are active on unsubstituted HG polymers; therefore section pretreatment with alkali to remove esters may optimize enzyme action and subsequent HG removal. Masking of cell wall polysaccharides may be a general feature of cell wall architecture and so pretreatment of sections with a range of glycosyl hydrolases against the constituent polymer classes may reveal subtleties in wall composition and architectures and give insight into polymer associations *in muro*.
4. For larger samples or in cases where sample processing produces poor quality blocks with insufficient infiltration of resin, incubation of the sample at each dehydration and infiltration step may be extended to 24 h.
5. Although large gelatin capsules are available from Agar Scientific that enable embedding of materials with ~10 mm diameter, it should be noted that longer incubation times must be employed during sample preparation to ensure proper

buffer exchange and infiltration of resin. For these reasons it may be preferable to excise a range of smaller-sized samples containing features of interest from an organ or tissue, rather than embedding the entire tissue.

6. A perfect ribbon can only be obtained from a well-trimmed block. A good quality, well-polymerized acrylic block should have a glassy surface when cut. It can be difficult to cut sections from wide blocks as this increases the cutting pressure and may lead to chatter (visible as lines on the cut face of the block). Compression of the sample may also become an issue as the face of the block increases in size. If this is the case, unwanted regions of sample should be trimmed away to leave a specific region of interest. An equivalent block may then be trimmed in a complementary fashion to give full coverage of the sample. If one encounters problems with material being pulled out of the resin during sectioning, microwave infiltration and pretreatment with (3-glycidoxypropyl) trimethoxysilane may address this issue [10].
7. Indirect immunofluorescence labelling of cell walls is a widely used technique that can accommodate several antibodies in the same protocol and also allows assessments of nonspecific binding and sample autofluorescence. The principles in the immunolabelling procedures are the same for whole-mount labelling of intact materials and hand-cut sections. Antibody incubations can be performed in tubes or plates, depending on the size of the material under study. Direct immunolabelling procedures, requiring just one step, are rapid, are highly effective, and may be combined with indirect immunolabelling to enable dual localization of cell wall epitopes in a single section.
8. The recommended dilution of an antibody is the highest dilution that results in a strong specific signal. Manufacturers of secondary reagents provide good guidance for dilution factors. For primary MABs, a 5–10-fold dilution of cell culture supernatants is often used; however, in some cases, up to a 200-fold dilution can be highly effective in terms of both analyses and costs.
9. Calcofluor White is used as a counter stain as it binds widely to β-glycans, including cellulose, and fluoresces under UV excitation and therefore can indicate all cell walls in sections and is useful for orientation and identification of immunolabelling in relation to organ and tissue anatomy. If sample autofluorescence is a problem, equivalent sections of the sample can be labelled with either MAB and Calcofluor White or MAB and Toluidine Blue O to allow visualization of all cell walls and MAB fluorescence without the contribution of sample autofluorescence.

10. Sample autofluorescence from, e.g., lignins and chlorophyll or induced by aldehyde fixation can be assessed by examining the no-primary-antibody-control in the FITC and TRITC channels of the microscope. Although usually an obscuring phenomenon for epitope localization, the intrinsic fluorescence of a tissue can be used as a diagnostic tool for the identification of organ anatomy and cell constituents [11].

Acknowledgment

We acknowledge the funding from the UK Biotechnology & Biological Sciences Research Council.

References

1. Caffall KH, Mohnen D (2009) The structure, function, and biosynthesis of plant cell wall pectic polysaccharides. Carbohydr Res 344: 1879–1900
2. Burton RA, Gidley MJ, Fincher GB (2010) Heterogeneity in the chemistry, structure and function of plant cell walls. Nat Chem Biol 6:724–732
3. Marcus SE, Verhertbruggen Y, Hervé C et al (2008) Pectic homogalacturonan masks abundant sets of xyloglucan epitopes in plant cell walls. BMC Plant Biol 8:60
4. Hervé C, Rogowski A, Gilbert HJ et al (2009) Enzymatic treatments reveal differential capacities for xylan recognition and degradation in primary and secondary plant cell walls. Plant J 58:413–422
5. Hervé C, Marcus SE, Knox JP (2011) Monoclonal antibodies, carbohydrate-binding modules, and the detection of polysaccharides in plant cell walls. In: Popper ZA (ed) Methods in molecular biology, vol 715, The plant cell wall—methods and protocols. Springer/Humana, New York, pp 103–113
6. Marcus SE, Blake AW, Benians TAS et al (2010) Restricted access of proteins to mannan polysaccharides in intact plant cell walls. Plant J 64:191–203
7. Meloche CG, Knox JP, Vaughn KC (2007) A cortical band of gelatinous fibers causes the coiling of redvine tendrils: a model based upon cytochemical and immunocytochemical studies. Planta 225:485–498
8. Davies LJ, Lilley CJ, Knox JP et al (2012) Syncytia formed by adult female *Heterodera schachtii* in *Arabidopsis thaliana* roots have a distinct cell wall molecular architecture. New Phytol 196:238–246
9. Willats WGT, Steele-King CG, Marcus SE et al (2002) Antibody techniques. In: Gilmartin PM, Bowler C (eds) Molecular plant biology: volume two—a practical approach. Oxford University Press, Oxford, UK, pp 199–219
10. Lindley VA (1992) A new procedure for handling impervious biological specimens. Microsc Res Tech 21:355–360
11. Razin SE (1999) Plant microtechnique and microscopy. Oxford University Press, Oxford, UK

Chapter 4

Automated Microscopy in Forward Genetic Screening of Arabidopsis

Tereza Dobisová and Jan Hejátko

Abstract

Tightly controlled spatiotemporal specificity of gene expression is intrinsic to developmental and adaptation responses of living systems throughout the kingdoms. Forward genetic screens employing well-characterized reporter lines can be used to identify as yet unknown genetic factors driving specific aspects of individual regulatory pathways. However, such screens are demanding with respect to data acquisition and analysis from thousands of mutant lines. Here, we describe a method that allows screening of a mutagenized GUS reporter line in Arabidopsis using an automated microscopy imaging system as a tool for rapid and efficient identification of mutants with modified expression profile for a gene of interest.

Key words Automated microscopy, Imaging, Forward genetics, Screening, GUS staining, .slide, Arabidopsis

1 Introduction

Most genes isolated to date have been identified through diverse forward genetic screens based upon identification of morphological or physiological malfunctions (e.g., refs. [1–3]). The tremendous progress in deciphering the molecular mechanism underlying particular responses permits screening for more subtle changes that are specific to particular developmental and adaptation pathways. Furthermore, many if not all of the developmental processes throughout the kingdoms are associated with the tightly regulated expression of regulatory genes. Consequently, many of the isolated mutants are affected in the spatiotemporal specificity as to the expression of important developmental regulators (e.g., refs. [4–7]). Thus, forward genetic screening that specifically allows to identify factors controlling the expression of important developmental regulators is of great interest.

Viktor Žárský and Fatima Cvrčková (eds.), *Plant Cell Morphogenesis: Methods and Protocols*, Methods in Molecular Biology, vol. 1080, DOI 10.1007/978-1-62703-643-6_4, © Springer Science+Business Media New York 2014

We adapted a "classical" EMS mutagenesis approach in combination with automated microscopy screening for isolation of mutations affecting spatiotemporal expression of genes of interest (GOI). The method is based on the identification of mutants in the expression profile of a transgenic line carrying a reporter under control of a GOI promoter. Alternatively, the translational fusion of a reporter with GOI coding sequence under control of the GOI promoter could be used.

As an example, we report a mutant screen directed to identifying factors regulating signaling via multistep phosphorelay (MSP) in Arabidopsis (reviewed in ref. [8]). To identify factors modulating expression of *CKI1* encoding one of the sensor histidine kinases initiating MSP signaling, we performed a forward genetic screen using the *ProCKI1:GUS* reporter line in a combination with EMS mutagenesis. This approach would be difficult without employing automated microscopy and innovative high-throughput imaging technology, thus enabling rapid acquisition and analysis of large datasets.

Here, we present a use of the "dot slide" (.slide) system for digital virtual microscopy (Olympus, http://www.olympus-europa.com) that was developed primarily for automated imaging of tissue sections in biomedical applications. We adapted this technology for effective scanning of specimens with very low contrast, i.e., Arabidopsis seedlings after GUS staining and tissue clearing [9]. This experimental setup has substantially accelerated screening the M2 population of an EMS-mutagenized *ProCKI1:GUS* transgenic line. In our hands, the entire process of screening approximately 2,000 M2 families by a single researcher can be completed within 6 months. Together with the following rescreen of interesting mutant lines, it is possible to begin mapping the mutation of interest after 1 year of work. The use of automated microscopy not only facilitates the data acquisition but also makes the subsequent data analysis less time-consuming. In any case, of course, the specimen preparation remains the most time- and labor-intensive part of the work.

This technology is suitable for use also with other types of reporters, including fluorescent proteins that can be visualized using additional, optional hardware. Finally, in addition to Arabidopsis seedlings, the automated microscopy can be used for detailed analysis of diverse tissue types and identification of rather moderate anatomical changes at microscopic level while screening for morphology mutants even in reporter-less lines.

The automated microscopy could be performed with other recently commercially available systems. However, we find the .slide system as the most open and therefore as the most suitable for diverse research applications. We are open to share the .slide system in our lab on either collaborative or commercial basis.

2 Materials

Unless otherwise specified, solutions are prepared using ultrapure double-distilled water (ddH_2O). The chemical compounds used were of *pro analysi* (p.a.) purity or higher.

2.1 Plant Seeds and Sterilization

1. Seeds of individual M2 families (progeny of self-pollinated M1 plants, grown from EMS-mutagenized M1 seeds) harvested separately and seeds from a non-mutagenized control (*see* **Note 1**).
2. 70 % (v/v) ethanol: mix 70 ml ethanol and 30 ml water.
3. Sterile filter papers, 1.5 ml tubes, gas-permeable (textile or paper) tape such as a surgical paper tape, and sterile Petri dishes.

2.2 Plant Cultivation

1. MS media (pH 5.7–5.9): Murashige and Skoog medium 4.3 g/l (Duchefa Biochemie); MES monohydrate 2.35 mM (Duchefa Biochemie), sucrose 1 %, plant agar 1 % (Duchefa Biochemie). Weigh 4.3 g of MS salts, 0.5 g of MES monohydrate, and 10 g of sucrose, and then adjust volume to 1,000 ml using ddH_2O. Adjust pH to 5.7–5.9 using 1 M KOH (approximately 1.25 ml).
2. Petri dishes: cylindrical, with diameter of 9 cm (or of similar size).
3. Cultivation growth chamber: e.g., Percival Scientific, Inc. or similar, allowing plant cultivation under selected growth conditions (*see* **Note 2**).

2.3 GUS Staining

1. 0.1 M Pi buffer: for preparation of 500 ml of 0.5 M $NaH_2PO_4{\cdot}H_2O$, weigh 34.49 g of $NaH_2PO_4{\cdot}H_2O$ and adjust volume to 500 ml by adding ddH_2O. For preparation of 500 ml of 0.5 M $Na_2HPO_4{\cdot}2H_2O$, weigh 44.99 g of $Na_2HPO_4{\cdot}2H_2O$ and adjust volume to 500 ml using ddH_2O. Mix 39 ml of 0.5 M $NaH_2PO_4{\cdot}H_2O$ and 61 ml 0.5 M $Na_2HPO_4{\cdot}2H_2O$ and adjust volume to 400 ml by ddH_2O.
2. 10 % (w/v) X-GlcA/DMF: weigh 100 mg of X-GlcA, sodium trihydrate (Duchefa Biochemie), and dissolve it in 1 ml DMF (N,N-dimethylformamide). 100 mg of X-GlcA is sufficient for staining approximately 100 samples (*see* **Note 3**).
3. 10 % (v/v) Triton X-100: measure 10 ml of Triton X-100 and adjust to 100 ml using ddH_2O (*see* **Note 4**).
4. 50 mM Fe salts: prepare 10 ml of 50 mM salts of $K_3[Fe(CN)_6]$ and of $K_4[Fe(CN)_6]$ in ddH_2O. Weigh 164.6 mg of $K_3[Fe(CN)_6]$ and dissolve in 10 ml of ddH_2O. Weigh 211.2 mg of $K_4[Fe(CN)_6]$ and dissolve in 10 ml of ddH_2O (*see* **Notes 5** and **6**).
5. GUS staining buffer: 0.1 M Pi buffer, pH 7, 0.1 % X-GlcA in DMF, 0.05 % Triton X-100, 0.1 mM Fe salts. To prepare

100 ml of GUS staining buffer, mix 100 ml of 0.1 M Pi buffer, 1 ml of 10 % (w/v) X-GlcA/DMF, 0.5 ml of 10 % Triton X-100, and 200 μL of 50 mM Fe salts. This volume is sufficient for staining approximately 100 samples (*see* **Notes 3** and **5**).

6. Staining plate: e.g., a multiwell cell culture plate (*see* **Note 7**).
7. Tweezers.
8. Incubator (37 °C).
9. Water vacuum pump (or other vacuum source of corresponding strength).
10. Desiccator.

2.4 Clearing of Samples

1. Ethanol series: prepare 100 ml of 80 % ethanol, 100 ml of 40 % ethanol, 100 ml of 20 % ethanol, 100 ml of 10 % ethanol. To prepare 100 ml of x % ethanol, mix x ml of 96 % ethanol and (100-x) ml of ddH_2O.
2. 5 % ethanol/50 % glycerol: mix 50 ml of 10 % ethanol and 50 ml of 50 % (v/v) glycerol in ddH_2O (*see* Subheading 2.5, **item 4**).
3. 0.25 M HCl/20 % (v/v) methanol: measure 457 ml ddH_2O and add 43 ml of 37 % HCl to obtain 500 ml of 1 M HCl solution. Prepare mixture of 125 ml 1 M HCl, 100 ml methanol, and 275 ml ddH_2O.
4. 7 % NaOH/60 % (v/v) ethanol: weigh 7 g of NaOH, dissolve it in 40 ml of ddH_2O, and finally add 60 ml of ethanol (*see* **Note 8**).
5. Orbital shaker, incubator (55 °C).

2.5 Specimen Preparation

1. Soft tweezers.
2. Microscope slides: size 76.2 (l) × 25.4 (w) × 1.0–1.2 (thickness) mm.
3. Cover slips: size 22 (l) mm × 50 (w) mm (*see* **Note 9**).
4. Mounting media: 50 % (v/v) glycerol in ddH_2O. Measure 50 ml of glycerol and add 50 ml of ddH_2O.

2.6 Microscope System Components

Soft imaging system ".slide" (Olympus Company). The system consists of hardware (Fig. 1) plus software that drives the system during image acquisition and for image analysis. Hardware includes BX 51 microscope with motorized high-precision X/Y stage and Z-drive, automatic slide loader with capacity of up to 50 slides, high-resolution cooled digital color CCD camera (1376 × 1032 pixel, pixel size 6.45 × 4.65 μm), very fast computer card for stage control, high-end workstation with TFT monitor (1,600 × 1,200 dpi), and set of super apochromatic objectives with magnifications 2×, 4×, 10×, 20×, and 40×.

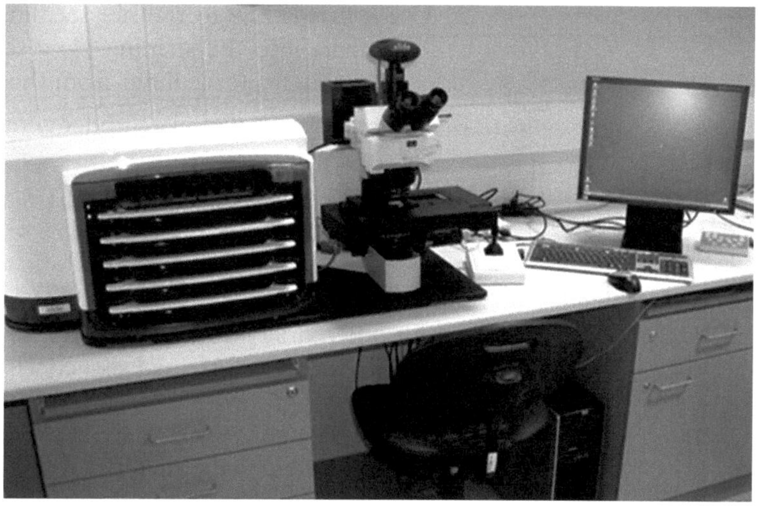

Fig. 1 The .slide automated microscopy system consisting of slide loader for 50 slides (*left*), BX51 microscope (*middle*), and the control unit with computer (*right*)

3 Methods

Unless otherwise specified, all procedures are performed at laboratory temperature.

3.1 Plant Cultivation

1. Temper freshly autoclaved MS media to 50–70 °C in a prewarmed water bath, then pour approximately 25 ml of the media into cylindrical Petri dishes. Work in a flow box under sterile conditions.
2. Place approximately 50 seeds from each M2 family separately into a 1.5 ml tube. Add 1 ml of 70 % ethanol per tube. Shake the tube with seeds several times for maximally 5 min (*see* **Note 10**). Transfer the seeds by pipetting (1,000 ml pipette) onto sterile filter papers placed in the flow box and let them dry. Transfer dry seeds by "dusting" (use sterile tweezers and a glass stick) onto prepared Petri dishes with MS media and seal the lid using air-permeable tape. For each biological replica and/or separate experiment, use the non-mutagenized background line (WT) as a control.
3. Store dishes with seeds at 4 °C in darkness for 2 days (*see* **Note 11**).
4. Place dishes into a growth chamber and cultivate for 7 days in vertical orientation (*see* **Note 12**).

3.2 GUS Staining

1. Always prepare fresh GUS staining buffer immediately before use. Add 1 ml of the GUS staining buffer to each well of the staining plate (*see* **Notes 3**, **5**, and **6**). This amount is sufficient for staining about 50 Arabidopsis seedlings 7 days old.

2. Use soft tweezers to transfer seedlings from a Petri dish with MS media into the staining plate so that each well contains approximately 30 seedlings from the particular M2 family.
3. Place the cultivation plate under gentle vacuum in the desiccator for 10 min (*see* **Note 13**).
4. Incubate the seedlings in the GUS staining buffer for 8 h at 37 °C in darkness (*see* **Note 14**).

3.3 Sample Clearing

This process allows obtaining a transparent specimen that is suitable for differential interference contrast (DIC) microscopy and is compatible with GUS-stained tissue [9]. In addition to seedlings, it can be used for other tissues from the later stages of development (e.g., flowers or inflorescence stems). In the first step, the chlorophyll is removed and the material is fixed. In the following steps, the sample is rehydrated in a decreasing ethanol series and saturated by 50 % glycerol. Unless otherwise specified, use laboratory temperature for the individual incubations.

1. Replace GUS staining buffer with 1 ml of 80 % ethanol by pipetting. Shake gently on an orbital shaker or rocking plate for 12 h (*see* **Note 15**).
2. Change the 80 % ethanol and continue incubation until the plant material turns yellowish white (usually, a further 12 h should be sufficient).
3. Replace the 80 % ethanol by pipetting with 1 ml of 0.25 M HCl/20 % methanol. Incubate for 15 min at 55 °C and remove the solution by pipetting (*see* **Note 16**).
4. Add 1 ml of 7 % NaOH/60 % ethanol. Incubate for 15 min, then remove the solution by pipetting (*see* **Notes 7**and **17**).
5. Add 1 ml of 40 % ethanol. Incubate for 10 min, then remove the solution by pipetting.
6. Add 1 ml of 20 % ethanol. Incubate for 10 min, then remove the solution by pipetting.
7. Add 1 ml of 10 % ethanol. Incubate for 10 min, then remove the solution by pipetting.
8. Add 1 ml of 5 % ethanol/50 % glycerol. Incubate for 30 min, then remove the solution by pipetting.
9. Add 1 ml of 50 % glycerol.
10. Store at 4 °C or continue to process as described below (Subheading 3.4, *see* **Note 18**).

3.4 Sample Preparation

For genetic screen of individual M2 families, use at least 20–30 seedlings per family and per slide (*see* **Note 19**).

1. Carefully place the cleared seedlings (or other plant tissues) onto the microscopic slide and mount them into a thin layer of 50 % glycerol (*see* **Note 20**).

2. Store the specimen at 4 °C or proceed directly with sample scanning via automated microscopy (*see* **Note 21**).

3.5 Automated Microscopy Procedure

The automated microscope system ".slide" allows digitalizing a large number of slides with minimal human effort during the scanning process (a video recording showing this procedure can be downloaded from the authors' website at http://www.ceitec.eu/functional-genomics-and-proteomics-of-plants/rg46#vybaveni).

1. Before the scanning is initiated, up to 50 slides are loaded into the slide loader. The robotic arm scans the slides via bar code reader (*see* **Note 22**). If the slides are not designated by bar codes (or the codes are not detected), the reader mechanically detects the presence of a slide in the respective loader position. After all the slides in the slide holder are successfully detected, the scanning process can be initiated.
2. The next step is to set up several scanning parameters: (a) information and designation as to the type of scanned slide (*see* **Note 23**), (b) slide positions in the loader that will be scanned (in most cases, all slides are selected), (c) magnification at which the slides will be scanned (*see* **Note 24**), (d) position of image storage area on disk, (e) acquisition properties (*see* **Note 25**), (f) specification of the slide scanning area (*see* **Note 26**), and (g) optionally, in a case of difficult specimens, image quality can be improved by using the extended focal imaging (EFI) scanning mode (*see* **Note 27**).
3. The scanning process can now be started simply by pushing the START button.
4. Slides are subsequently loaded from the slide loader by robotic arm (via a vacuum-based adaptor) and placed into the motorized stage (*see* **Notes 8** and **20**).
5. As the first step, the system creates an overview of the slide at low magnification (2×). In the slide overview, contrasting objects are automatically recognized. Slides are scanned according to the user settings (*see* **step 2**).
6. If the manual adjusting mode of the scanning area was selected (*see* **step 2**, f), then the slide scanning area must now be set. This scanning area will be used for all slides in the loader that are selected to be processed.
7. In the next step, a focusing map is automatically set up (*see* **Note 28**) and the scanning process initiates. The final scanned image is composed of small overlaid frames, each corresponding to the field of view of the respective objective. The image processing and creation of the final image (virtual slide) runs in real time and can be observed on the computer screen.

8. The technology is working with a huge amount of data, and therefore, it is very demanding of data storage capacity (*see* **Note 29**).
9. To use the outputs in other applications, the native ".vsi" data format can be converted to the required format (e.g., ".jpg" or ".tiff" files).

3.6 Specific Considerations for Arabidopsis Seedling Slides

1. Place slides with seedlings of an individual M2 family into the slide loader (*see* **Note 30**).
2. For scanning of highly transparent objects such as are the cleared Arabidopsis seedlings, it is necessary to set up particular parameters as follows: (1) completely close the aperture stop and (2) set the "shade avoidance" and "white balance" in the calibration settings (*see* **Note 31**).
3. Initiate the reading of slides in the slide loader by selecting the "batch scan" mode.
4. Set up the sample information and then set the options for detailed scanning as follows: In the "more options" menu, select acquisition to the highest sensitivity of sample detection, "focusing sample only," and check off the "auto adjustment of scanning area" option. Specify the location at the storage media where the image files will be saved and the file designations. Choose those slides that you want to scan and set the scanning of slides with objective 10× (Fig. 2).
5. Start scan.
6. After the overview scan is finished, specify manually the whole slide as the scanning area (*see* **Notes 26** and **32**) and proceed with the scanning (*see* **Note 33**).
7. Every slide in the loader will be processed according to the aforementioned settings. The average time when using these settings is approximately 15–20 min per slide, and the final file size is approximately 300 MB.

3.7 Image Analysis

Based upon our experience, we recommend manual image analysis to identify mutants with changes in the expression profile of the gene of interest (*see* **Note 34**).

1. All loaded samples are by default saved in the native ".vsi" format.
2. The ".vsi" format allows reliable viewing of the virtual slide using your computer's mouse wheel to change the magnification in a manner similarly to that of the Google Maps application. A sample file of the entire virtual slide of 7 days old ProCKI1:GUS M2 family seedlings scanned using a 10× objective can be downloaded from http://www.ceitec.eu/programs/genomics-and-proteomics-of-plant-systems/functional-genomics-and-proteomics-of-plants/. To view the

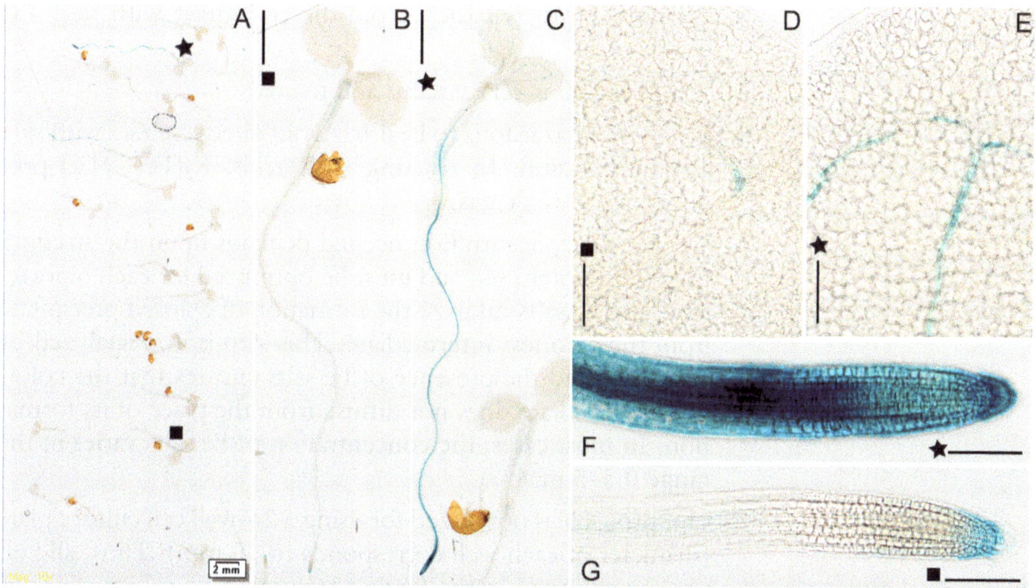

Fig. 2 Example of automated microscopy imaging using the .slide automated microscopy system (for the original file in the native ".vsi" format, *see* http://www.ceitec.eu/functional-genomics-and-proteomics-of-plants/rg46#vybaveni). Shown is an illustrative result of the scanned slide with sample specimen from the *ProCKI1*:GUS M2 family after EMS mutagenesis. Imaging of the whole slide of seedlings 7 days old was performed using a 10× objective. All of the presented figures represent visualizations from a single virtual slide obtained via changes in the virtual slide magnification. (**a**) Whole slide. In 2 out of 18 seedlings (designated by *star*), the abnormal GUS pattern was identified in comparison to the WT seedlings (designated by *square*). This implies potential occurrence of a single-point mutation in the gene-regulating expression of *CKI1*. Scale bar 2 mm. (**b**, **c**) Closer view of the WT seedling (**b**) and the mutant (**c**). Scale bars 1 mm. Detail of cotyledons (**d**, **e**) and root tip (**f**, **g**) imaging of the selected WT and mutant seedlings. Scale bars 100 μm

image in the native ".vsi" format, download the software OlyVIA (olyVIA_v2.4.exe file) that is freely available at http://downloads.olympus-europa.com/microscopy/software/.

3. For viewing details on the low-magnification background, the electronic lens tool can be used.
4. A particular area of interest can be cropped and saved in various commonly used image format types (e.g., ".bmp", ".jpg", ".tiff") for visualization in other applications (Fig. 2).

4 Notes

1. Harvesting seeds from each M2 family separately facilitates the identification of individual mutation events in each family and enables evaluating their segregation ratio.
2. We used an 8/16 h (light/dark) photoperiod, with light intensity 150 μmol/m^{2}s, 50 % humidity, and temperature 21 °C/19 °C (light/dark).

3. Caution, DMF is toxic. Avoid direct contact with skin and perspiration.
4. Triton X-100 is very viscous and foams.
5. Fe salts are hazardous to health. Avoid direct contact with skin and perspiration. In reacting with acids, $K_3[Fe(CN)_6]$ produces highly toxic gas.
6. The Fe-salt concentration needed depends upon the strength of the promoter used and must be optimized for each reporter line. The Fe salts catalyze the formation of colored precipitate from the colorless intermediate. This step is not catalyzed by the GUS, and the presence of Fe salts ensures that the colorless intermediate does not diffuse from the place of its formation. In most cases, the concentration of Fe salts varies in the range 0.5–5 mM.
7. Our protocol is optimized for using a 24-well cell culture plate (diameter of each well corresponds to 15 mm). Thus, all volumes mentioned in the further text could be subject to optimization if a staining plate of different dimensions were used.
8. Be careful during preparation of clearing solutions. Protect yourself (wear gloves and goggles) and work in a fume hood. Caution, both reactions are exothermic! Keep the 7 % NaOH/60 % ethanol solution at 4 °C in an opaque container for no more than 1 month or until the yellowish color of the solution is detectable.
9. Use a cover slip of maximum length 50 mm. The loader in the automated microscope uses the slide label area to handle the specimen, and longer cover slips would interfere with that. This could cause damage to your preparation and/or the automated microscopy system.
10. Alternatively, a motorized lab tube rotator could be used.
11. The purpose of this cold treatment is to break seed dormancy (vernalization). Even in the case of using nondormant ecotypes (e.g., ecotype Col-0), however, the cultivation of seeds at 4 °C synchronizes germination and improves its frequency.
12. Use plastic stands to avoid leakage of condensate from the Petri dishes during cultivation. The condensate contains sucrose that would result in heavy contamination of the growth chamber by fungi.
13. Do not use a strong vacuum source (e.g., an oil pump), as strong vacuum could damage the specimen.
14. Incubation time is dependent on the strength of the promoter used in the respective reporter line. It can differ among individual reporter lines, growth conditions, tissue types, etc., and must therefore be optimized. Plants are not fixed during staining and in cases of highly sensitive promoters and/or genes,

non-reproducible inconsistencies in the GUS patterning could sometimes be observed during repetitions.

15. The 80 % ethanol will stop the GUS staining, fix the plant material, and begin its bleaching by chlorophyll extraction.
16. The solution containing HCl digests cell walls and ensures elasticity and transparency of the plant tissues. The samples will start to become sticky and might adhere to the pipette tip during removal of the HCl solution. Be careful, therefore, to avoid damaging the specimen; use tweezers whenever necessary. Work carefully and use gloves when handling clearing solutions.
17. NaOH solution is used to stop the HCl-mediated degradation process by neutralization. Work carefully and use gloves when handling clearing solutions.
18. The protocol can be interrupted here and samples stored in 50 % glycerol for up to 6 months in parafilm-sealed staining plates. Check the amount of 50 % glycerol during storage to avoid samples' drying during storage.
19. According to the expected segregation ratios, recessive homozygous mutation appears at frequency of approximately 1:7 (assuming only one of the two initials in the SAM is mutated and the recessive mutation segregates according to Mendelian rules).
20. Use a small volume of mounting media, as this can increase the quality of the resulting virtual slide because all the seedlings will be in the same plane as defined by the thin layer of mounting media. We routinely use 100 μl of 50 % glycerol for 22 mm × 50 mm cover slips. Using a standardized volume of mounting media ensures the comparable quality of all samples. Avoid excessive spattering of the slide label area, as this commonly is responsible for any inaccurate handling (loading problems) by the automated microscope and for damage to specimens and/or the automated microscopy system. When following the instructions stated in our protocol, fixing the cover slip using tape or nail polish is not necessary. Avoid the presence of bubbles during the specimen preparation. Use mounting media (50 % glycerol) prepared separately, as using 50 % glycerol from the staining plate will ultimately increase the amount of contaminants and bubbles in the specimen and can significantly decrease the quality of the final image.
21. Samples can be stored for a long time in a refrigerator (at 4 °C). According to our experience, storing the samples for up to 6 months in a horizontal position and in a plastic slide holder does not negatively affect the specimen quality. Avoid storing the samples in paper slide holders or in a non-horizontal position. Both lead to leakage of glycerol and specimen

drying. Before scanning stored samples, allow them to acclimate to room temperature so as to avoid water condensation on slides and cover slips.

22. Bar code-labeled slides are saved after scanning using the bar code information implemented in the resulting file. Any slides without bar code are saved numerically in the default settings, which also can be changed (*see* **Note 23**). The system also allows scanning the slide label area and saving this information as an integral part of the virtual slide. This, however, increases the file size.
23. Optionally, the following information can be added manually for each slide to the synoptic file table: bar code, name, information, specimen, species, tissue, preparation, staining, tray number, slide number, and tray bar code.
24. Magnification depends on screening type. For the screening of seedlings 7 days old, a 10× objective is sufficient (*see* Fig. 2 and sample data online).
25. Quality of the automatic scanning process depends greatly on the setting of the scanning properties. In the acquisition mode, sample detection sensitivity (high or low) during the scanning process can be set. In the highest-sensitivity mode, even a low-contrast specimen, but also most of the impurities on the slide, is detected by the software. At the lowest-sensitivity setting, on the other hand, you will be able to detect and automatically focus upon only strongly contrasting objects. Finally, for autonomous slide scanning, it is important to set the density of the focusing map (*see* **Note 28**).
26. For scanning of transparent samples, we recommend to adjust the scan area manually to take in the whole slide. Scanning of the entire slide ensures that all of the specimens will be scanned at an acceptable quality, even if the automated recognition on the slide overview fails to identify all specimens or their parts. This can happen quite often, particularly in the case of transparent and tiny Arabidopsis root.
27. EFI staining combines multiple images taken in different planes of focus into a single, layered image. This ensures the best image quality (by increasing the depth of focus) of the specimen. However, using EFI dramatically increases both scanning time and file size.
28. The focusing map is generated to avoid the need for focusing the respective objective in each field of view on the slide and thereby to accelerate imaging. After the specimen is identified in the image overview, the microscope focuses upon the specimen in a pattern, the density of which was set up as a scanning parameter (*see* Subheading 3.5, **step 2**). The position of focusing

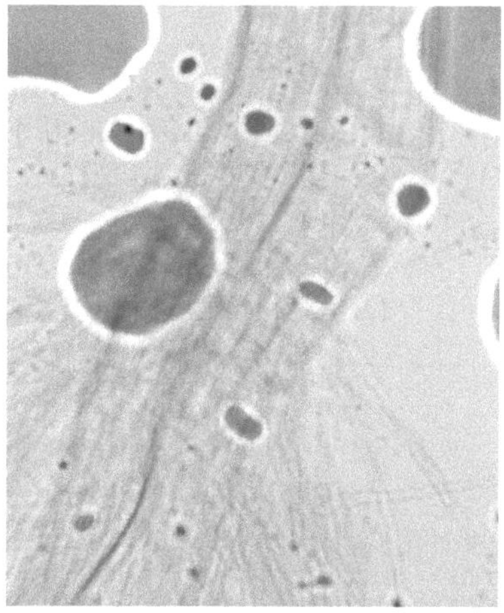

Fig. 3 Detailed snapshot from a virtual slide obtained by imaging of a slide spattered with glycerol on the cover slip. This is an example of how incorrectly prepared slides can negatively influence automatic sample recognition. This is problematic especially in case of low-contrast samples, such as cleared Arabidopsis seedlings

points is shown by green rectangles. During slide scanning, the focus plane is determined by an imaginary line that connects the neighboring focusing points.

29. In our case, we use a data server equipped with disk array of raw capacity 6.0 TB.
30. The maximum number of samples that can be loaded depends upon the size of the slide loader. Our system is designed for a maximum 50 slides.
31. The proper setting of the shade avoidance and white balance individually for each scan ensures the highest possible quality and helps in recognizing the transparent objects, thereby ensuring the specimen is optimally in focus.
32. This setting ensures that whole Arabidopsis seedlings—including their most transparent portions (in particular, the tiny root parts)—will be scanned. Avoid scanning of untidy slides, which leads to poor recognition of samples and ultimately to images that are not sharp (Fig. 3). According to our experience, scanning with a 10× objective is sufficient for reliable signal detection in the tiny structures of young developing Arabidopsis seedlings.
33. Usually we allow the scanning of slides to run overnight.
34. Automated image analysis also can be applied, but an application suitable for that purposes is not a part of the software package delivered by Olympus.

Acknowledgments

The work was supported by the Czech Science Foundation (grants 13-25280S and P501/11/1150). This work was supported by the project "CEITEC—Central European Institute of Technology" (CZ.1.05/1.1.00/02.0068) from the European Regional Development Fund and by the European Social Fund (CZ.1.07/2.3.00/20.0189).

References

1. Scheres B, Dilaurenzio L, Willemsen V et al (1995) Mutations affecting the radial organization of the Arabidopsis root display specific defects throughout the embryonic axis. Development 121:53–62
2. Torres-Ruiz RA, Jurgens G (1994) Mutations in the FASS gene uncouple pattern formation and morphogenesis in Arabidopsis development. Development 120:2967–2978
3. Mayer KF, Schoof H, Haecker A et al (1998) Role of WUSCHEL in regulating stem cell fate in the Arabidopsis shoot meristem. Cell 95: 805–815
4. Donner TJ, Sherr I, Scarpella E (2009) Regulation of preprocambial cell state acquisition by auxin signaling in Arabidopsis leaves. Development 136:3235–3246
5. Muller B, Sheen J (2008) Cytokinin and auxin interaction in root stem-cell specification during early embryogenesis. Nature 453:1094–1097
6. Palmeirim I, Henrique D, Ish-Horowicz D et al (1997) Avian hairy gene expression identifies a molecular clock linked to vertebrate segmentation and somitogenesis. Cell 91:639–648
7. Kumar SV, Wigge PA (2010) H2A.Z-containing nucleosomes mediate the thermosensory response in Arabidopsis. Cell 140:136–147
8. Schaller GE, Shiu SH, Armitage JP (2011) Two-component systems and their co-option for eukaryotic signal transduction. Curr Biol 21:R320–R330
9. Malamy JE, Benfey PN (1997) Organization and cell differentiation in lateral roots of *Arabidopsis thaliana*. Development 124:33–44

Chapter 5

Image Analysis: Basic Procedures for Description of Plant Structures

Jana Albrechtová, Zuzana Kubínová, Aleš Soukup, and Jiří Janáček

Abstract

This chapter gives examples of basic procedures of quantification of plant structures with the use of image analysis, which are commonly employed to describe differences among experimental treatments or phenotypes of plant material. Tasks are demonstrated with the use of ImageJ, a widely used public domain Java image processing program. Principles of sampling design based on systematic uniform random sampling for quantitative studies of anatomical parameters are given to obtain their unbiased estimations and simplified "rules of thumb" are presented. The basic procedures mentioned in the text are (1) sampling, (2) calibration, (3) manual length measurement, (4) leaf surface area measurement, (5) estimation of particle density demonstrated on an example of stomatal density, and (6) analysis of epidermal cell shape.

Key words Counting frame, Experiment design, Image analysis, Image calibration, Length estimation, Number estimation, Stereology, Systematic uniform random sampling, Unbiased estimation

1 Introduction

Image analysis has become a powerful tool for quantification of plant structures—macroscopic or microscopic ones. The quality of captured digital signal (microscopic image) is essential for further processing and quantitative analysis. Image spatial resolution is number of picture elements (pixels, pxl in 2D) per length or area unit, which determines distinction of structures within the image. Resolution in color or gray scale (bit depth) is another important parameter. Standard 8-bit images allow for 256 (2^8) levels while 16-bit images provide more detailed scale of 65536 levels to choose from. Colors of the image are most commonly encoded by three principal color channels—R (red), G (green), and B (blue) of the additive color mixing model (RGB) or HSB (or HSI or HSL) model, which uses hue, saturation, and brightness to specify colors. During image processing it is thus possible to work with individual channels of RGB or HSB containing different information.

Viktor Žárský and Fatima Cvrčková (eds.), *Plant Cell Morphogenesis: Methods and Protocols*, Methods in Molecular Biology, vol. 1080, DOI 10.1007/978-1-62703-643-6_5, © Springer Science+Business Media New York 2014

Objects in the plane perpendicular to the objective axis could be acquired without distortion (*see* **Note 1**) and their physical properties might be analyzed, if the size of the image pixels is known. Dimensions of pixels are determined during calibration (*see* **Note 2**).

Selection of proper image format affects quality of the image data. Some formats (e.g., JPG) use lossy compression and might cause some data degradation. That is why lossless formats (e.g., TIFF) are far more suitable for primary data storage. The images intended for image analysis should have sufficient both image resolution and bit depth. The size of the pixel shall be at least one half of the least requested detail in the image as defined by the Nyquist sampling theorem. Higher bit depth (12 bits and more) enables more accurate image processing and segmentation.

Image preprocessing can be used to correct some deficiencies (background gradients, dust, etc.) and to improve visual image quality. Unlike human the computer does not identify the object of interest, so image segmentation should be used to separate those from background. The most common and simplest method of segmentation is thresholding, which results in binary image (binary values per pixel; 0/1; black/white) definitions of the objects. Subsequent adjustments of binary image based on mathematic morphology—operations of dilation, erosion, closing, and opening—are used to improve object representation. Objects, once defined, can be measured and classified using various geometrical parameters (*see* **Note 3**). Subset of objects might be selected in specified region of interest (ROI), with superimposed mask or a counting frame [1–3].

The majority of plant structures exhibits gradients [4] in quantitative anatomical parameters, reflecting polarity of plant structures established during early plant ontogenetic development and organogenesis. A known example is the dependence of some anatomical parameters of leaf on its insertion—i.e., distance from the root system. Because of the spatial heterogeneity of values of anatomical parameters in biological structures, it might be difficult to specify average values of a particular anatomical parameters within an organ or tissue. Proper sampling designs based on a systematic uniform random (SUR) sampling (e.g., ref. [2]) are necessary to gain reproducible values. Unbiased estimation of a measured parameter can be achieved by SUR sampling, which ensures the prerequisite of unbiased estimation, i.e., that each particle or object has the same probability to be selected by the sampling system or rule. This is a very important issue, often neglected in formation of sampling design of biological studies. Quite often the sampling design and the method of quantification are not fully described, and thus the results can be difficult to compare and discuss with other results as a consequence of biological treatment or processes.

Stereological methods might be also employed to analyze an image. Stereological methods provide 3D characteristics of objects based on measurement of parameters in 2D or 3D image, which is done by application of some test system (composed of points, lines, quarter-circles, cycloids, etc.) and counting the intersections of the structure under study and parts of a test system (e.g., refs. [2, 3]); for introduction to stereology see also C. V. Howard's web pages (http://www.liv.ac.uk/fetoxpath/quantoxpath/stereol.htm). Stereological methods along with uniform random sampling can be very efficiently used to get unbiased estimations of quantitative anatomical parameters of plant organs. For more details on principles of stereological method application see the source studies [5–9].

In this chapter, several examples of basic procedures for quantification of plant structures are presented. We focused on parameters which can be quantified from 2D images: length, surface area in 2D, particle density of particles laying in one plain, and analysis of epidermal cell shape.

The very important rules for obtaining reproducible results of image analysis are the following: (1) object sampling should be done on the base of uniform random sampling to achieve unbiased estimation, (2) right calibration has to be set to obtain results in right units, (3) proper statistical rules and methods have to be applied during results processing and interpretation, and (4) all used methods have to be thoroughly described with enough details to enable reproduction or comparison of obtained results by other researchers.

2 Materials

There are many commercial image analysis programs, which can be successfully used for the following tasks. All examples in this chapter are described on the public domain program ImageJ [10]. The ImageJ program can be downloaded from http://imagej.nih.gov/ij/, which also provides a detailed installation manual, documentation, user manual, and many plug-in modules for various tasks. The program is based on Java and runs on Windows, Mac OS, Mac OS X, and Linux. Its source code is freely available and anyone can program further plug-ins to be added into the program (*see* **Notes 4** and **5**). The program window consists of three lines: Menu Commands ("Menu" in the subsequent text), Toolbar ("Tools" in the subsequent text), and Status and Progress Bar (Fig. 1).

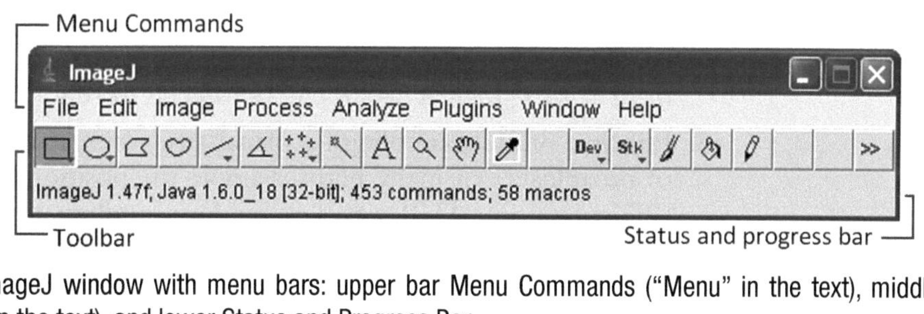

Fig. 1 ImageJ window with menu bars: upper bar Menu Commands ("Menu" in the text), middle Toolbar ("Tools" in the text), and lower Status and Progress Bar

3 Methods

3.1 Sampling

To get unbiased estimation of quantitative parameters of studied objects, it is necessary to sample the specimens in a SUR way, so each item has the same probability to be sampled. The more heterogeneous the structure is, the more segments, samples, and sampling windows are needed [11]. The aim of good experimental design is to quantify natural variability of data due to biological reasons (treatment, or biological process). To characterize natural variability of data, the other, artificial sources of the error have to be minimized—such as error due to wrong sampling and design experimental error (observer, devices, etc.) or sample processing (*see* **Note 1**). SUR sampling principles applied in design lead to unbiased estimations of followed structural parameters.

Answer the following questions and modify the design based on your answers:

1. What and where you have to sample? Are there known gradients of the parameter you are going to measure? Sample the same/comparable parts of plants to exclude natural variation in anatomical gradients.
2. Do you need the average value for the whole plant/organ/tissues? In that case you have to apply the SUR principle for sampling, e.g., segments of an organ to be further processed to get sections, to which any measurement procedure is applied. For example, for study on leaves you have to sample several segments in SUR way.
3. Are you interested just in comparison of some specific parts of the organ/plant? For example, for more extensive study on leaves you can compromise to comparison of well-defined part within the leaf blade, e.g., middle of the coniferous needle.
4. Principle of SUR sampling for elongated plant structures, e.g., coniferous needles: Sample the positions of segments along longitudinal organ axis. Determine the interval (distance T) between the segments, in which segments will be sampled along the organ. The value of T can be derived from the average length of the axis, divided by 5–10 (you want to sample between 5 and 10 segments). Get a random number from the set $\{0, 1, 2,\ldots, T\text{-}1\}$ from a table of random numbers or from some generator (e.g., http://www.random.org) to determine the position of first segment—z. Cut the initial segment at distance z, next segment at $z+T$, then at $z+2T$, etc. (for details *see* refs. [7, 8]).
5. Principle of SUR sampling for planar structures, such as blade of broadleaves of many herbs. Sample the positions in a flat (2D) organ in a SUR way. Get random numbers for coordinates x, y from a table of random numbers or from some generator (see above)—for the horizontal coordinate x from the set $\{0, 1, 2, \ldots, a-1\}$ and for vertical coordinate y from the set

{0, 1, 2, …, *b*–1}, where *a* is a distance in *x*-axis direction and *b* is a distance in *y*-axis direction to determine the position of the initial point of the superimposed point grid (square or rectangular), the points of which denote the same corner of a sampled segment (for details *see* refs. [5, 7–9]).

6. Apply the principle of SUR sampling design on all hierarchical levels of your design, i.e., segments, sections, sampling windows, or measured objects.
7. Calculate variability from the results obtained in the pilot study. The convenient measure of how precise the estimate is is called the *coefficient of error*, or *CE*. The coefficient of error is a statistical value used extensively in the stereological literature (defined as the standard deviation divided by mean of the sample). Coefficient of error depends on the number of samples and corresponds to proportional variability of the estimate; its value should be lower than 0.05. If *CE* is higher, modify the sampling design by adding more segments, sections, sampling windows, or measured objects.
8. In the majority of cases it is possible to apply just a few basic rules: (1) Five individuals per group are usually enough for the parameter estimation [12]; (2) when sampling sections along the longitudinal axis, cover the whole object systematically, in such a manner, that 5–10 segments or sections per organ are sampled; (3) when using sampling windows, 5–10 of them should be superimposed on each section; and (4) in most cases it is not necessary to count more than 200 points or intersections with stereological probe per organ in each compartment of interest [11].

3.2 Calibration: Manual Setting of Microscopic Image Calibration

Correct calibration is a must to measure absolute values of a measured parameter in real units. The calibration is specific for particular acquisition settings. Objective magnification, tubus factor, camera model (size and resolution of chip), and settings of pixel binning on camera are particularly important while acquiring the image with microscope. Calibrate the pixel size with stage micrometer or reticle of known dimensions as follows:

1. Acquire the image of stage micrometer or reticle with particular optical setup (objective and other components affecting magnification).
2. Open the image of stage micrometer acquired with the same objective (and other acquisition settings such as image size in pixels and zoom) as the image (or images) you would like to analyze. For the instructions to manual setting of known calibration; *see* **Note 6**.
3. Draw a line (Tools/Straight line) along the side of a line whose size is known. Consider the thickness of the lines of the stage micrometer—include it only once (Fig. 2).

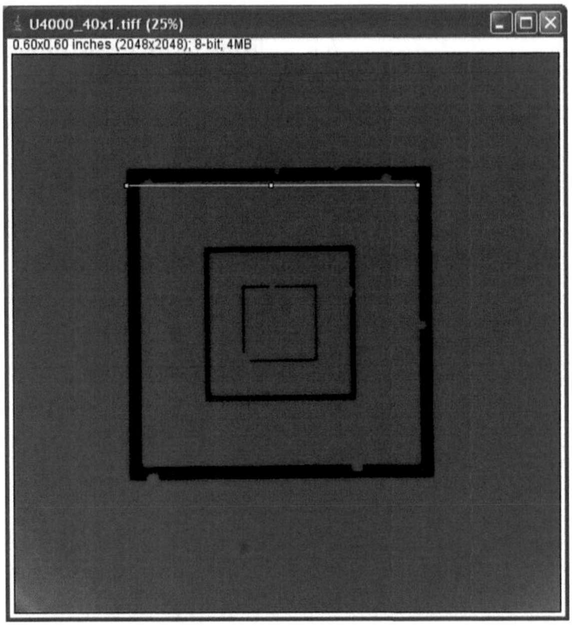

Fig. 2 Calibrating square of known length of a square side

4. Fill in the known length and units of length of the drawn line in the Set Scale window (Menu/Analyze/Set Scale; *see* **Note 2**). If you tick "Global," the calibration is saved and used for all subsequently opened images until the calibration is changed again.
5. Check the calibration by measuring the same line (Menu/Analyze/Measure).

3.3 Manual Length Measurement

1. Open your image and set right calibration (*see* Subheading 3.2 and **Note 6**).
2. For objects under study draw curves using the Tools, most often Straight line tool (Tools/Straight line), possibly other convenient tools (Tools/Oval or Rectangular), or trace the boundaries of the object (Tools/Polygon selection or Freehand selection) and measure their lengths. The length of the line appears in the dialog box.
3. To record your measurements, use the command Measure (Menu/Analyze/Measure) after every measurement. The results will appear in the Results window.
4. Save the acquired results (File/Save as in the Results window).

3.4 Measurement of Leaf Surface Area and Other Geometric Characteristics of the Leaf

1. Open your image of scanned leaves (or other objects, whose area you want to determine) and set right calibration (*see* **Note 6**).
2. Threshold the image (Menu/Image/Adjust/Threshold; Ctrl + Shift + T) to specify area of leaves within the image. Depending on the color of the background tick or uncheck "Dark background" in the dialog box. The threshold defines

brightness values by the ruler under the histogram. Automatic segmentation is offered. Several values can be sampled from the image and used for segmentation. At the end the leaves shall be masked with color, while the background shall be unchanged. When you have set the threshold, press "Select" and then "Sample" to get the binary image.

3. Specify which geometrical characteristics you intend to measure (Analyze/Set Measurements). For their explanation, see the manual parameter description. This applies particularly, for example, to the parameter of circularity, formula of which can vary in different image analysis software.
4. To exclude thresholded tiny objects in the background, set a minimal area of a measured object (Analyze/Set Measurements).
5. Measure the objects by command Analyze Particles (Menu/Analyze/Analyze Particles).
6. Save the acquired results (File/Save as in the Results window).

3.5 Estimation of Stomatal Density

1. Install the Sampling_Window plug-in to ImageJ: Download the file "Sampling_Window.class" from http://rsbweb.nih.gov/ij/plugins/sampling-window/index.html and copy it to the folder (Program files\...) ImageJ\plugins. Restart ImageJ.
2. Open your image of an epidermal peel, which was sampled in a uniform random way, and set right calibration.
3. Superimpose the counting frame on the image (Menu/Plugins/Sampling_Window). Set the size, color, line width, and position of the frame in the dialog box. Note the size of the frame and calculate its area.
4. Count the stomata (*see* **Note 7**) which are selected by the counting frame according to the following rules: stomata fully inside, stomata laying partly inside, and intersecting the dashed line of the sampling frame. Do not count stomata, which lay partly inside and simultaneously intersect full exclusion lines of the frame (Fig. 3c). You can use the Multipoint selection tool (Tools/Multipoint selection tool) to mark the counted stomata by a point, which are then numbered. You can undo the selection by Alt + left click.
5. Relate the recorded number of stomata to the area of the counting frame and get the estimation of stomatal density.

3.6 Analysis of Epidermal Cell Shape

1. Install the Sampling_Window plug-in to ImageJ (*see* Subheading 3.5, **step 1**).
2. Open your image of an epidermal peel, which was sampled in a uniform random way and acquired in such way to get a high contrast of cell walls (e.g., using cell wall polyphenolic fluorescence or stained cell walls with toluidine blue). Set right calibration.

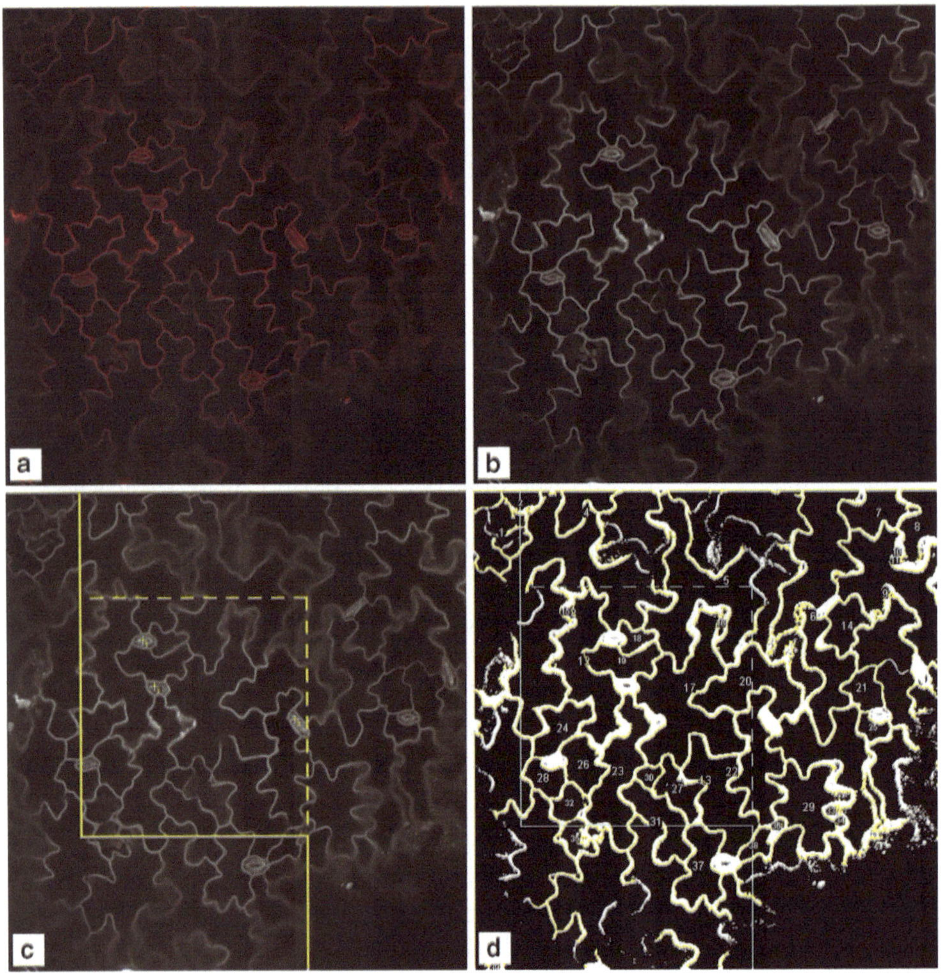

Fig. 3 Analysis of epidermis—estimation of stomatal density and analysis of epidermal cell shape: (**a**) Original image of epidermis acquired by fluorescence microscope. (**b**) Red channel separated from the original image (Menu/Image/Color/Split Channels). (**c**) Sampling window superimposed on the image. Three stomata are manually selected, because they are lying at least partly in the frame and are not intersected by the (full-drawn) exclusion line. (**d**) Sampling window superimposed on the adjusted image. Cells number 5, 12, 16–20, 22, 24, 26–28, 30 and 32 are selected, because they do not intersect the frame rectangle (or are inside the rectangle) and simultaneously do not intersect the exclusion lines. Results for each cell appear in the Results table

3. If your image is a multichannel image, split the channels (Image/Color/Split Channels) and for further analysis choose the most contrasting channel.
4. Smooth the image (Process/Smooth).
5. Enhance the contrast (Process/Enhance contrast).
6. Set the threshold (Image/Adjust/Threshold).
7. Dilate the image (Menu/Process/Binary/Dilate).
8. If needed, edit the image manually with the pencil or paint brush tool (More Tools/Drawing Tools).

9. Superimpose the sampling window on the image (Menu/Plugins/Sampling_Window) at random position. Set the size, color, line width and position of the frame in the dialog box.
10. Measurement: Set the parameters for analysis in Menu/Analyze/Set measurements (e.g., Area, Shape descriptors, Perimeter, Feret's diameter). Specify the minimal size of analyzed particles in Menu/Analyze/Analyze Particles (e.g., 5-Infinity), tick Display results, Clear results, and Add to manager.
11. Save results for the cells selected by the counting frame—you can browse from cell to cell in the ROI manager window to see which cell corresponds to which number. Other possibility is to measure the cells selected by the counting frame by the Wand (tracing) tool and command Ctrl + M (or Analyze/Measure). The selected cells are those which intersect the frame rectangle (or are inside the rectangle) and simultaneously do not intersect the exclusion lines (Fig. 3d).
12. Calculate the shape complexity of the object under study (ratio of its perimeter to square root of its area: $b/\sqrt{a}$, where b is perimeter and a is area of the object, [11]. $b/\sqrt{a}$ is minimal for circle: $2\pi.\ \sqrt{\pi} = 3.54$, for highly structured objects, such as endoplasmic reticulum, it can exceed the value of 30.

4 Notes

1. The real object size could change in consequence of specimen processing. Therefore possible size artifacts, e.g., in structure shrinking, collapsing, and distortion, should be identified to get proper information about real dimensions (*see,* e.g., ref. [5]).
2. The pixels are commonly square shaped but some chips have rectangular sensor elements. In case of non-square rectangular pixels, the pixel aspect ratio (width/height).
3. Some parameters, such as circularity, can be defined in different ways—thus it is recommended to check the formula in the program manual.
4. Sequence of commands can be recorded as macro, saved to text file, and repeatedly run. This can speed up routine analyses of many images substantially. Saving the image analysis steps in macro is also useful for documentation purposes. This technique is accessible also to non-experts as a sequence can be in most programs "recorded" during work.
5. The new plug-ins can be written in JAVA using ImageJ application interface. Extensive documentation as well as a lot of source files of plug-ins published on ImageJ website can be used for study of programming the image analysis modules. Some programming experience and knowledge of JAVA and ImageJ API is necessary.

6. Open the image you would like to analyze. Fill in the known information in the Set Scale window (Menu/Analyze/Set Scale). If the known scanner resolution is in pixels per inch and you need your results in mm, consider the unit transfer (1 in. = 25.4 mm). The ImageJ works with decimal point, not with decimal comma. If you tick "Global", the calibration is saved and used for all subsequently opened images until the calibration is changed again. You can also fill the calibration in the Image Properties (Menu/Image/Properties). Check the calibration by measuring a line length (Menu/Analyze/Measure).
7. The same method can be used for estimation of density of any object in 2D, such as epidermal cells. The particles must lay in one plane, otherwise it is not possible to estimate their density properly from 2D image. Detailed methodology of estimation of the number and sizes of stomata is described in the literature [8].

Acknowledgments

The authors wish to acknowledge the funding by the project GACR P501/10/0340, by the Charles University in Prague, SVV 265203 and institutional support RVO:67985823.

References

1. Sterio DC (1984) The unbiased estimation of number and sizes of arbitrary particles using the disector. J Microsc 134:127–136
2. Howard CV, Reed MG (1998) Unbiased stereology. BIOS, Oxford
3. Weibel ER (1979) Stereological methods—vol 1: practical methods for biological morphometry. Academic, London
4. Pazourek J (1966) Anatomical gradients. Acta Univ Carol Biol Suppl. 1/2:19
5. Lhotáková Z, Albrechtová J, Janáček J et al (2008) Advantages and pitfalls of using freehand sections of frozen needles for three-dimensional analysis of mesophyll by stereology and confocal microscopy. J Microsc 232: 56–63
6. Albrechtová J, Kubínová L (1991) Quantitative analysis of the structure of etiolated barley leaf using stereological methods. J Exp Bot 42: 1311–1314
7. Kubínová L (1993) Recent stereological methods for the measurement of leaf anatomical characteristics: estimation of volume density, volume and surface area. J Exp Bot 44: 165–173
8. Kubínová L (1994) Recent stereological methods for measuring leaf anatomical characteristics: estimation of the number and sizes of stomata and mesophyll cells. J Exp Bot 45: 119–127
9. Albrechtová J, Janáček J, Lhotáková Z et al (2007) Novel efficient methods for measuring mesophyll anatomical characteristics from fresh thick sections using stereology and confocal microscopy: application on acid rain-treated Norway spruce needles. J Exp Bot 58:1451–1461
10. Schneider CA, Rasband WS, Eliceiri KW (2012) NIH Image to ImageJ: 25 years of image analysis. Nat Methods 9:671–675
11. Gundersen HJG, Jensen EB (1987) The efficiency of systematic sampling in stereology and its prediction. J Microsc 147:229–263
12. Cruz-Orive LM, Weibel ER (1990) Recent stereological methods for cell biology: a brief survey. Am J Physiol 258:L148–L156

Chapter 6

Identifying Subcellular Protein Localization with Fluorescent Protein Fusions After Transient Expression in Onion Epidermal Cells

Andreas Nebenführ

Abstract

Most biochemical functions of plant cells are carried out by proteins which act at very specific places within these cells, for example, within different organelles. Identifying the subcellular localization of proteins is therefore a useful tool to narrow down the possible functions that a novel or unknown protein may carry out. The discovery of genetically encoded fluorescent markers has made it possible to tag specific proteins and visualize them in vivo under a variety of conditions. This chapter describes a simple method to use transient expression of such fluorescently tagged proteins in onion epidermal cells to determine their subcellular localization relative to known markers.

Key words Protein localization, Transient expression, Particle bombardment, Fluorescence microscopy, Onion epidermis, Organelle markers

1 Introduction

Plant cells carry out a wide variety of functions, ranging from photosynthesis and basic metabolism over secretion and cytoplasmic streaming to environmental and pathogen responses. These functions depend on the proper distribution and interaction of a large number of proteins within the cells. In recent years, it has become evident that the dynamics of these protein distributions and interactions are essential for their function which makes it imperative to develop methods to identify these dynamic events in living cells.

Detection of subcellular localization and dynamics of proteins is usually achieved by creating genetically encoded fluorescent derivatives of the proteins of interest, for example, by fusing them with green fluorescent protein (GFP, ref. [1]). The discovery of red fluorescent proteins [2] combined with the targeted modification of fluorescent protein (FP) genes to create brighter varieties or different colors (e.g., refs. [3, 4]) allows for the direct comparison of two or more proteins within the same cell, thus greatly facilitating

Viktor Žárský and Fatima Cvrčková (eds.), *Plant Cell Morphogenesis: Methods and Protocols*, Methods in Molecular Biology, vol. 1080, DOI 10.1007/978-1-62703-643-6_6, © Springer Science+Business Media New York 2014

localization and protein–protein interaction studies. In fact, a number of collections of organelle marker constructs are available from stock centers (e.g., refs. [5, 6]) that make identification of subcellular localization studies relatively straightforward.

Stable transformation of plants with genes encoding fluorescently tagged proteins, preferably under control of their native promoters, is clearly desirable since this approach will allow for observation of long-term effects and, ideally, complementation of mutant phenotypes. This approach, however, requires a substantial investment of time. Transient expression approaches, on the other hand, can already yield important insights into protein localization and function and can be achieved without much technical effort by *Agrobacterium* infiltration [7–9]. In this chapter, however, we describe the use of transient expression by means of particle bombardment. As long as the necessary equipment is available, particle bombardment is usually faster than Agrobacterium-mediated approaches since it does not require integration of the gene construct into a binary plasmid and its mobilization into an appropriate *Agrobacterium* strain. The second part of the protocol describes basic epifluorescence techniques to visualize fluorescent proteins in living plant tissues which apply to all transformation techniques. The methods presented here can also be modified to accommodate more complex microscopy techniques such as laser scanning or spinning disk confocal microscopy.

2 Materials

2.1 Particle Bombardment

This protocol is based on the PDS1000/He system (BioRad):

1. For onion tissues, M17 tungsten particles (1.0 μm, BioRad) work best.
2. A supply of macrocarriers, stopping screens, and rupture disks.
3. Agar plates with standard growth medium (for example, 1/2× Murashige–Skoog medium, 1 % sucrose, pH 6.0); one per sample.
4. 2.5 M $MgCl_2$.
5. 200 mM spermidine (store in −20 °C freezer).
6. 70 % and 100 % ethanol.
7. Purified plasmids encoding the expression constructs (approx. 100 ng/μl). Miniprep DNA is usually sufficient.
8. Fresh onion.
9. Fine forceps.
10. Razor blades or scalpel.

Table 1
Filters recommended for visualizing common fluorescent proteins

Fluorescent protein	Excitation	Dichroic	Emission
CFP	BP 436/25[a]	455	BP 480/40
GFP	BP 470/40	495	BP 525/50
YFP	BP 500/25	515	BP 535/30
RFP (mCherry)	BP 572/25	590	BP 629/62
Triple cube (CFP + YFP + mCherry)	BP 430/24 BP 500/20 BP 577/25	Multiple transmission windows	TBP 470/24 TBP 537/30 TBP 635/65

[a]BP 436/25 = bandpass filter centered around 436 nm with a transmission window width of 25 nm at half-maximal height

2.2 Microscopy

1. Microscope slides and cover slips.
2. Overview objective (10× or 20×), high-magnification objective (63×/1.4 NA, oil immersion).
3. Appropriate filters for fluorescence (*see* Table 1). As an alternative to the individual filter cubes for specific fluorescent proteins, a "triple cube" (for example, Chroma set no. 69308) with separate excitation filters can be used. In this setup, specific fluorescent proteins can be excited by mounting the excitation filters in a separate filter wheel (e.g., Lambda 10-2, Sutter Instruments) or a wavelength switcher (DG-4, Sutter Instruments) while the filter cube with the dichroic and emissions filters does not have to be changed. This setup allows for faster image capture but increases the risk of bleed-through between channels (*see* Subheading 3.7). A conventional "triple cube" that does not allow separate excitation of the fluorophores combined with a color camera is not advisable as it is virtually impossible to separate the different signals after capture for quantitative image analysis.

3 Methods

3.1 Preparation of Particles

1. Weigh out 30 mg of M17 tungsten particles (1.0 μm; BioRad) in a microcentrifuge tube (*see* **Note 1**) and add 500 μl 70 % ethanol (freshly prepared).
2. Vortex at half-maximal speed for at least 10 min to suspend particles. Pellet particles in microcentrifuge for less than 5 s (*see* **Note 2**).

3. Remove the supernatant with pipette and wash three times with 500 μl sterile water. At every wash step, vortex the particles for about 1 min, let them settle for 1 min, and pellet them in a microfuge for less than 2 s (*see* **Note 2**).
4. Resuspend particles in 500 μl sterile 50 % glycerol by vortexing. Particles are stable at room temperature for at least 1 month.

3.2 Preparation of Onion Tissue

1. Cut a fresh onion (*see* **Note 3**) into quarters. Remove the innermost leaves since they are usually too highly curved.
2. Gently cut the epidermis on the adaxial (concave) surface into small strips of approximately 0.5 mm × 2 mm (*see* **Note 4**).
3. Using fine forceps, peel off the epidermal strips and lay them, with the outer surface down, in the center of an agar plate. Collect enough epidermal strips to cover an area of about 3 cm in diameter. Prepare one petri dish per sample.

3.3 Preparation of Macrocarriers

1. Combine expression plasmids in microfuge tube with a final volume of 10 μl (*see* **Note 5**). Mix well.
2. Add 25 μl M17 particles from Subheading 3.1 (thoroughly suspended by vortexing), 25 μl 2.5 M $MgCl_2$, 5 μl 200 mM spermidine.
3. Vortex mixture for 15 min at half-maximal speed. Let particles settle for 1 min before pelleting them for less than 2 s in a microcentrifuge (*see* **Note 2**).
4. Remove the supernatant without disturbing the pellet. Add 100 μl of 70 % ethanol (freshly prepared), wait 30 s, and remove supernatant (*see* **Note 6**).
5. Repeat wash steps three times with 100 % ethanol to remove residual water.
6. Resuspend particles in 25 μl 100 % ethanol. Vortex particles for at least 1 min. Set pipettor to 15 μl and pipette particles up and down for a few times to break up larger clumps. Vortex particles again for 1 min.
7. Set down two macrocarrier disks (BioRad) on filter paper in an empty petri dish (*see* **Note** 7). Place 8 μl of finely suspended particles into the center of each macrocarrier disk and put entire petri dish in 37 °C incubator for 5 min to evaporate the ethanol (*see* **Note 8**).

3.4 Bombardment

1. Turn on particle gun and vacuum pump.
2. Load rupture disk (1,100 psi; BioRad) and securely tighten holder (*see* **Note 9**). Put stopping screen in macrocarrier assembly. Place the macrocarrier with the particles facing down in the holder and place on top of macrocarrier assembly. Slide macrocarrier assembly on top shelf.

3. Put open agar plate with onion peels on second shelf from the bottom.
4. Close particle gun and pull vacuum to about 27 mmHg. Press "Fire" switch until rupture disk breaks. Vent the chamber and remove the agar plate.
5. Repeat the procedure (**steps 2–4**) with the same agar plate for the second macrocarrier disk of this sample.
6. After the second shot with the same sample, wrap agar plate with parafilm and store in a dark place at room temperature until the next day (*see* **Note 10**).
7. Repeat as needed for all samples.

3.5 Mounting Epidermal Peels for Microscopy

1. Place clean cover slip on a paper towel and add small drops of water all over its surface.
2. With fine forceps, pick up the epidermal peel and place it on the cover slip in same orientation as on the agar (in other words, the outer surface that was down on the agar plate should also be down on the cover slip). Do this with a slight rolling movement (similar to rolling out a rug) to prevent the formation of air bubbles between the cover slip and the tissue. Put as many epidermal strips on the cover slip as possible.
3. Put a few drops of water on the back of the epidermal peels. Gently lay a microscope slide on the cover slip and pick it up immediately; the cover slip will adhere to the slide.

3.6 Microscopy: Identifying Transformed Cells

1. Orient yourself with a low-magnification objective (10× or 20×) under bright-field illumination and focus on the epidermal cell layer (*see* **Note 11**). Move sample so that the lower left corner of the tissue piece is in view.
2. Switch over to fluorescent illumination (*see* **Note 12**) and scan over the tissue piece to identify transformed cells (*see* **Note 13**). On an inverted microscope, the position of transformed cells can be marked on the upper side of the slide with a small dot from a felt-tip marker to make it easier to return to them later. On upright microscopes, the coordinates can be noted down from the stage markings (*see* **Note 14**).

3.7 Microscopy: High-Magnification Imaging

1. Switch to a high-magnification objective suitable for observation of subcellular structures (e.g., 63×/1.4 oil immersion). Focus on the marker signal. The highest quality images can be obtained in the cytoplasm right behind the outer plasma membrane (*see* **Note 15**). This area also has the advantage that many organelles can be observed at the same time.
2. Switch the fluorescent filters to observe the signal from the unknown protein. Ideally, the signal of the marker and the unknown should be of roughly equal intensity to avoid

bleed-through (*see* **Note 16**). As a general rule, the exposure setting should be adjusted such that the signal brightness becomes maximal without saturating any pixels (*see* **Note 17**). Since most organelles show rapid movements, it may be necessary to limit the exposure time to prevent streaking of the organelles (*see* **Note 18**). The two images for the unknown protein and the organelle marker have to be taken in close succession to minimize movement of the organelles between images (*see* **Note 19**).

3. To remove background noise, a second set of images should be taken with the same exposure settings, but the excitation light shutter closed (*see* **Note 20**).

3.8 Image Analysis

After background subtraction (*see* **Note 20**) the images can be analyzed. To detect colocalization, the two images can be false-colored and superimposed to reveal overlapping signals. The best colors to use are red and green since their overlap will result in a different color, yellow, that can easily be detected (*see* **Note 21**). This can be achieved by creating a new RGB image in ImageJ, converting this into an RGB stack (with "Image > Type") and pasting the two images into the first and second frame of the stack. Converting this image back to an RGB image will complete the procedure.

4 Notes

1. The tungsten particles can be hazardous because of their small size. Wear gloves and possibly also a respirator.
2. Keep the centrifugal force to a minimum to prevent clumping of the particles since this makes it difficult to resuspend them later.
3. Commercial onions usually work well. Avoid previously frozen onions and do not keep cut onions more than a few days.
4. The smaller size of the strips makes it easier to spread the curved epidermal peels on the flat agar surface.
5. Anywhere from 100 ng to 1 μg of DNA per plasmid is usually sufficient for clearly detectable expression of fusion proteins. The precise amount depends on the size of the plasmid, the promoter, and nature of the fusion protein. Organelle markers tend to express very well and require little DNA (50–100 ng).
6. There is no need to resuspend the particles by vortexing. Several tubes can be prepared in parallel.
7. There should be enough particles for three disks, but two shots per sample normally yield sufficient numbers of cells to determine protein localization. The second shot can be omitted too, but is handy if something goes wrong with the first.

8. The particles should be finely dispersed on the macrocarrier. Clumps will lead to cell death and lower transformation rates. Presence of small amounts of water in the resuspended particles will lead to clumping during drying. It is not necessary to put the macrocarriers in a 37 °C incubator when the relative humidity in the lab is low.
9. If the rupture disk holder is not securely tightened, the disk will slip out at low pressure. This may still result in transformation, but the yield of expressing cells may be lower.
10. Optimal incubation time depends on the proteins that need to be expressed. First signals can usually be observed after a few hours. Expression can be stable for several days, but fungal growth often appears on the second day of incubations. A typical incubation is between 16 and 24 h, in other words from the afternoon of the first day to the morning or afternoon of the next.
11. The epidermal cells can be identified by their well-defined outlines. Further into the sample are the mesophyll cells which were typically broken during the peeling process. It is important that the epidermis is closest to the cover slip to ensure best image quality.
12. It is usually best to perform the initial scan with illumination for an organelle marker since they tend to express very well and result in bright fluorescent images. For example, a CFP marker is easy to detect while the weak autofluorescence of the cell walls will allow for orientation during the scanning process. Scanning for RFP is usually more difficult since the tissue emits very little signal at this wavelength and makes orientation difficult. Autofluorescence of the tissue, e.g., resulting from damage during preparation, can sometimes be confused with actual signal from the FPs. However, most autofluorescence appears with several fluorescence filters, whereas FP fluorescence can be detected only by the appropriate filter set.
13. It is best to follow a regular path to ensure that all cells are examined. For example, move the stage up in a straight line to observe the left edge of the sample, shift over to the right by approximately one field of view, and move down until the lower edge of the tissue is reached.
14. In most cases, distribution of transformed cells is not uniform but occurs in patches. It is often possible to identify patches with 10–20 or more cells in a small area. These patches are very convenient since it is easy to move to neighboring cells even with a high-magnification objective.
15. Depending on the intracellular distribution of the fluorescent signal, it may be necessary to focus further into the cell. For example, the nucleus of these epidermal cells is usually found attached to the back wall, about 70 μm into the cell. Due to

diffraction of light by the anticlinal walls (parallel to the incident light), however, the image quality in this part of the cell is not as high as closer to the cover slip.

16. FPs have both broad absorption and emission peaks. As a result, we often observe bleed-through of one fluorophore into the filter set of another fluorophore. For example, GFP will be visible with both CFP and YFP filters. Similarly, it is possible to excite RFP with GFP or YFP filters. This problem becomes more noticeable with multiwavelength filters, but it can also occur with dedicated filter sets when one fluorophore is very bright and the other is very dim. In this case, the long exposure times necessary to detect the weak second signal may make it possible to pick up the small amount of bleed-through of the brighter fluorophore. To test whether this problem may occur, perform a transient expression with a single marker only and collect an image with optimal settings for this fluorophore. Then switch to another filter set and take an image with identical settings. Repeat taking images with increasing exposure times (twice as long, four times as long, etc.) until the signal becomes visible with this "wrong" filter. Depending on the filters and the fluorophores, a two- to fourfold difference in exposure time may result in negligible bleed-through, but this has to be determined for every microscope setup independently. Also note that contrast enhancement may bring out weak signals that may go unnoticed at first. Once these "bleed-through limits" have been established, it is possible to detect this problem simply by comparing exposure settings for marker and unknown. Should this be a problem, it may be possible to reduce the amount of DNA for the marker construct to reduce its signal.

17. The maximal signal intensities per image pixel should be about 10 % below the maximum that can be handled by the camera. This would be an intensity of about 3,700 for 12-bit cameras, 15,000 for 14-bit cameras, and 59,000 for 16-bit cameras. Bright signals ensure highest signal-to-noise ratios and hence best image quality.

18. Organelles can often move with speeds exceeding 2 μm/s and reaching as high as 8 or 10 μm/s. In this case, it is necessary to reduce exposure times to 250 ms or shorter. This may require increasing the electronic gain setting of the camera, even though this tends to increase the noise more than longer exposure times.

19. If possible, computer automation should be used to capture the two images in close succession. Typically, the exposure settings are first determined and stored in the computer which then controls the microscope to expose the camera with the appropriate filter sets. With this approach, the delay between the images is only limited by the exposure times and the speed with which the filter sets can be switched. For this reason, a

multiwavelength filter cube combined with an external excitation filter switcher (*see* Subheading 2.2) is of advantage. In this case, the chance of the bleed-through between channels is increased (*see* **Note 16**).

20. All camera images bring a certain level of background noise with them, which is evident in pixel values significantly above zero. For quantitative image analysis, it is necessary to remove this background in order to obtain accurate measurements of fluorescent signal intensity. The simplest way to achieve this is to subtract the pixel intensities of the background images from the experimental images. Most microscope software programs have this command built-in. The same effect can be achieved with the "Process > Image Calculator..." command in ImageJ.
21. Journals often recommend using magenta and green to accommodate red–green colorblind people. In this case, the red channel is duplicated in the blue channel which is visible to colorblind people. While this often achieves the same effect as red–green images, it tends to be more difficult to identify weakly colocalizing signals. Under no circumstances should the "true" colors be used, since the overlap, for example, of cyan (for CFP) and yellowish-green (for YFP) does not lead to a distinct signal and therefore cannot be clearly distinguished from the individual channels.

Acknowledgments

I thank the members of my lab for numerous improvements of the procedure. Work in my lab is supported by the National Science Foundation (MCB-0822111).

References

1. Chalfie M, Tu Y, Euskirchen G et al (1994) Green fluorescent protein as a marker for gene expression. Science 263:802–805
2. Matz MV, Fradkov AF, Labas YA et al (1999) Fluorescent proteins from nonbioluminescent Anthozoa species. Nat Biotechnol 17:969–973
3. Rizzo MA, Springer GH, Granada B et al (2004) An improved cyan fluorescent protein variant useful for FRET. Nat Biotechnol 22: 445–449
4. Shaner NC, Campbell RE, Steinbach PA et al (2004) Improved monomeric red, orange and yellow fluorescent proteins derived from *Discosoma* sp. red fluorescent protein. Nat Biotechnol 22:1567–1572
5. Geldner N, Dénervaud-Tendon V, Hyman DL et al (2009) Rapid, combinatorial analysis of membrane compartments in intact plants with a multicolor marker set. Plant J 59:169–178
6. Nelson BK, Cai X, Nebenführ A (2007) A multicolor set of in vivo organelle markers for colocalization studies in *Arabidopsis* and other plants. Plant J 51:1126–1136
7. Batoko H, Zheng H-Q, Hawes C et al (2000) A Rab1 GTPase is required for transport between the endoplasmic reticulum and Golgi apparatus and for normal Golgi movement in plants. Plant Cell 12:2201–2218
8. Li J-F, Park E, von Arnim AG et al (2009) The FAST technique: a simplified Agrobacterium-based transformation method for transient gene expression analysis in seedlings of Arabidopsis and other plant species. Plant Methods 5:6
9. Marion J, Bach L, Bellec Y et al (2008) Systematic analysis of protein subcellular localization and interaction using high-throughput transient transformation of Arabidopsis seedlings. Plant J 56:169–179

Chapter 7

Visualizing and Quantifying the In Vivo Structure and Dynamics of the *Arabidopsis* Cortical Cytoskeleton Using CLSM and VAEM

Amparo Rosero, Viktor Žárský, and Fatima Cvrčková

Abstract

The cortical microtubules, and to some extent also the actin meshwork, play a central role in the shaping of plant cells. Transgenic plants expressing fluorescent protein markers specifically tagging the two main cytoskeletal systems are available, allowing noninvasive in vivo studies. Advanced microscopy techniques, in particular confocal laser scanning microscopy (CLSM) and variable angle epifluorescence microscopy (VAEM), can be nowadays used for imaging the cortical cytoskeleton of living cells with unprecedented spatial and temporal resolution. With the aid of suitable computing techniques, quantitative information can be extracted from microscopic images and video sequences, providing insight into both architecture and dynamics of the cortical cytoskeleton.

Key words Actin, Microtubules, Fluorescent proteins, CLSM, VAEM, Image analysis

1 Introduction

Cortical microtubules are long known to play a major part in the morphogenesis of plant cells, in particular due to their intimate relationship with the biosynthesis of the cellulosic cell wall microfibrils (*see* ref. [1]). However, the actin cytoskeleton, which undergoes constant dynamic remodeling [2], is crucial for processes such as trichome morphogenesis [3], tip growth in root hairs [4], or development of epidermal cell lobes [5] and apparently contributes to the localization of exocytosis, affecting also the positioning of cellulose synthase complexes [1, 6]. Detailed characterization of the spatial structure and temporal behavior of the two main cytoskeletal systems in vivo may thus substantially contribute to our understanding of plant cell shaping.

Such studies depend on several prerequisites. Sufficiently specific and nondisruptive fluorescent cytoskeletal markers must be introduced into the tissues of interest, and suitable high-resolution imaging technology must be available, even if the aim of the study

Viktor Žárský and Fatima Cvrčková (eds.), *Plant Cell Morphogenesis: Methods and Protocols*, Methods in Molecular Biology, vol. 1080, DOI 10.1007/978-1-62703-643-6_7, © Springer Science+Business Media New York 2014

was a "mere" morphological characterization. In addition, if quantitative information is to be extracted from image data, appropriate protocols and software are needed.

A variety of fluorescent protein-based markers has been used to trace cytoskeletal structures in living cells, including those of plants. Plant microtubules have been successfully visualized using both GFP-tubulin fusions [7, 8] and GFP-tagged ortho- or heterologous microtubule-associated proteins (MAPs) such as mammalian MAP4 [9] or several isoforms of Arabidopsis MAP65 [10]. In addition to labeling microtubules along their whole length, microtubule ends can be specifically marked by tags based on end-binding proteins such as EB1 preferring minus ends [11] or mammalian CLIP170 for plus ends [12]. For actin visualization, fluorescent protein-tagged mammalian talin [13] or constructs based on the C-terminal actin-binding domain of Arabidopsis fimbrin (FABD, refs. [14, 15]) can be used. A very promising actin marker is the 17 amino acid actin-binding peptide known as LifeAct, which has been successfully used to target fluorescent proteins to actin filaments also in plant cells [16].

It has to be stressed that any experiments including (over)expression of tagged (i.e., modified) and possibly heterologous proteins have to be interpreted with caution, as (1) only a subset of the relevant cytoskeletal structures may be labeled, as shown, e.g., for the various MAP65 isoforms [10], and (2) the tag itself may affect cytoskeletal structure and dynamics. Both talin and MAP4-based markers cause visible phenotypic alteration on the whole plant level [17], and in particular GFP-tagged talin was shown to interfere with actin dynamics and aggravate the effects of some treatments and mutations affecting the actin cytoskeleton [18, 19].

A suitable high-resolution fluorescence microscopy and microphotography equipment is required to make full advantage of in vivo cytoskeletal labeling. Conventional fluorescence microscopy, although useful, is limited by spatial resolution, interfering background (auto)fluorescence, and usually also by long exposure times. However, advanced microscopy techniques, such as confocal laser scanning microscopy (CLSM), can be used to improve spatial resolution. Very thin samples can be observed with supreme spatial and temporal resolution using the total internal reflection microscopy (TIRFM) technique; however, TIRFM can only reach up to some 200 nm from the cover slip. Nevertheless, TIRFM hardware can also be used in variable angle epifluorescence microscopy (VAEM) mode with a reasonable trade-off between lateral resolution and imaging depth, allowing thus visualization of a thin cortical layer of the cytoplasm through the cell wall [20–25]. However, it is possible that in plant cells, the evanescent wave might be initiated between the cell wall and plasmalemma, the cell wall thus being a part of the optical system, and even true TIRFM may thus work [22].

A variety of computational techniques can be used to analyze high-resolution images of the plant cortical cytoskeleton and quantify their biologically relevant parameters. With a bit of exaggeration, there may be as many, or even more, image analysis methods as there are publications devoted to the topic, which often hampers comparison of data from different laboratories. Here we are presenting the protocols currently used in our laboratory [25], but based to a large extent on previous work published by others [26–29], with the hope to contribute to the standardization of basic approaches. Some of the quantification methods presented here can be used also for evaluation of images obtained from fixed material, e.g., after antibody staining.

2 Materials and Equipment

Besides specialized equipment and materials listed below, standard equipment, tools, and consumables for plant in vitro culture will be required.

2.1 Plants

The fluorescent markers listed above are likely to be available upon request from the authors who published them (*see* also Chapter 6), either in the form of a plasmid suitable for transformation (which may be useful for introducing the marker into mutants) or in the form of seeds of stable transgenic lines. Transgenic *A. thaliana* lines carrying GFP-tagged tubulin markers, GFP-TUB6 and GFP-TUA6, can be obtained also from the public Arabidopsis stock collections—NASC (http://arabidopsis.info) and ABRC (http://www.arabidopsis.org)—under stock codes N6550 and N6551 (NASC) or CS6550 and CS6551 (ABRC), respectively. Stable transgenic plants carrying the marker of interest can be then used to introduce the markers into different genetic backgrounds (e.g., various mutants) by crossing.

While, in principle, any fluorescent cytoskeletal marker can be used for in vivo imaging, our experience is based mainly on observations in plant lines carrying two marker constructs expressed under the viral 35S promoter—GFP-MAP4 [9] and GFP-FABD [30].

We usually observe roots and cotyledons of young seedlings (5–8 days after germination) grown on vertical MS plates at 22 °C with a 16 h-light/8 h-dark cycle (*see* **Note 1**). Pharmacological treatments may be included during cultivation, and seedlings may be alternatively grown in the dark to achieve etiolation, as etiolated hypocotyls provide another interesting model especially for TIRF observation ([20]; *see* **Note 2**). Appropriate controls (e.g., wild type for mutants, or non-treated plants for pharmacological studies) have to be included at the same time, since all measurements can be interpreted only in comparison with data from simultaneously grown control plants (*see* **Note 3**).

2.2 Microscopy and Image Processing

For both CLSM and VAEM, we provide information on instrument configuration we are using, as well as basic settings (in Subheading 3.2) as a guide, albeit modifications and some experimenting will be necessary with different hardware (*see* **Note 4**):

1. CLSM: Leica TCS SP2 confocal laser scanning microscope equipped with a 63×/1.2 water-immersion objective and 488-nm argon laser for excitation.
2. VAEM: Leica AF6000 LX fluorescence platform with integrated TIRF module, HCX PL APO 100×/1.46 oil immersion objective, equipped with the Leica DFC350FXR2 digital camera for recording.
3. Microscopy slides, cover slips (preferentially larger size to accommodate the whole length of a stretched seedling), chambered slides (Nunc Lab-Tek II, 1 well, catalogue number 154453), sterile water, immersion oil, tweezers, sterile toothpick, paper tissues, and nail polish (optional).
4. Personal computer with the Windows operation system (XP or higher) with the microscope manufacturer's image processing software installed (LCS Lite for CLSM, Leica Application Suite AF Lite for VAEM).
5. On the same or another computer (*see* **Note 5**), ImageJ ([31]; http://rsbweb.nih.gov/ij/) or its distribution Fiji (http://fiji.sc) should be installed, with the following plug-ins: the MBF plug-in collection to open file formats provided by the microscope from McMaster Biophotonics Facility ([32]; http://rsbweb.nih.gov/ij/plugins/mbf-collection.html), the KashiwaBioImaging plug-in collection (KBI ImageJ Plugins and the Scala runtime library, available from http://hasezawa.ib.k.u-tokyo.ac.jp/zp/Kbi/ImageJKbiPlugins), the Higaki Laboratory macros hig_skewness.txt and hig_255counts.txt (from http://hasezawa.ib.k.u-tokyo.ac.jp/zp/Kbi/HigStomata), and Multiple Kymograph from European Molecular Biology Laboratory (http://www.embl.de/eamnet/html/kymograph.html; not required if using Fiji as it is already contained within the package). A table calculator (such as Microsoft Excel or Libre Office Calc) will be also required.

3 Methods

3.1 Preparing Plant Materials for Visualization

1. For CLSM, place a seedling (collected off the agar plate using sterile toothpick) into a drop of water or cultivation medium on a microscope slide and cover with a cover slip, avoiding bubbles as far as possible. Remove excess water at the slide edges with a torn bit of paper tissue (*see* **Notes 6** and 7).

2. For VAEM, cut a piece of agar containing the seedlings (width 1.5 cm, length according to the seedling size). We usually observe two or three 5 days old plants per agar piece. Put a drop of water on the chambered slide and put the piece of agar placing the seedlings in contact with the glass, avoiding bubbles (the piece of agar helps to press the seedling tissues in contact with the slide). Remove excess water by gently touching the edges of agar with paper tissue.

3.2 Image Acquisition

Follow the recommended standard procedures for your microscope.

1. In case of CLSM, we record single slices (capturing an area of the cell cortex adjacent to the cover slip) and Z series using the following settings: excitation laser (488-nm argon) intensity 25 mW, detector window using the GFP preset values (*see* **Note 8**), XY field size 1,024 × 1,024 pixels, line averaging of 4–8 times, Z series interval 0.7–1 μm, and color depth of 12 bits.
2. In case of VAEM, we use 400 nm peak excitation for GFP constructs, 150–210 ms exposure time, frames taken in 0.5 s intervals over the course of 2 min, and color depth of 8 bits.

3.3 Measuring Cytoskeletal Network Density on CLSM Stacks

1. Obtain serial optical sections (XYZ, i.e., Z-stack) of the cortical cytoplasm of a cell expressing a suitable marker by CLSM. In general, we aim towards imaging about 7–10 plants per sample, with 5–10 cells per plant evaluated (*see* **Note 9**).
2. Open the stack by dragging the microscope-generated *.lei file onto the ImageJ window; use the "open as hyperstack" option in the dialog box. Skeletonize the original serial optical sections (Fig. 1a) using Plugins > kbi > Kbi_Filter2d (set filter: lineFilters and parms linemode: thinLine; Fig. 1b). Generate a Z projection (Image > Stacks > Zproject) using the maximum intensity option and save the resulting image as a new 8 bits *.tif file (Image > Type > 8bits; Fig. 1c).
3. Select the area to be analyzed (a whole cell or a well-focused region of the image) manually by ROI selection (Plugins > ROI > Specify ROI). To specify multiple ROIs of the same size and shape within an image, you may duplicate the selected ROI (right mouse click > duplicate).
4. Evaluate the filament density within the ROI by estimation of the GFP signal occupancy, i.e., the fraction of pixels constituting the skeletonized filaments relative to the total pixel number of the ROI. Count pixel number of selected ROI using the macro: hig_255counts.txt (Plugins > hig_255counts). The occupancy value is proportional to the overall filament density in the cell region of interest and was shown to serve as a useful indicator to evaluate, e.g., the changes in the microfilament organization induced by physiological processes, treatments

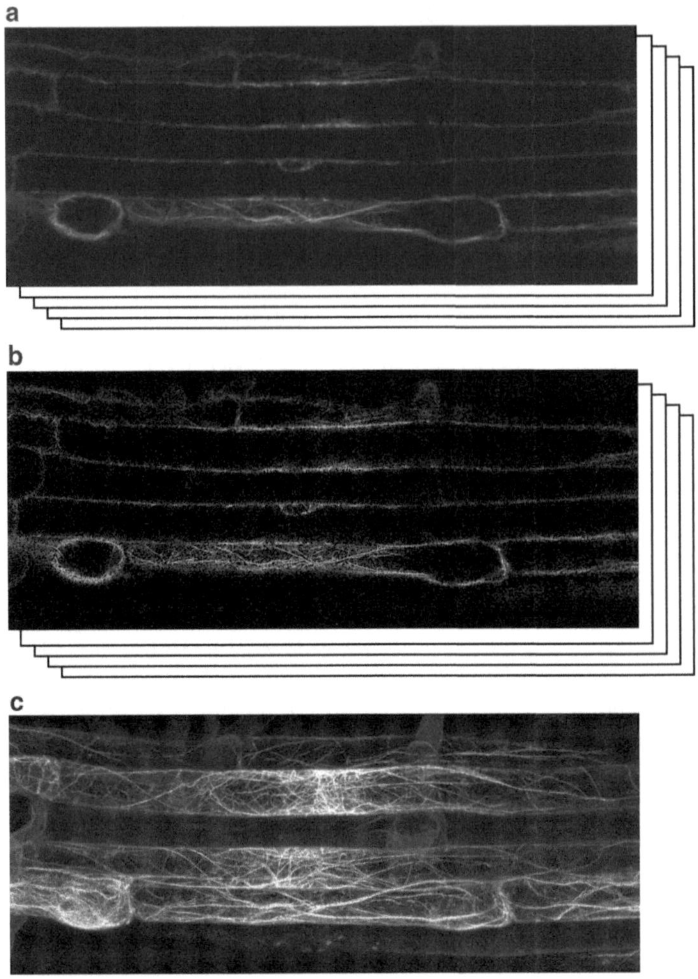

Fig. 1 Stages of image processing prior to determining actin density and bundling using ImageJ. (a) Serial optical sections from CSLM. (b) Skeletonization of serial optical sections. (c) Single image from maximum intensity projections

with inhibitors, such as latrunculin B [28] or by gene mutations ([25, 33, 34]; *see* **Note 10**).

5. The size and organization of microtubules make their analysis easier than in the case of actin; microtubule density can be estimated in small specific area also by direct manual counting [25].

3.4 Evaluating Actin Bundle Thickness by Measuring the Skewness of Fluorescence Distribution

1. Record an image stack and prepare a skeletonized image of a maximum intensity projection as described in Subheading 3.3, **steps 1** and **2**.
2. Open the image in ImageJ and measure the skewness of the fluorescence intensity distribution (a measure of the degree of asymmetry of a distribution, correlated with microfilament

bundling because bundles exhibit brighter fluorescence) in the microfilament-containing pixels using the macro: hig_skewness.txt (Plugins > hig skewness).

3.5 Evaluating Actin Bundle Thickness Using the Histogram Method

1. From original stack of optical sections obtained by CLSM, prepare a maximum intensity projection (Image > Stacks > Zproject) and save it as an 8 bits *.tif image (Image > type > 8bits).
2. Define a specific line length by ROI selection (Plugins > ROI > SpecifyLine); put the line across a representative area of the image (across a well-focused part of a cell; *see* **Note 11**).
3. Generate a profile of GFP fluorescence intensity (Analyze > Plot profile) and record the brightness values of all peaks corresponding to microfilament bundles crossed by the line (values appear on mouse over or generate a list of values by pressing "List"). Record also background values in an area devoid of actin filaments.
4. Using the table calculator, subtract the average background value from the peak values and generate a histogram of the distribution of the resulting net peak values into three or four equally broad classes of gray level (in arbitrary units). The resulting plot documents microfilament bundling, as low intensity represents weakly labeled bundles or single filaments and high intensity corresponds to brightly labeled bundles.

3.6 Quantifying Filament Dynamics from VAEM Image Series

1. Acquire temporal series of single-plane optical sections (XYT) of the cortical cytoplasm of a cell expressing a suitable marker by VAEM. We aim towards imaging at least five plants per sample, with 15–20 movies per sample evaluated (*see* **Note 9**).
2. To measure microfilament dynamics, select randomly ten actin bundles per sample and measure their pause duration (for monitoring over time, use multipoint selection tool in ImageJ and register manually the time when the filament end shows a change in behavior). Values and distribution of pause duration can serve, e.g., as an indicator of differences either in bundle size or in the degree of actin cross-linking ([25]; *see* **Note 12**).
3. To quantify microtubule turnover, select randomly 10–20 microtubule ends per sample and monitor their behavior over time (2 min); use the pen or brush tool in ImageJ to mark the already evaluated ends. Count microtubules in the four distinct phases (growing, shrinking, pausing, and alternating between growth and shrinkage).
4. To estimate of microtubule growth and shrinkage rates, select randomly 5–10 microtubule ends per cell and measure their distances from the starting position during specific time using ImageJ.

3.7 Kymograph Construction from VAEM Image Series

Kymographs can be used to visualize aspects of microfilament and microtubule dynamics that are not easily observed in the video sequences.

1. Open the *.lif file generated by VAEM by dragging it onto the ImageJ window; use the "open as hyperstack" option in the dialog box, and select the desired image to evaluate.
2. Define a specific line length by ROI selection (Plugins > ROI > SpecifyLine) and locate the line across a representative area of the image (across a well-focused part of a cell).
3. Generate the kymographs using the plug-in Multiple Kymograph (Plugins > MultipleKymograph with linewidth: 3).
4. The image generated shows velocity, movement, and different phases of microfilament or microtubule turnover (*see* **Note 13**).

4 Notes

1. Use the culture media and protocols established in the laboratory; any medium and culture setup that allows easy removal of intact seedlings from plates should work. Alternatively, seedlings may be grown on a medium-covered slide surface in situ to avoid disturbance of, e.g., root hairs.
2. Any tissue that can be positioned flat towards the cover slip surface ought to be accessible to CLSM and VAEM; especially for the latter, tight contact with the cover slip is critical. Leaves of glabrous mutants and petals may be especially worth exploring.
3. Ideally, the measurements should be done at least in a single-blind manner to eliminate observer bias (i.e., the person performing quantitative image analysis should not know which image series belongs to which genotype or experimental treatment).
4. In our case, both microscopes are in the inverted configuration. For an upright microscope, sample preparation may have to be modified.
5. This software exists also in versions for other operation systems such as Linux.
6. Take care to treat all the seedlings equally, since mechanical stress may elicit modification of cytoskeletal organization and dynamics during the plant manipulation, media exchange, or even cover slip placement. For longer observation, edges of the cover slip may be sealed with nail polish, but this is usually not necessary. Do not use too much water, as the cover slip should be held in place by capillary forces rather than move around on excess liquid (this is easier to achieve with large cover slips).

7. If working carefully, live seedlings can be recovered from the slide after observation and transferred ex vitro for further cultivation, but do not expect 100 % survival.
8. In case of high autofluorescence background, the window should be narrowed (i.e., longer wavelengths should be cut off in the Beam Path settings).
9. Cytoskeletal structure and dynamics varies dramatically with anatomical location; therefore, imaged cells should be located consistently (e.g., at the bottom of the root tip elongation zone or in the middle of the cotyledon).
10. Maximum intensity projections from the serial optical sections can serve also for determining cytoskeletal filament orientation. Reference [28] describes a procedure employing noise reduction, conversion of the images to binary, and skeletonization. The resulting skeletonized image is used to evaluate cytoskeletal architecture in guard cells of the stomata. Mean angular difference between microfilament pixel pairs and the nearest pixel pairs of a specific cell edge (the stomatal pore) is used as a measure of microfilament orientation. The procedure can be checked by obtaining synthesized images. Analogously, microtubule orientation can be determined with respect to the cell's specific axis, e.g., the longitudinal axis of the hypocotyl [35].
11. It is recommendable to maintain a constant location/direction of the sampling line within a cell, e.g., along the longitudinal axis in case of rhizodermis.
12. Additional data about actin assembly rates, filament origin, and severing frequency can be obtained by the analysis of stochastic dynamics as described in ref. [20], where actin filaments are tracked manually through the time-lapse stack of images and different actin dynamic parameters are estimated by overlapping images or monitoring breaks along the filament over time.
13. Choose the length and location of the sampling line consistently (e.g., parallel to the longitudinal axis in roots and hypocotyls; *see* ref. [29]). While 1 min is usually enough to document microfilament dynamics, in the case of the less mobile microtubules 2 min provides a more informative result.

Acknowledgments

This work has been supported by the GAČR P305/10/0433 project. We thank Boris Voigt and Richard Cyr for transgenic *Arabidopsis* lines; Ondřej Šebesta, Ondřej Horváth, and Aleš Soukup for expert microscopy advice; and Marta Čadyová for technical assistance.

References

1. Crowell EF, Gonneau M, Vernhettes S et al (2010) Regulation of anisotropic cell expansion in higher plants. C R Biol 333(4):320–324
2. Blanchoin L, Boujemaa-Paterski R, Henty JL et al (2010) Actin dynamics in plant cells: a team effort from multiple proteins orchestrates this very fast-paced game. Curr Opin Plant Biol 13:714–723
3. Szymanski DB (2005) Breaking the WAVE complex: the point of Arabidopsis trichomes. Curr Opin Plant Biol 8:103–112
4. Pei W, Du F, Zhang Y et al (2012) Control of the actin cytoskeleton in root hair development. Plant Sci 197:10–18
5. Fu Y, Gu Y, Zheng Z et al (2005) Arabidopsis interdigitating cell growth requires two antagonistic pathways with opposing action on cell morphogenesis. Cell 120:687–700
6. Žárský V, Cvrčková F, Potocký M et al (2009) Exocytosis and cell polarity in plants: exocyst and recycling domains. New Phytol 183:255–272
7. Ueda K, Matsuyama T, Hashimoto T (1999) Visualization of microtubules in living cells of transgenic *Arabidopsis thaliana*. Protoplasma 206:201–206
8. Nakamura M, Naoi K, Shoji T et al (2004) Low concentrations of propyzamide and oryzalin alter microtubule dynamics in Arabidopsis epidermal cells. Plant Cell Physiol 45:1330–1334
9. Marc J, Granger CL, Brincat J et al (1998) A GFP-MAP4 reporter gene for visualizing cortical microtubule rearrangements in living epidermal cells. Plant Cell 10:1927–1939
10. Van Damme D, Van Poucke K, Boutant E et al (2004) In vivo dynamics and differential microtubule-binding activities of MAP65 proteins. Plant Physiol 136:3956–3967
11. Chan J, Calder G, Fox S et al (2005) Localization of the microtubule end binding protein EB1 reveals alternative pathways of spindle development in Arabidopsis suspension cells. Plant Cell 17:1737–1748
12. Dhonukshe P, Gadella TWJ (2003) Alteration of microtubule dynamic instability during preprophase band formation revealed by yellow fluorescent protein-CLIP170 microtubule plus-end labeling. Plant Cell 15:597–611
13. Kost B, Spielhofer P, Chua NH (1998) A GFP-mouse talin fusion protein labels plant actin filaments in vivo and visualizes the actin cytoskeleton in growing pollen tubes. Plant J 16:393–401
14. Sheahan MB, Staiger CJ, Rose RJ et al (2004) A green fluorescent protein fusion to actin-binding domain 2 of Arabidopsis fimbrin highlights new features of a dynamic actin cytoskeleton in live plant cells. Plant Physiol 136:3968–3978
15. Voigt B, Timmers ACJ, Šamaj J et al (2005) GFP-FABD2 fusion construct allows in vivo visualization of the dynamic actin cytoskeleton in all cells of Arabidopsis seedlings. Eur J Cell Biol 84(6):595–608
16. Era A, Tominaga M, Ebine K et al (2009) Application of Lifeact reveals F-actin dynamics in *Arabidopsis thaliana* and the liverwort, Marchantia polymorpha. Plant Cell Physiol 50:1041–1048
17. Hashimoto T (2002) Molecular genetic analysis of left-right handedness in plants. Philos Trans R Soc Lond B Biol Sci 357:799–808
18. Ketelaar T, Anthony RG, Hussey PJ (2004) Green fluorescent protein-mTalin causes defects in actin organization and cell expansion in Arabidopsis and inhibits actin depolymerizing factor's actin depolymerizing activity in vitro. Plant Physiol 136:3990–3998
19. Cvrčková F, Grunt M, Žárský V (2012) Expression of GFP-mTalin reveals an actin-related role for the Arabidopsis Class II formin AtFH12. Biologia Plantarum 56:431–440
20. Staiger CJ, Sheahan MB, Khurana P et al (2009) Actin filament dynamics are dominated by rapid growth and severing activity in the Arabidopsis cortical array. J Cell Biol 184: 269–280
21. Smertenko A, Deeks MJ, Hussey P (2010) Strategies of actin reorganisation in plant cells. J Cell Sci 123:3019–3028
22. Vizcay-Barrena G, Webb S, Martin-Fernandez M et al (2011) Subcellular and single-molecule imaging of plant fluorescent proteins using total internal reflection fluorescent microscopy (TIRFM). J Exp Bot 62:5419–5428
23. Wan Y, Ash WM 3rd, Fan L et al (2011) Variable-angle total internal reflection fluorescence microscopy of intact cells of *Arabidopsis thaliana*. Plant Methods 7:27
24. Sparkes I, Graumann K, Martiniere A et al (2011) Bleach it, switch it, bounce it, pull it: using laser to reveal plant cell dynamics. J Exp Bot 62:1–7
25. Rosero A, Žárský V, Cvrčková F (2013) AtFH1 formin mutation affects actin filament and microtubule dynamics in *Arabidopsis thaliana*. J Exp Bot 64:585–597
26. van der Honing H, Kieft H, Emons A et al (2012) Arabidopsis VILLIN2 and VILLIN3 are required for the generation of thick actin filament bundles and for directional organ growth. Plant Physiol 58:1426–1438
27. Higaki T, Kutsuna N, Sano T et al (2008) Quantitative analysis of changes in actin microfilament contribution to cell plate development in plant cytokinesis. BMC Plant Biol 8:80

28. Higaki T, Kutsuna N, Sano T et al (2010) Quantification and cluster analysis of actin cytoskeletal structures in plant cells: role of actin bundling in stomatal movement during diurnal cycles in Arabidopsis guard cells. Plant J 61:156–165
29. Sampathkumar A, Lindeboom J, Debolt S et al (2011) Live cell imaging reveals structural associations between the actin and microtubule cytoskeleton in Arabidopsis. Plant Cell 23:2302–2313
30. Ketelaar T, Allwood EG, Anthony RG et al (2004) The actin-interacting protein AIP is essential for actin organization and plant development. Curr Biol 14:149
31. Abramoff MD, Magelhaes PJ, Ram SJ (2004) Image processing with ImageJ. Biophotonics International 11:36–42
32. Collins T (2007) ImageJ for microscopy. Biotechniques 43:S25–S30
33. Henty JL, Bledsoe S, Khurana P et al (2011) Arabidopsis actin depolymerizing factor 4 modulates the stochastic dynamic behavior of actin filaments in the cortical array of epidermal cells. Plant Cell 23:3711–3726
34. Li J, Henty JL, Huang S et al (2012) Capping protein modulates the dynamic behavior of actin filaments in response to phosphatidic acid in Arabidopsis. Plant Cell 24:3742–3754
35. Yao M, Wakamatsu Y, Itoh T et al (2008) Arabidopsis SPIRAL2 promotes uninterrupted microtubule growth by suppressing the pause state of microtubule dynamics. J Cell Sci 121:2372–2381

Chapter 8

Sequential Replicas for In Vivo Imaging of Growing Organ Surfaces

Dorota Kwiatkowska and Agata Burian

Abstract

Sequential replica method facilitates in vivo imaging of plant surface and provides data sufficient for detailed computation of geometry and growth. It enables obtaining a series of high-resolution images visualizing details of the examined surface. Series of molds, made in dental polymer, representing the examined surface are used to obtain casts in epoxy resin, which are in turn observed by scanning electron microscopy, while the structure itself remains intact. Images obtained from casts can be further used for data extraction, comprising 3D reconstruction and computation of local geometry and cell growth parameters. The sequential replica method is a universal method and can be applied to image complex shapes of a range of structures, like meristems, flowers, stems, leaves, or various types of trichomes. Different plant species growing in various conditions can be studied.

Key words In vivo imaging, Organ geometry, Cell growth, Epidermis, Shoot apical meristem

1 Introduction

Plant morphogenesis comprises maintenance and changes of organ shape (geometry) that can be realized by various growth patterns [1, 2]. In order to understand how morphogenesis is regulated, one usually needs to perform detailed quantification of geometry and growth of the developing organ. This quantification is often a challenge in technical and computational terms because, in most organs, geometry and growth patterns are complex and changing in time. If changes of a complex organ shape are to be quantified, 3D reconstruction of the organ surface is required. Moreover, if growth pattern is changing in time, as in majority of developing organs, sequential imaging in vivo is required for growth quantification. Noteworthy, there are only few cases when temporal changes of growth do not have to be accounted for; accordingly one-time examination of the organ is enough for growth quantification. These exceptions are root apices [3] or shoot apical meristems with relatively small primordia [4] considered within a short time frame.

Viktor Žárský and Fatima Cvrčková (eds.), *Plant Cell Morphogenesis: Methods and Protocols*, Methods in Molecular Biology, vol. 1080, DOI 10.1007/978-1-62703-643-6_8, © Springer Science+Business Media New York 2014

Sequential replica method, developed by Paul B. Green and collaborators [5, 6], is one of the in vivo imaging methods that provide data sufficient for detailed computation of geometry and growth. It enables obtaining a series of high-resolution images visualizing details (superficial cell outlines, trichomes, etc.) of a surface of individual plant organs. The organ surface has to be only partially exposed for the method application. Series of molds, made in dental polymers, representing the growing organ surface are used to obtain casts in epoxy resin, which are in turn observed in scanning electron microscopy (SEM), while the organ itself remains intact. Images obtained from casts can be further used for data extraction, 3D reconstruction, and computation of local geometry (curvature) and cell growth parameters for which software are available [7, 8]. Furthermore, the replica method can be combined with in vivo confocal laser scanning microscopy (ref. [9]; *see* also Chapter 9) to complement growth and geometry data with distribution of a reporter gene signal in the organ. Also, the specimens can be fixed after the last replica has been taken and used for various histological procedures. The replica method is a universal method and can be applied to image complex shapes of a range of structures (meristems, flowers, stems, leaves, or various types of trichomes) of different plant species (*see* Fig. 1), growing in various conditions (outdoors, indoors, in in vitro culture, etc.).

2 Materials

2.1 Plant Material and Growth Conditions

Sequential replicas can be obtained from virtually every organ surface of almost every plant species as long as this surface can be kept in a dry state for the time of mold taking (but *see* **Note 1**) and it is not in contact with any other organ or a solid object that cannot be temporarily moved away. Plants growing in a wide scope of conditions can be used, including potted plants, plants growing outdoors, or on solid media in in vitro culture (*see* **Note 2**). The main limitation is imposed by the conditions necessary during the time of mold taking itself.

2.2 Exposing the Organ Surface

1. Scaffolding: small disposable polystyrene petri dishes, disposable syringes (dish and syringe size depends on the plant size), thin bamboo sticks, waterproof tape, and waterproof glue.
2. Ultrafine threads used in ophthalmologic surgery. We use ultrafine monofilament nylon threads (manufactured by Ethicon Corp., Somerville, USA).
3. Epoxy gel (not a regular resin!). After setting, the gel is not brittle and has lower stiffness than a regular resin. We use Devcon high strength 5 min epoxy gel. Since you will need only small equal-volume amounts of the two epoxy gel components

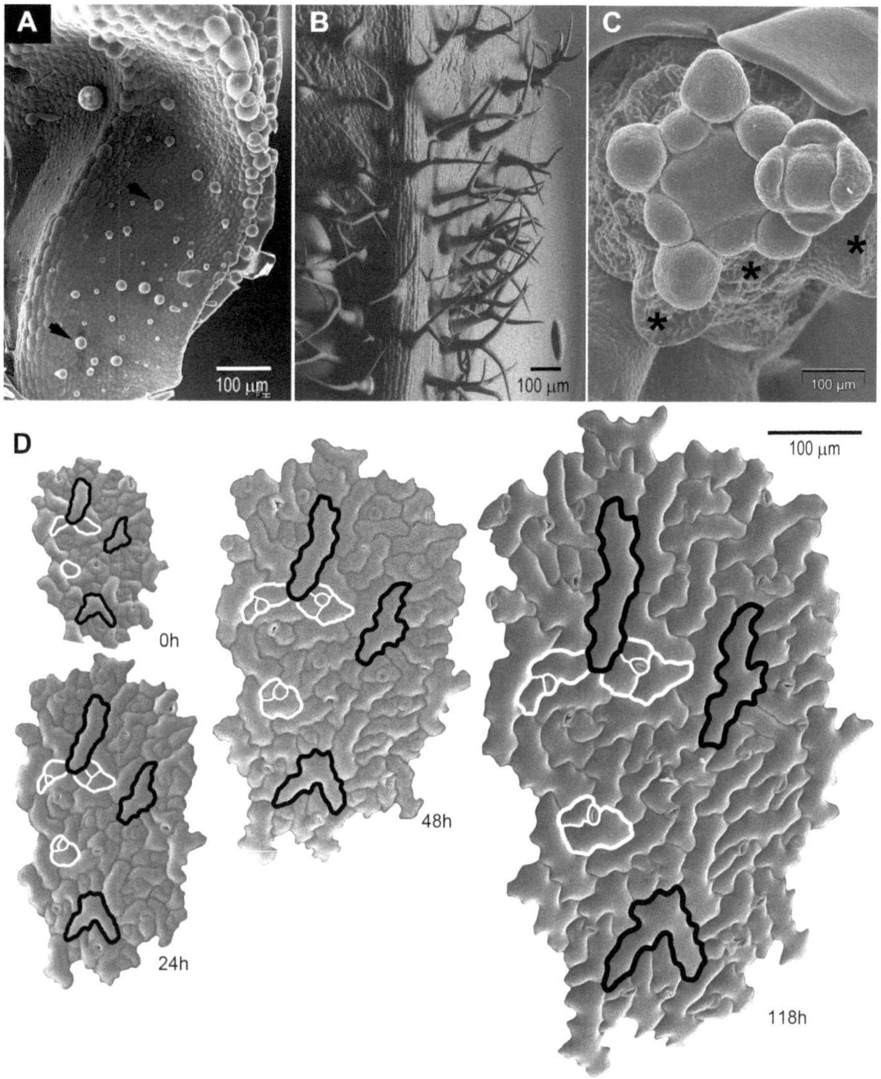

Fig. 1 SEM micrographs showing exemplary casts of various organs of *Anagallis arvensis* (**a**) and *A. thaliana* (**b–d**). (**a**) Abaxial surface of a young expanding leaf with grandular trichomes (*arrows*). Note the complex shape of the leaf in this developmental stage. (**b**) Basal portion of a rosette leaf, whose abaxial surface is covered by large, branched trichomes. Note how well the complex trichome shapes are reproduced in the cast. (**c**) Inflorescence shoot apex grown in in vitro culture on agar-solidified medium. The same specimen can be observed in vivo in CLSM (*see* Chapter 9). Exemplary places from where older organs (flower buds, cauline leaves) surrounding the apex have been removed are pointed by *asterisks*. (**d**) Sequence of replicas taken from abaxial leaf surface of *transparent testa glabra1* mutant plant. The time at which replicas were taken is given in lower right corner of each image. Exemplary cells or cell packets are outlined: in *black* if no cell division took place or in white if the cells are still dividing. Images shown in (**a**, **d**) were taken with the aid of SEM machine LEO435VP; (**b**)—Philips XL 30 TMP ESEN; (**c**)—Hitachi S-800

each time, it is convenient to put a small amount of each in two disposable syringes (10–15 ml volume) and close syringes by original caps or short injection needles.

4. Any type of dental silicon polymer.
5. Wooden toothpicks and fine forceps.

2.3 Mold Preparation

1. Dental silicon (polyvinyl) impression material dedicated for obtaining patient's mouth impression of the finest details, with work time circa 2 min and total set time 5 min. It is better to use less hydrophilic materials (*see* **Note 3**). We recommend Take 1 (the hydrophilic vinyl; wash material; regular set) manufactured by Kerr Corporation. Original product is available as a pair of cartridges with two differently colored pastes. You will need only small equal-volume amounts of the pastes each time. Thus, it is convenient to put a small amount of the two pastes in two disposable syringes (5–10 ml volume) and close syringes by original caps or short injection needles.
2. Silicon sealant (clear, used for plumbing).
3. Wooden toothpicks, dental filling instruments or excavators, fine forceps, glass slides, and disposable petri dishes.

2.4 Cast Preparation and Observation by Scanning Electron Microscopy

1. Epoxy resin (transparent with long setting time, preferably setting overnight). We use Devcon 2 t epoxy, transparent and long setting.
2. Epoxy gel (*see* Subheading 2.2, **item 3**).
3. Scanning electron microscopy (SEM) stubs.
4. Wooden toothpicks, fine forceps, razor blades, injection needles, thin glass tubes with ends stretched and thinned in a flame, dental filling instruments or excavators, and disposable petri dishes.

In case of **items 1** and **2** put some of the two resin components in two disposable syringes (10–15 ml volume), and close them by original caps or short injection needles (only a small amount of a resin is used each time).

3 Methods

Carry out all procedures at room temperature, not below 16 °C and not exceeding 24 °C (*see* **Note 4**). High air humidity is preferable. The spectrum of light and its intensity during the mold taking (steps described in Subheadings 3.1 and 3.2.) are limited mainly by the type of organ and process under investigation that may require usage of stereomicroscope, monochromatic light, etc.

3.1 Exposing the Examined Organ Surface

The surface of interest has to be at least partly and temporally exposed so that the dental silicon polymer, which is fluid during application, may flow onto the surface (*see* **Note 5**). The described

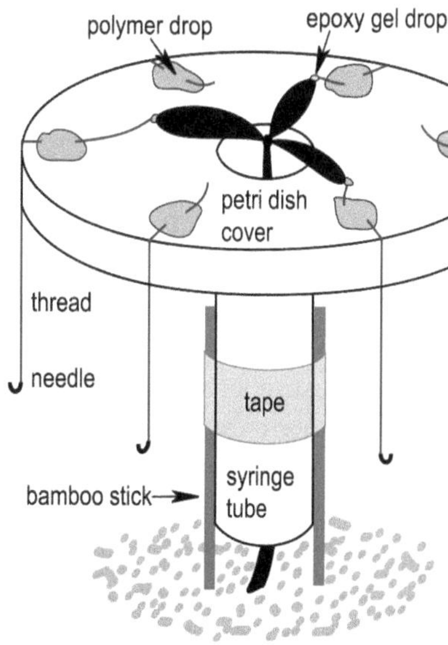

Fig. 2 Scaffolding assembly that can be used to expose the apex of the shoot with elongating internodes (shown in *black*)

procedure of surface exposing applies to apices (or apical meristems) of shoots with elongating internodes (not rosettes of leaves) from which leaves or flower buds have to be bent away for the time of mold taking; plants are growing in soil [10].

1. Prepare a scaffolding (*see* Fig. 2), gluing a cover of a small petri dish, with a hole cut in its center, to the container part of the disposable syringe from which the needle end was cut off. Attach two thin bamboo sticks (as those used for barbeque) to the syringe tube, using a waterproof tape. Attach threads (*see* Subheading 2.2, **item 2**) to the petri dish upper surface, placing the middle part of each thread in a freshly mixed drop of dental silicon polymer (*see* Subheading 2.2, **item 4**), each thread in a separate drop. Needle ends of the threads should hang free from the dish rim.
2. Insert the shoot in the syringe tube so that the shoot tip is just above the petri dish surface at the same time driving the sticks into the soil.
3. Mix for a few minutes two small equal-volume drops of epoxy gel, using a toothpick.
4. Attach needle-free ends of threads to tips of leaves or other organs that you would like to bend away, placing them in tiny drops of freshly mixed epoxy gel, applied to the organ surface.
5. When the epoxy gel sets, after circa 15 min, bend away the organs gently pulling the needle ends of threads by forceps.

3.2 Obtaining the Mold

1. Mix rapidly, for circa 10 s, two small equal-volume drops of the dental polymer pastes on a lipid-free smooth surface (e.g., petri dish cover) using a toothpick. Mix the polymer directly before usage, preferably when the organ surface is already exposed. Prepare the mixture of the dental polymer separately for each mold to ensure that it is fluid enough during application.
2. Quickly (as soon as possible), during less than half a minute, apply a small amount of the polymer on the studied surface using a toothpick with a sharpened tip or a dental filling instrument or excavator. The size of toothpick or the instrument endings needs to be adjusted to the size of the surface of interest.
3. After a few minutes check with your finger whether the remnant of mixed polymer on the petri dish is no longer sticky. If it is not, check gently the mixture applied on the organ surface—it may set for a bit longer time than the one on the dish. As soon as the polymer is not sticky slowly remove the mold with forceps pulling gently one of its margins from a side so that the mold is removed gradually, not from the whole surface at once. The latter may lead to organ breakage. Put the mold in a safe place (no dust or rapid air movement), e.g., in a petri dish (*see* **Note 6**).
4. If you were using a scaffolding, allow the organs, which were bent away, to move back to their original position, releasing the threads with forceps.
5. If you are making a sequence of replicas, put a drop of distilled water on the organ surface from which replica has been taken or put a transparent cover over the plants to ensure a high humidity.
6. Attach the mold to a glass slide, putting a drop of a silicon sealant on a lipid-free glass slide and placing the mold in the drop. Make sure that the mold is in such position that the orientation of the surface of interest is more or less horizontal and that the later-applied resin (*see* Subheading 3.3, **step 2**) will not all flow away. Be careful since the sealant sticks really well to the dental polymer, also to the surface of interest (!). Because the resin will not set in presence of acetic acid vapor coming from the sealant, leave the glass slide in an open space (never in a closed small container) overnight before filling with resin.
7. Repeat **steps 1–6** to obtain the sequence of molds from the surface of interest. The number of repetitions and the time interval between consecutive molds depend on the organ studied (*see* **Note 7**).

3.3 Preparing the Cast

1. Mix rapidly two small equal-volume drops of the two epoxy resin components (resin and hardener) with a toothpick for circa 5 min.
2. Fill the molds with the resin (*see* **Note 8**). Use stretched end of thin glass tubes, toothpick with sharpened ending, or a dental instrument with fine ending to apply the resin. If air bubbles appear in the resin contacting the surface of interest, the glass tubes or toothpicks can be used to remove them gently. It is crucial to touch only the bubbles, not the polymer surface, since it could damage the mold. Do not mind small bubbles in the cast interior (not contacting the surface of interest)—they will not be visible in SEM. The resin mixture cannot be used after it became apparently more viscous (it takes usually circa 15 min) than immediately after mixing in order to ensure proper penetration of the mold. Make also sure that the cast is not very thin, applying additional amount of the already setting resin mixture if necessary. Leave the molds filled with resin overnight in an open petri dish.
3. Remove gently the casts from molds using forceps (*see* **Note 9**) and place them in drops of freshly mixed epoxy gel on a SEM stub. The surface of cast in contact with the epoxy gel has to be large and provide the hold strong enough for later cast trimming. The epoxy gel mixture cannot be used after it became viscous (circa 5 min from mixing), so that long threads are drawn during application. The threads may stick to the cast surface of interest.
4. Pay attention that the surface of interests is fully uncovered and thus accessible for the SEM observation. This can be achieved trimming the casts. After the epoxy gel sets strongly (it takes circa 1 h), the excess of resin or casts of some organs that obstruct the view (*see* Fig. 3a, b) can be removed with a razor blade or an injection needle under stereomicroscope. This is often tricky and you may damage the surface of interest (*see* Fig. 3a) but the casts can be obtained several times from the same mold and trimmed again (*see* **Note 10**). Before SEM examination, get rid of all epoxy resin debris using a compressed gas duster.

3.4 Scanning Electron Microscopy Examination

1. The requirements of the available SEM machine have to be followed, keeping in mind that the epoxy resin casts do not require drying but need only to be sputter-coated. The vacuum mode of SEM operation always gives satisfactory results (*see* **Note 11**). Although in principle the environment mode could be also used, it usually gives blurred cell outlines; thus, we recommend the vacuum mode with sputter-coated casts.
2. If the SEM images are to be used for a 3D reconstruction, which is crucial in the case of organs of complex geometry,

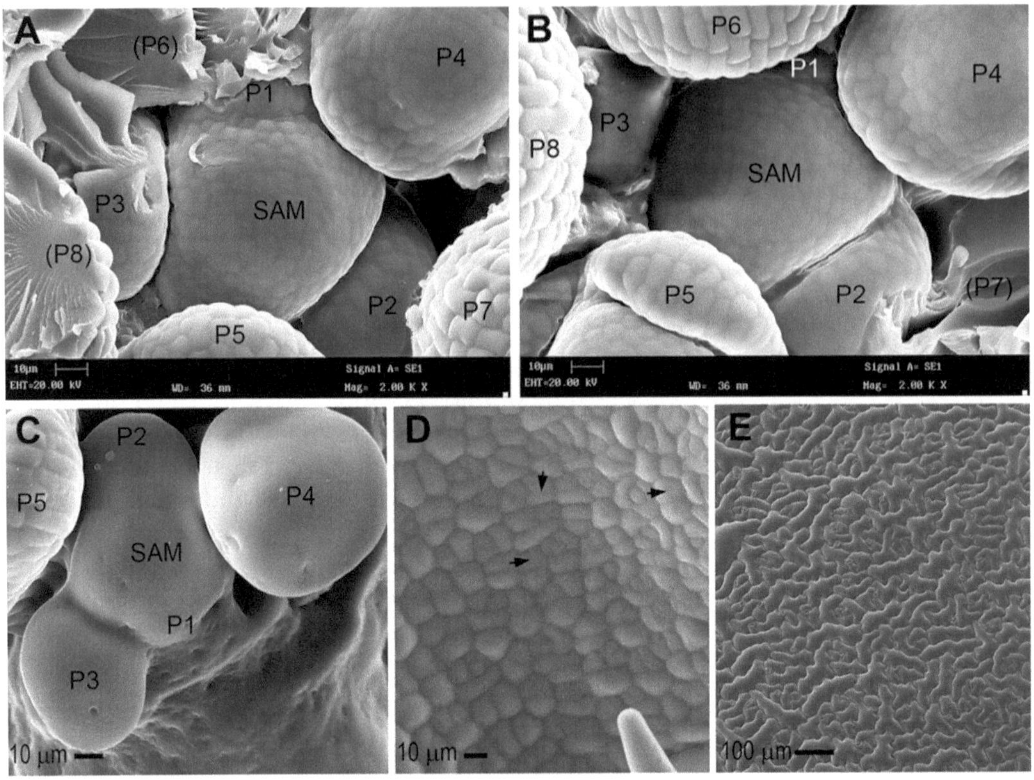

Fig. 3 SEM micrographs of casts representing apices of inflorescence shoots (**a–c**) or leaf epidermis (**d**, **e**) of *A. thaliana*. (**a**, **b**) Two casts obtained from the same, high-quality mold, which were trimmed in a different way. Flower primordia are labelled in both casts with the same symbols (*P* and the number increasing with the primordium age). (**a**) Cast trimming enables better visualization of the shoot apical meristem (SAM): the cell outlines are more apparent and more of the SAM surface can be examined than in the other cast obtained from this mold, because all the older primordia were cut off with razor blade (their symbols are in *parentheses*). However, the surface of the SAM is locally damaged between primordia P1 and P3. (**b**) The specimen is charged because of a deep grove between primordia and the SAM. Parts of the SAM periphery, as well as the primordium P1, are hidden behind surrounding primordia. (**c**) The cast obtained from a mold of insufficient quality, in which the dental polymer has not set properly, presumably because of earlier plant reaction to aphids. Before the mold was taken, some older primordia were removed (*lower right* part of the image), and the released cell sap probably affected the polymer setting. Unlike this apex portion, the surface of P5 is rather well represented. (**d**, **e**) Casts from molds taken from the adaxial epidermis of the same leaf at 6 days time interval. In the younger epidermis (**d**) cell outlines are less prominent than in older (**e**). Moreover, in the younger epidermis (**d**), cells are still dividing and the younger anticlinal walls (e.g., those pointed by *arrows*) are much less distinct than older walls, unlike the walls in older nearly differentiated epidermis (**e**). All the images were taken with the aid of SEM machine LEO435VP

a stereopair of images has to be taken [7, 8]. These are two images of the same region taken at different SEM stage tilt angles (we use a 10° difference in tilt for shoot apices or leaves). After taking one image, tilt the stage precisely with respect to the *X*-axis (*see* **Note 12**) and then move the specimen along the *Y*-axis to come back to the region of interest. During this operation you may decrease the magnification temporarily but you cannot use rotation with respect to the Z-axis. Images with a scale bar pasted are preferable. Information about the exact tilt angle, magnification, and work distance is required for each image for further analysis, i.e., 3D reconstruction, curvature, and growth computation [8].

4 Notes

1. The surface to which the polymer is applied must be really dry. If there is even a small amount of water or other liquid on the examined surface, the polymer will not reproduce the surface details (no visible cells) or may not set. According to product information and our own experience, dental silicon polymers used for molds do not set in the presence of sulfur compounds. In such cases often an oily and sticky film is formed on the organ surface that cannot be removed, and thus the specimen is lost. This can be a serious problem in case of species that produce sulfur-containing exudates on the organ surface or with sulfur compounds in a cell sap, since it is often unavoidable that some tissue in the vicinity of examined region is damaged during mold taking. It may also be the reason why it is often virtually impossible to obtain good molds from *Arabidopsis thaliana* plants that have been attacked by aphids or other pests (*see* Fig. 3c) because sulfur compounds can be produced in reaction to pests in myrosin cells of plants from Brassicaceae family [11, 12].
2. The sequential replica method can be combined with in vivo confocal laser scanning microscopy (CLSM) (*see* Chapter 9). In this case plants are grown in in vitro culture on agar-solidified medium. Replica method can be combined also with other methods. For example, the specimen can be fixed immediately after the last replica has been taken and used for β-glucuronidase reaction, nucleus staining [13], or analysis of inner cell wall pattern. Although it may be time-consuming, it is possible to recognize the same cells in the replica and fixed material (*see* also **Note 11**).
3. The composition of products is changing and the manufacturing companies adjust them to dentist/patient requirements that are not always the same as those for replica taking from plant material. Therefore, we recommend checking the product before usage, e.g., applying it to an easily accessible shoot surface, obtaining a mold, and checking for epidermal cells viability next day. The disadvantage of the new products is in the addition of components that make them more hydrophilic, which may possibly enhance plant cell desiccation.

4. The time for the mold setting depends on temperature: the higher the temperature, the shorter the time, and we advise to minimize the time of organ surface contact with the dental polymer (we observe surface damage after leaving the polymer on the organ overnight; also, the presence of set polymer on the organ surface is in fact a mechanical factor). However, too high temperature may cause organ desiccation and makes the time window, when the dental polymer is fluid enough to be applied on the surface of interest, too short. Thus, the temperature cannot be too high. On the other hand, a lower temperature (below 16 °C) may increase the setting time to several hours or even prevent the setting of the dental polymer.
5. Using the scaffolding is necessary especially when apices or apical meristems of elongated shoots in the vegetative phase of development are studied [10, 14]. In case of some organs, whose surface is easier accessible, this step can be much simplified. For example, in the case of the inflorescence shoot apex (or apical meristem) of *Arabidopsis thaliana* [15], it is enough to bend away young flower primordia with a sharpened end of a toothpick—they often remain in such a changed position for a moment long enough to apply the dental polymer. Young leaves or stems of many species, in turn, do not require any additional operation at all.
6. It is good to check the appearance of the mold surface immediately after taking (e.g., using a stereomicroscope). Sometimes the polymer has not reached the surface of interest or there is an air bubble on this surface. This can be recognized already in the mold and the next trial to obtain the mold can be made before water is applied to the organ surface. Such repetition should, of course, be avoided if possible.
7. We observed that growth of shoot apices [14–16] and young leaves, e.g., 1–5 mm long leaves of *Arabidopsis thaliana* [13] or *Anagallis arvensis*, apparently slows down or even ceases after four successive replicas have been taken. The minimal time interval that we recommend for shoot apices is 10 h [15] and for expanding leaves, 48 h. If the interval is too long, the recognition of the same region (individual cells and their progeny) may be difficult or even impossible in consecutive replicas because too many cell divisions take place during that time. On the other hand if the interval is too short, differences in the cell size between the consecutive time points may be too small for growth computation.
8. If a surface studied is flat, the cast can be made also from nail polish and observed under light microscope in a drop of water or 50 % glycerol solution. The nail polish has to be transparent [13, 17].

9. It is worth time to realize that newly set epoxy resin casts are rather elastic. Therefore, if you use a strong grip with the forceps a nearly transparent cast may be "catapulted," jumping away far enough to be lost.
10. It is usually best to trim the specimen next day after filling the mold with resin, because later on the resin becomes more and more brittle and it may break during trimming forming cracks in the cast portions of interest. Nevertheless, the molds can be used several times to obtain the casts. If one is careful, it is only rarely that the mold is damaged when filling with resin or removing the casts. The molds can be stored in room temperature in closed containers to avoid dust, although we do not recommend to store them for longer than half a year because slow deformation of molds cannot be excluded. This time, however, is long enough to make first casts, check them in SEM, and repeat if necessary.
11. Cell outlines can be recognized in casts as groves in organ surface that are formed above junctions of anticlinal walls to the outer periclinal wall. In case of meristematic tissue, these groves are rather shallow (*see* Fig. 3d), while in case of differentiated cells like those of a young leaf epidermis (*see* Fig. 3e), the groves are prominent. It is important to realize that not all the anticlinal walls are apparent in replicas if cells are still dividing. This is because newly formed walls are not immediately visible—groves appear only after the new wall has shrunk a bit, forming a grove on a surface. Therefore, in case of meristematic tissues, some recently divided cells look as if the division has not yet taken place.
12. It is crucial for the stereoscopic reconstruction of the cast surface that the tilt is precisely with respect to *X*- (or *Y*-) axis of the image. Otherwise, the reconstruction with available software is impossible. Better results can be obtained if single-stub SEM stages are used (even with a few casts on it) instead of multi-stub [8].

Acknowledgments

The sequential replica method has been developed by the late Paul B. Green. While writing this chapter, we have used his numerous indispensable advices that we have learned from Dr. Jacques Dumais, the last graduate student of Paul. We would like to thank Drs. Zofia Czarna and Krystyna Heller (Electron Microscopy Laboratory, Wrocław University of Agricultural Sciences, Poland) and Ewa Teper (Laboratory of Scanning Electron Microscopy, Faculty of Earth Sciences, University of Silesia) for their help with scanning electron microscopy and Dr. Joanna Elsner (University of Silesia) for providing SEM micrographs of leaf epidermis. The work in D.K. lab is financially supported by Polish Ministry of Science and Higher Education and by the MAESTRO research grant No 2011/02/A/NZ3/00079 from the National Science Centre, Poland.

References

1. Green PB, Erickson RO, Richmond PA (1970) On the physical basis of cell morphogenesis. Ann N Y Acad Sci 175:712–731
2. Hejnowicz Z, Nakielski J (1979) Modeling of growth in shoot apical dome. Acta Soc Bot Pol 48:423–442
3. Silk WK, Lord EM, Eckard KJ (1989) Growth patterns inferred from anatomical records: empirical tests using longisections of roots of *Zea mays* L. Plant Physiol 90:708–713
4. Hejnowicz Z, Nakielski J, Włoch W et al (1988) Growth and development of the shoot apex of barley III. Study of growth rate variation by means of the growth tensor. Acta Soc Bot Pol 57:31–50
5. Williams MH (1991) A sequential study of cell divisions and expansion patterns on a single developing shoot apex of *Vinca major*. Ann Bot 68:541–546
6. Williams MH, Green PB (1988) Sequential scanning electron microscopy of a growing plant meristem. Protoplasma 147:77–79
7. Dumais J, Kwiatkowska D (2002) Analysis of surface growth in shoot apices. Plant J 31: 229–241
8. Routier-Kierzkowska A-L, Kwiatkowska D (2008) New stereoscopic reconstruction protocol for scanning electron microscope images and its application to *in vivo* replicas of the shoot apical meristem. Funct Plant Biol 35:1034–1046
9. Uyttewaal M, Burian A, Alim K et al (2012) Mechanical stress acts via katanin to amplify differences in growth rate between adjacent cells in Arabidopsis. Cell 149:439–451
10. Green PB, Havelange A, Bernier G (1991) Floral morphogenesis in *Anagallis*: scanning-electron-micrograph sequences from individual growing meristems before, during, and after the transition to flowering. Planta 185: 502–512
11. Andréasson E, Jørgensen LB, Höglund A-S et al (2001) Different myrosinase and idioblast distribution in Arabidopsis and *Brassica napus*. Plant Physiol 127:1750–1763
12. Bones A, Rossiter JT (1996) The myrosinase-glucosinolate system, its organization and biochemistry. Physiol Plant 97:194–208
13. Elsner J, Michalski M, Kwiatkowska D (2012) Spatiotemporal variation of leaf epidermal cell growth: a quantitative analysis of *Arabidopsis thaliana* wild-type and triple *cyclinD3* mutant plants. Ann Bot 109:897–910
14. Kwiatkowska D, Routier-Kierzkowska A-L (2009) Morphogenesis at the inflorescence shoot apex of *Anagallis arvensis*: surface geometry and growth in comparison with the vegetative shoot. J Exp Bot 60:3407–3418
15. Kwiatkowska D (2006) Flower primordium formation at the Arabidopsis shoot apex: quantitative analysis of surface geometry and growth. J Exp Bot 57:571–580
16. Kwiatkowska D (2004) Surface growth at the reproductive shoot apex of *Arabidopsis thaliana*: *pin-formed 1* and wild type. J Exp Bot 55:1021–1032
17. Geisler MJ, Sack FD (2002) Variable timing of developmental progression in the stomatal pathway in Arabidopsis cotyledons. New Phytol 153:469–476

Chapter 9

Time-Lapse Imaging of Developing Meristems Using Confocal Laser Scanning Microscope

Olivier Hamant, Pradeep Das, and Agata Burian

Abstract

Analysis of shoot meristem shape and gene expression pattern has been conducted in many species over the past decades. Recent live imaging techniques have allowed an unprecedented accumulation of data on the biology of meristematic cells, as well as a better understanding of the molecular and biophysical mechanisms behind shape changes in this tissue. Here we describe in detail how to prepare shoot apices of both Arabidopsis and tomato, in order to image them over time using a confocal microscope equipped with a long-distance water-dipping lens.

Key words Confocal laser scanning microscopy, Time-lapse and live imaging, Shoot apical meristem, Development, Cell growth

1 Introduction

Meristems are groups of dividing cells that are responsible for the generation of all the aerial organs of the plant. Their identification, and ongoing characterization, has always been associated with technological imaging improvements. Meristems have been formally identified for the first time in 1759 by Caspar Friedrich Wolff using a rudimentary microscope, but it is only in 1858 that Karl Nageli could observe meristematic cells using a true microscope. This started a long series of histological analysis, with three essential steps: Hanstein in 1868 found that the meristem is organized in three layers, Schmidt in 1924 observed that these layers can be associated with anticlinal and periclinal cell division planes, and Plantefol and Buvat in the 1950s identified different zones in the meristem with more (peripheral zone) or less (central zone) activity. Various staining (e.g., with ^{3}H thymidine) confirmed the higher metabolic activity (e.g., DNA synthesis) in the peripheral zone of

Viktor Žárský and Fatima Cvrčková (eds.), *Plant Cell Morphogenesis: Methods and Protocols*, Methods in Molecular Biology, vol. 1080, DOI 10.1007/978-1-62703-643-6_9, © Springer Science+Business Media New York 2014

the meristem [1]. The genetic basis of meristem function was then analyzed: many mutants with an altered number of organs or with defective organ identities were isolated from the 1980s onwards (e.g., ref. [2]). The observation of mutant meristems was conducted, and in parallel, gene expression was monitored with reporter gene expression (e.g., promoter activity with GUS activity), in situ hybridization (mRNA accumulation in meristem sections), or immunolocalization (e.g., ref. [3]). The impact of mutations on meristem shape was analyzed, using in particular scanning electron microscopy (e.g., ref. [4]). The generalization of the use of confocal laser scanning microscopes (CLSM) in the 1990s allowed a more precise analysis of cell behavior and meristem shape in 3D while being still mostly limited to fixed tissues (e.g., ref. [5]). The localization of certain GFP-fused proteins in the meristem was initiated, notably to show the existence of transport mechanisms across meristem layers (e.g., ref. [6]). In 2004, two studies were published, in which whole meristems were observed over time [7, 8]. Today, this switch to live imaging still represents a major step forward in the analysis of meristem biology and this is what this protocol paper is about.

Two plant model systems have received more attention for time-lapse imaging, namely, the inflorescence meristem of *Arabidopsis thaliana* and the vegetative meristem of *Lycopersicon esculentum* (tomato). As meristems are usually covered by young organs, this hinders their analysis under a microscope. Young organs are thus dissected out, and cut stems can be grown in vitro [7, 9–12]. Alternatively, whole plantlets can be grown in vitro in the presence of the auxin transport inhibitor (and flower formation inhibitor) NPA to generate naked meristems [8, 13]. Meristems can be observed with a CLSM equipped with a water-dipping lens in water at room temperature over time, which makes them ideal systems to study the cellular basis of morphogenesis in the aerial part of the plant. In particular, using transgenic lines, one can follow gene expression or protein dynamics, such as the polarity of the auxin efflux carrier PIN1 or cortical microtubule orientations, and associate these behaviors with shape changes in the developing meristems (e.g., refs. [14, 15]). Last, these living meristems can be treated with drugs—e.g., the microtubule-depolymerizing drug oryzalin [8, 16]—and hormones, e.g., auxin [17], and the tissue can be mechanically perturbed by local compressions or laser-induced cell ablations [13, 18]. Other imaging techniques can be used sequentially to obtain additional information on the meristems, such as high-resolution growth quantifications with the replica method [19–21] or with other 3D reconstruction methods such as MARS-ALT [12] or MorphoGraphX [22], as well as local mechanical properties with indentation techniques [23, 24].

2 Materials

2.1 Plant Material and Growth Conditions

1. Inflorescence meristems of *Arabidopsis thaliana*: seeds are sowed on soil and plants are usually kept at 22 °C in short day conditions (8 h light/16 h dark) for at least 3 weeks to increase the size of the meristem. Plants are then transferred to long day conditions (16 h light/8 h dark; *see* **Note 1**). For light, we use a combination of white and gro-lux neons.
2. NPA-treated Arabidopsis seedlings: sterile seeds are sowed on the Arabidopsis medium supplemented with 10 μM NPA (*see* below). Seedlings are grown in tall petri dishes, in long day conditions, at 22 °C.
3. Vegetative meristems: 11- to 12-day-old tomato (*Solanum lycopersicum/Lycopersicon esculentum,* e.g., *cv M82 or Moneymaker)* seedlings are used. Seedlings are grown in soil in long day conditions (16 h light per day, 110 μEm^2/s) in 65 ± 10 % humidity, at 20 ± 2 °C.

2.2 Media

1. Medium used to grow Arabidopsis shoot apices (Arabidopsis apex culture medium, ACM): dissolve 2.2 g/l MS medium without vitamins (Duchefa Biochemie—MS basal salt mixture), add 1 % sucrose, and adjust pH to 5.8 with KOH. The solid medium is obtained by adding 0.8 % agarose. Vitamins are added to the lukewarm solution from the stock listed below.
2. 1,000× vitamin stock solution: for 50 ml, dissolve 5 g myoinositol, 0.05 g nicotinic acid, 0.05 g pyridoxine hydrochloride, 0.05 g thiamine hydrochloride, 0.1 g glycine in millipore water. Filter the solution at 22 μm under laminar hood, aliquot, and store at −20 °C.
3. Medium for growing NPA-treated seedlings (whole plantlets with naked meristems): we use the Duchefa "Arabidopsis" medium (DU0742.0025). Dissolve 47.28 g in 4 l of water, add 8 ml of 1 M $Ca(NO3)_2$ $4H_2O$, adjust pH to 5.8 with KOH, and add 1 % of agar (Merck). We germinate the seeds on 10 μM NPA (*N*-(1-naphthyl) phthalamic acid) using a 0.1 M stock solution (10,000×) in DMSO and stored for no more than 3 months at −20 °C. There is more chance to get naked stems if seedlings are densely sowed.
4. Medium used to grow tomato shoot apices (tomato apex culture medium): dissolve 1× MS minimal organic powder medium; add 2 % (final conc.) sucrose, 1.5 % (final conc.) agarose; adjust pH to 5.8 with KOH; and add hormones—gibberellic acid A3 (0.01 μM) and kinetin (0.01 μM). To avoid medium contamination, add preservative for plant tissue culture; we use 1 μl/ml PPM (Preservative for Plant Tissue Culture Media, Plant Cell Tech).

2.3 Tools and Microscopes

1. To dissect out old flowers, buds, or leaves, tweezers, injection needles, razor blades, or scalpels can be used. A good binocular, with increasing magnification, is necessary: an Arabidopsis meristem is between 100 and 300 μm wide.
2. To image the meristems, we use small disposable petri dishes or higher tissue culture dishes, which can be put under the microscope objectives (we use a 20×, 40× or 63× water immersion achroplan lens with a long working distance—40× is the most versatile). We use an upright confocal laser scanning microscope (e.g., Zeiss LSM510, 700, 710, 780 or Leica SP5). An inverted CLSM is also usable, if the meristem is placed deep into the medium, so that the meristem surface is close to the medium surface, thus allowing the water column between the lens and the meristem to be maintained over the time of image acquisition.

2.4 Dyes

1. To visualize cell outlines, the FM4-64 lipophilic dye can be used. A stock solution is prepared in water at a concentration of 330 μg/ml and stored at −20 °C. This stock solution can be considered to be 1× to 10× depending on the permeability of the sample to the dye. Once diluted and in use, it can be stored at 4 °C.
2. Propidium iodide (PI) can be used to stain cell walls. A 0.1 % PI solution is prepared by mixing the PI powder in water, filtering and aliquoting in tubes. If not used, they can be stored at −20 °C. Note that PI can also be used to detect dead cells: PI does not go through an intact plasma membranes and binds to the cell wall, but in a damaged cells, it enters into the cell and binds to DNA [25].

3 Methods

3.1 Accessing the Meristem

1. In Arabidopsis, when plants on soil start to bolt (i.e., when the inflorescence stem is about 2 cm above the rosette leaves), the stem is cut out and placed vertically in box containing 1 % agar MS medium (*see* **Note 2**). Note that as the size of the Arabidopsis inflorescence meristem is decreasing as stem elongates, we recommend using short stems, where the apical meristem is easier to dissect and handle. Using a pair of forceps (that were previously sharpened as much as possible on a stone), floral buds are removed from the most aged to the youngest, by clipping them out. To remove the younger buds, a good binocular is absolutely necessary. In the end, a glossy dark green dome surrounded by light green young organs should be visible—it corresponds to the inflorescence meristem (Fig. 1; *see* **Note 3**). Organs that are not obstructing the

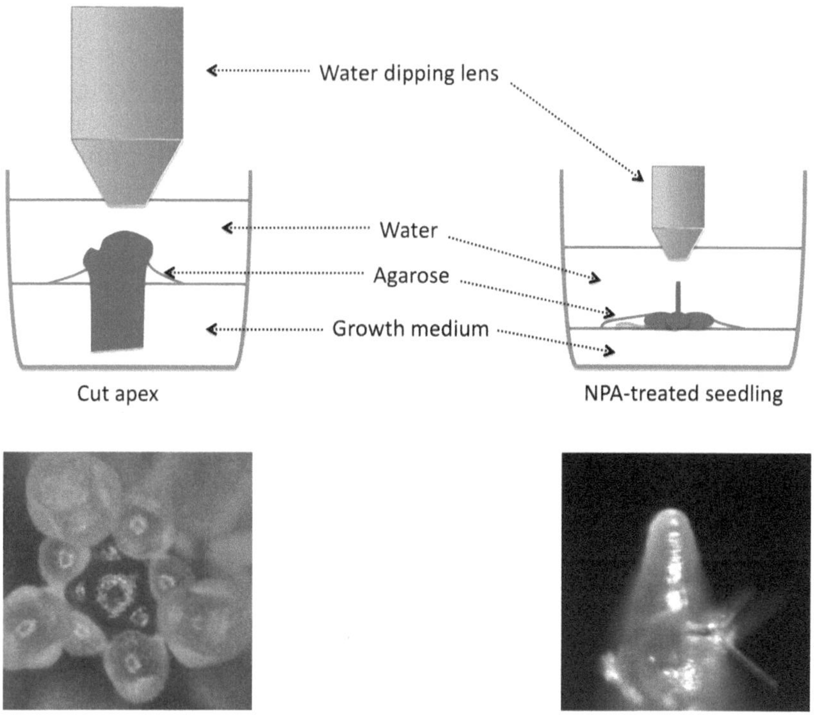

Fig. 1 Time-lapse imaging of developing meristems using confocal laser scanning microscope. Cut apices or NPA-treated seedlings can be used to access the meristem surface, using a water-dipping lens. The images below are taken from Arabidopsis and adapted from [28] and [8]

optical path should not be dissected out. Cut apices are then transferred to a fresh ACM medium in a clean box that will be used for imaging. Make sure to keep the apices at a distance from the boxes edges to allow the displacement of the lens along the *X*-, *Y*-, *Z*-axis while imaging the meristem. It is usually better to dissect the apex a few hours before imaging, to let the apex recover from the wound stress.

2. As soon as naked meristems can be observed in the NPA-grown conditions, seedlings are transferred to a box containing Arabidopsis medium without NPA. The transfer can be done at a later stage too, but we do not recommend it as the meristem is likely to become smaller as the stem elongates and as a longer stem will increase the chances of having random meristem movements while imaging it under the CLSM. The whole plantlets are stuck to the medium with a generous amount of lukewarm melted 0.5 % agarose (Fig. 1). Using a binocular, make sure that the meristem is properly oriented. In comparison with cut apices from soil grown plants, the integrity of the whole plant is preserved when using NPA-grown seedlings. However, NPA may induce artifacts and the size of the meristem is much smaller.

3. In tomato (*see* **Note 4**), 11–12-day-old seedlings are cut. The apex is put vertically in a petri dish filled with a solid agarose medium (e.g., 1 % agarose) to prevent drying of apex during the dissection. Most of the preexisting leaf primordia are removed, but it is reasonable to leave the youngest primordia. If organs are covering the meristem, their most apical part can be cut, although removing too much could damage the meristem or affect its development through the indirect effects of wounding. The dissected apices of a subapical region ca. 5 mm long are transferred to the tomato apex culture medium. It is recommended to put only one apex per dish, to minimize time that each the apex will be under the CLSM during imaging. It is usually better to dissect the apex a few hours before imaging, to let the apex recover from the wound stress.

3.2 Staining the Meristem Before Imaging

To reconstruct meristem geometry from CLSM stacks and/or segment the cell topology for further analysis, cell outlines must be visualized. The signal-to-noise ratio (from either plasma membranes or cell walls) must be high enough to ensure the best results from the segmentation tools. One can use a transgenic line, in which fluorescent protein is fused with plasma membranes (e.g., the Arabidopsis *p35S::GFP-LTI6b* line), or fluorescent dyes for either plasma membranes (FM4-64) or cell walls (PI).

1. Staining plasma membranes: before imaging, the meristem is immersed for at least 5 min in distilled water (*see* **Note 5**). This will notably fully hydrate the gel and prevent the stem from moving too much in the Z direction while imaging. This is notably crucial when stacks of images are obtained. The immersion in water may also hydrate the meristem surface and facilitate the uptake of dyes, like the membrane marker FM4-64. If membranes need to be stained, the water is removed, inflorescences are treated with about 2 μl of a 330 μg/ml stock in water of FM4-64 (Invitrogen) for 1–5 min (*see* **Note 6**) and the inflorescences are immersed in water again.
2. Staining cell walls: a 0.1 % PI solution is used. Generally, the meristem has to be stained each time immediately before the imaging. To stain the meristem, pour the PI solution to the petri dish or other culture dish, so that the whole apex is immersed in the solution. If the meristem is stained for the first time, incubate for ca. 5 min. For the following times 2–3 min is enough. After removing the PI solution (the solution can be reused again a few times), wash the apex in the water at least twice. Note that although PI is light sensitive, it can be handled under regular room light conditions.

3.3 Imaging the Meristem

1. To image the meristem with water-dipping lenses, pour distilled water into the box containing the cut apices or NPA-grown seedlings so that the meristems are fully covered with water (Fig. 1). Since the solid growth medium absorb the water and can swell, we recommend to keep the water in the dish for at least 10 min before image acquisition.
2. The stack of images is obtained with a resolution of 0.5–1 μm along the *Z*-axis (an Arabidopsis meristematic cell is about 5 μm long in width and height).
3. If the meristem is moving during image acquisition, because the medium does not adhere to the apex enough, it is useful to add a few drops of more concentrated agarose (e.g., 2 %) around the apex to immobilize it. To check if the initial position of the object has changed during the stack acquisition, restart an acquisition and check whether the position has changed. Although the signal would still be visible, make sure not to cover the meristem with agarose.
4. Confocal settings have to be adapted to the object and the software for data processing. For time-lapse imaging, it is better to reduce both time of scanning and laser irradiation as much as possible.
5. After imaging, water is removed and the boxes containing the meristems are returned to a growth chamber (*see* **Note 7**). Depending on the time interval between successive time points, restaining with FM4-64 is not always necessary.

4 Notes

1. Short day conditions are not absolutely required in this protocol, and live imaging of the meristem can be achieved from plants grown in long day and continuous white light conditions too.
2. MS medium or any other kind of standard medium can be used here. The most important thing is to have a relatively concentrated agar medium to maintain the stem in a vertical orientation.
3. Dissection can be done in air or in water. It is preferable to dissect in water, as the meristem may dry out fast. If the dissection only last for a few minutes, dissection in air can provide good results and can be easier to start with, as dissection of the meristem in water under the binocular can prove challenging because of optical aberrations.
4. In tomato, the transition from vegetative to reproductive phase is autonomous and the meristem stays vegetative until 6–12 leaves are formed [26]. The time of transition can be modulated by exogenous stimuli (light, temperature) and is different in different backgrounds. The transition from vegetative to inflorescence meristems can occur in vitro, when too old seedlings are used.

However, even when using 11–12 days old seedlings for time-lapse imaging (at 12 h interval), we occasionally observe the transition at the fourth or fifth time point. During the floral transition, the shape of meristem is changing, i.e., the meristem is more bulgy (and instead of leaf primordia, floral meristems are formed).

5. Immersion of the meristem can be done in any water solution, including tap water, although sterile water must be preferred when doing time-lapse imaging over several days, to prevent contaminations. While this may represent a hypotonic stress, in our hands, we found little impact on signals in Arabidopsis meristem. This is likely to depend on the species, as tomato meristems seem to have a higher isotonic point than Arabidopsis meristems (*see*, e.g., ref. [27]).
6. We find that the extent of FM4-64 staining can be unpredictable. Depending on the biological question, using membrane-bound fluorescent reporters (e.g., *p35S::GFP-LTI6b* line) can be preferable, notably when focusing on the epidermal layer of the tissue.
7. Many physiological processes are regulated by light; thus, light conditions in a growth chamber, in which in vitro culture is conducted, are very important. For example, to follow morphogenesis at the tomato meristem, we use a constant light for in vitro culture, as organogenesis can be controlled by light. In particular, light can change the distribution of auxin at the meristem and PIN1 localization at plasma membranes and affect cytokinin signaling and expression of key regulatory genes [11].

Acknowledgments

This work was supported by a bilateral grant from INRA, France, and Ministry of Science and Higher Education, Poland, and by a grant from Agence Nationale de la Recherche ANR-10-BLAN-1516 "Mechastem." We thank Marion Louveaux for helpful comments on this manuscript.

References

1. Bernier G, Jensen WA (1966) Pattern of DNA synthesis in the meristematic cells of sinapis. Histochemie 6:85–92
2. Hake S, Vollbrecht E, Freeling M (1989) Cloning knotted, the dominant morphological mutant in maize using Ds2 as a transposon tag. EMBO J 8:15–22
3. Jackson D, Veit B, Hake S (1994) Expression of maize knotted1 related homeobox genes in the shoot apical meristem predicts patterns of morphogenesis in the vegetative shoot. Development 120:405–413
4. Bowman JL, Smyth DR, Meyerowitz EM (1989) Genes directing flower development in Arabidopsis. Plant Cell 1:37–52
5. Laufs P, Grandjean O, Jonak C et al (1998) Cellular parameters of the shoot apical meristem in Arabidopsis. Plant Cell 10:1375–1390
6. Lucas WJ, Bouche-Pillon S, Jackson DP et al (1995) Selective trafficking of knotted1

homeodomain protein and its mRNA through plasmodesmata. Science 270:1980–1983
7. Reddy GV, Heisler MG, Ehrhardt DW et al (2004) Real-time lineage analysis reveals oriented cell divisions associated with morphogenesis at the shoot apex of *Arabidopsis thaliana*. Development 131:4225–4237
8. Grandjean O, Vernoux T, Laufs P et al (2004) In vivo analysis of cell division, cell growth, and differentiation at the shoot apical meristem in Arabidopsis. Plant Cell 16:74–87
9. Bayer EM, Smith RS, Mandel T et al (2009) Integration of transport-based models for phyllotaxis and midvein formation. Genes Dev 23:373–384
10. Fleming AJ, McQueen-Mason S, Mandel T et al (1997) Induction of leaf primordia by the cell wall protein expansion. Science 276: 1415–1418
11. Yoshida S, Mandel T, Kuhlemeier C (2011) Stem cell activation by light guides plant organogenesis. Genes Dev 25:1439–1450
12. Fernandez R, Das P, Mirabet V et al (2010) Imaging plant growth in 4D: robust tissue reconstruction and lineaging at cell resolution. Nat Methods 7:547–553
13. Hamant O, Heisler MG, Jonsson H et al (2008) Developmental patterning by mechanical signals in Arabidopsis. Science 322:1650–1655
14. Heisler MG, Ohno C, Das P et al (2005) Patterns of auxin transport and gene expression during primordium development revealed by live imaging of the Arabidopsis inflorescence meristem. Curr Biol 15:1899–1911
15. Uyttewaal M, Burian A, Alim K et al (2012) Mechanical stress acts via katanin to amplify differences in growth rate between adjacent cells in Arabidopsis. Cell 149:439–451
16. Corson F, Hamant O, Bohn S et al (2009) Turning a plant tissue into a living cell froth through isotropic growth. Proc Natl Acad Sci USA 106:8453–8458
17. Reinhardt D, Mandel T, Kuhlemeier C (2000) Auxin regulates the initiation and radial position of plant lateral organs. Plant Cell 12: 507–518
18. Heisler MG, Hamant O, Krupinski P et al (2010) Alignment between PIN1 polarity and microtubule orientation in the shoot apical meristem reveals a tight coupling between morphogenesis and auxin transport. PLoS Biol 8:e1000516
19. Williams MH, Green PB (1988) Sequential scanning electron microscopy of a growing plant meristem. Protoplasma 147:77–79
20. Dumais J, Kwiatkowska D (2002) Analysis of surface growth in shoot apices. Plant J 31: 229–241
21. Routier-Kierzkowska A-L, Kwiatkowska D (2008) New stereoscopic reconstruction protocol for scanning electron microscope images and its application to in vivo replicas of the shoot apical meristem. Funct Plant Biol 35: 1034–1046
22. Kierzkowski D, Nakayama N, Routier-Kierzkowska AL et al (2012) Elastic domains regulate growth and organogenesis in the plant shoot apical meristem. Science 335:1096–1099
23. Milani P, Gholamirad M, Traas J et al (2011) In vivo analysis of local wall stiffness at the shoot apical meristem in Arabidopsis using atomic force microscopy. Plant J 67: 1116–1123
24. Peaucelle A, Braybrook SA, Le Guillou L et al (2011) Pectin-induced changes in cell wall mechanics underlie organ initiation in Arabidopsis. Curr Biol 21:1720–1726
25. Rounds CM, Lubeck E, Hepler PK et al (2011) Propidium iodide competes with Ca(2+) to label pectin in pollen tubes and Arabidopsis root hairs. Plant Physiol 157: 175–187
26. Samach A, Lotan H (2007) The transition to flowering in tomato. Plant biotechnol 24: 71–82
27. Nakayama N, Smith RS, Mandel T et al (2012) Mechanical regulation of auxin-mediated growth. Curr Biol 22:1468–1476
28. Vernoux T, Besnard F, Traas J (2010) Auxin at the shoot apical meristem. Cold Spring Harb Perspect Biol 2:a001487

Chapter 10

Quantifying Cell Shape and Gene Expression in the Shoot Apical Meristem Using MorphoGraphX

Pierre Barbier de Reuille, Sarah Robinson, and Richard S. Smith

Abstract

Confocal microscopy is a technique widely used to live-image plant tissue. Cells can be visualized by using fluorescent probes that mark the cell wall or plasma membrane. This enables the confocal microscope to be used as a 3D scanner with submicron precision. Here we present a protocol using the 3D image processing software MorphoGraphX (http://www.MorphoGraphX.org) to extract the surface geometry and cell shapes in the shoot apex. By segmenting cells over consecutive time points, precise growth maps of the shoot apex can be produced. It is also possible to tag a protein of interest with a fluorescent marker and quantify protein expression at the cellular level.

Key words Shoot apex, Confocal microscopy, Image analysis, Segmentation, Watershed, Shape analysis, Expression pattern, 3D visualization

1 Introduction

MorphoGraphX is an open-source application designed for the visualization and processing of data in 3D either as volumes (i.e., 3D images) or as surfaces. In this chapter, we will present the protocol used to extract information from the shoot apical meristem of the tomato or *Arabidopsis thaliana*, as published previously [1, 2]. Also provided in the notes are variants of this protocol for use with other tissues, such as the Arabidopsis root [3].

MorphoGraphX can manipulate two kinds of data: *stacks* (3D volumetric images) and *meshes* (triangulated surfaces).

A *stack* corresponds to a specimen for a given time point and contains two images, called *stores*: the main store and the work store. When processing images, MorphoGraphX can use either store as the input but always writes to the work store. This allows the use of the main store as a checkpoint while optimizing parameters.

Viktor Žárský and Fatima Cvrčková (eds.), *Plant Cell Morphogenesis: Methods and Protocols*, Methods in Molecular Biology, vol. 1080, DOI 10.1007/978-1-62703-643-6_10, © Springer Science+Business Media New York 2014

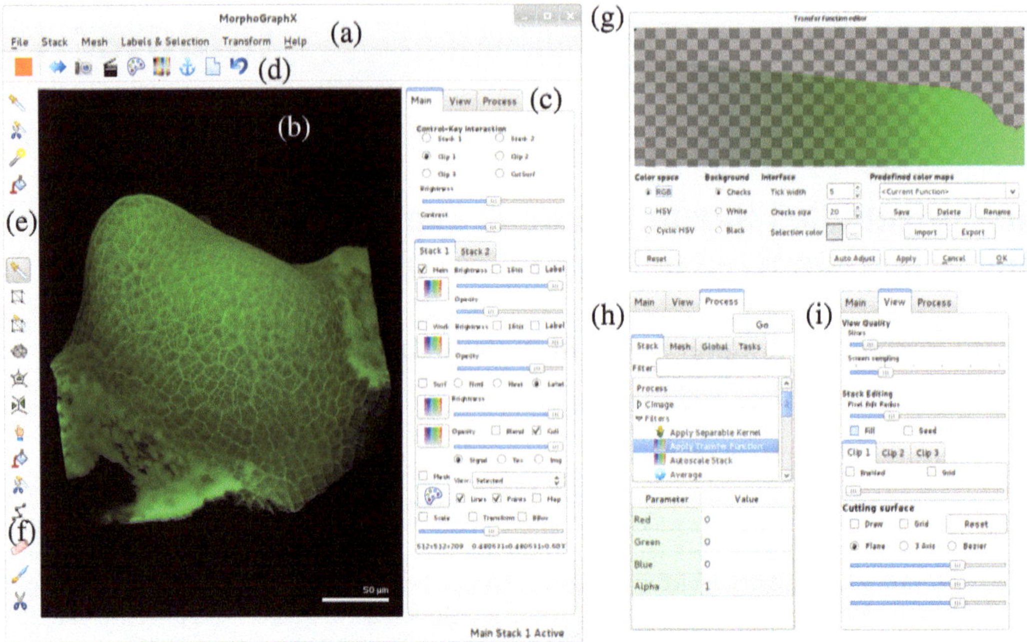

Fig. 1 MorphoGraphX main window: (**a**) menu bar; (**b**) visualization window; (**c**) tab panel; (**d**) global toolbar; (**e**) volume toolbar; (**f**) mesh toolbar; (**g**) transfer function dialog box; (**h**) "Process" tab; (**i**) "View" tab. The visualization panel shows the loaded stack

A *mesh* consists of vertices connected into triangles to define a surface. MorphoGraphX can perform image processing operations on this curved surface, with each vertex of the mesh corresponding to a pixel in a flat 2D image. It is also possible to modify the structure of the mesh, for example, by subdividing triangles or deleting vertices.

Each stack is associated with a single mesh. They share the same reference system and therefore cannot be displaced relative to each other. However, MorphoGraphX does allow two stacks and their associated meshes to be loaded at the same time. These stacks have their own coordinate systems and therefore can be positioned independently.

Any modification that does not require user interaction is achieved through the use of *processes*. The processes are grouped in three categories: (1) *stack processes*, which modify only the stacks and their positions; (2) *mesh processes*, which modify only the meshes (surfaces) and their properties; and (3) *global processes*, which may modify anything. The list of processes is located in the interface on the "Process" tab (Fig. 1h). In the box below the process list, the relevant parameters are displayed for the selected process. Once the parameters are entered, the process is executed either by pressing the "Go" button on the top-left corner of the tab or by double-clicking on the process name. The result of a process is either a modification of the state (visualization, file saved, etc.) or an error which will appear in a dialog box.

2 Materials

2.1 Confocal Images

For this protocol, we require confocal images of a shoot apical meristem with the cell wall or cell membrane labelled in some way. For example, the samples in the figures used propidium iodide (PI) staining. Optionally other markers, localized on the plasma membrane or inside the cell, may be collected in other channels [1, 2].

When acquiring images for the purpose of 3D reconstruction with MorphoGraphX, the following principles should be followed.

1. Images should be acquired in 16 bits per pixel: this is the native image format of MorphoGraphX. All the processing will be done in 16 bits, and the extra information is often crucial for correctly reconstructing areas of the image with low intensity. If 16 bit collection is not available on your microscope, use 12 bit or the highest available setting.
2. Use cubic voxels if possible (i.e., the z-step is equal to the resolution in x and y).
3. Add planes (i.e., reduce the z-step) rather than using line or frame averaging. For the same amount of exposure to the laser, you will get extra information on the z position.
4. Adjust the offset and gain of the sensor to maximize the dynamic range of your image (*see* **Note 1**).
5. Collect extra Z-slices above and below the area of interest of the meristem. The first and last slices should not contain any useful signal: they are used to limit the size of the meristem along the z-axis. This will ensure you have the correct shape for the whole meristem, including the tip.
6. Most microscopes save their images in a proprietary format. MorphoGraphX can read a set of standard file formats but works best with single TIFF stacks as written by ImageJ or FiJi [4–6]. Load the images using ImageJ with the LOCI plug-in [7] and save it as a single TIFF stack (*see* **Note 2**).

2.2 Computer

You will need a computer with a recent nVIDIA graphics card that supports CUDA. A card with 2 Gb or more dedicated video memory is recommended. The computer itself should have at least 8 Gb of main memory, and a multi-core CPU is beneficial for some processes.

3 Methods

3.1 Visualization and User Interaction

Once the image is acquired, the first step is to visualize it. The stack should be in a single TIFF file, with only one time point and channel per file.

1. Launch MorphoGraphX. You should see a single window, with a black central area (the visualization zone), some menus and toolbars, and a right panel with three tabs (Fig. 1). You will also see a terminal window opened with some text (*see* **Note 3**).
2. Drag the file containing the first time point of your meristem and drop it into the visualization area of MorphoGraphX. You should see the volumetric image appear in the visualization area (Fig. 1b, *see* **Note 4**). You should be able to rotate and translate the image by using the left and right mouse buttons. The wheel can be used to zoom. A complete description of camera manipulations can be found in the online help of the software, accessible either from the "Help" menu or by pressing the "h" key.
3. Assess the quality of the acquisition (*see* **Note 6**): the meristem should be "complete" (i.e., it must not be cropped); the cells of interest must be entirely visible; there should not be holes or spots in the image (i.e., no leak of the wall marker inside the cells and no large interruption of the marking of the wall).

3.2 Extracting the Meristem Surface

Since the meristem is a dome-shaped organ with considerable surface curvature, there is no flat projection of the cells that would conserve their geometry. Instead, the shape of the meristem's surface is extracted, and the signal projected onto this curved surface for further processing.

1. Use the stack process "Gaussian Blur Stack" in the "Filters" folder to smooth out the noise (*see* **Note 7**).
2. If required, remove any unwanted objects. Use the voxel editing tools in erase mode (*see* **Note 8**) to remove any object that is not part of the meristem.
3. Save the stack to a new file, and copy it to the main store with the stack process "Copy to Work to Main Stack" in the "Multi-stack" folder.
4. Use the stack process "Edge Detect" in the "Morphology" folder to obtain a filled-in stack of the meristem shape (*see* **Note 9**). The threshold determines what light intensity (ranging from 0 to 65,536) is considered to be the surface. If there are large holes in the shape, then repeat the edge detection with a lower threshold. If there are only small holes, use the stack process "Fill Holes" to fill them in. It is important that this step is done well. To check that the filled-in shape matches the original stack well, turn on the main store to make it visible, and then use the clipping plane to compare the two stores (*see* **Note 5**). The top of the data should match the top of the extracted shape (Fig. 2a). If there is not a good match, try again with a higher threshold. Too high a threshold will result in holes.

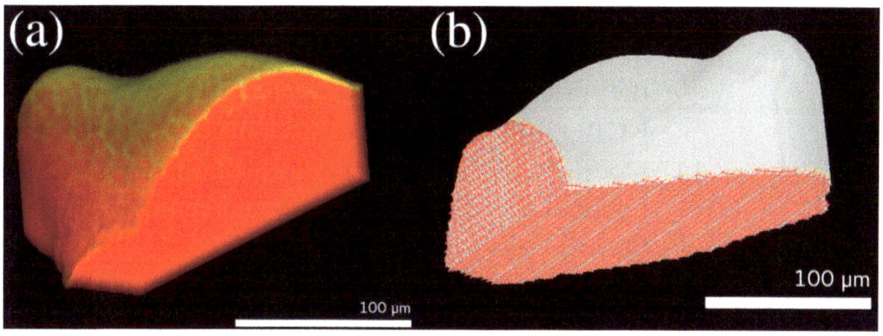

Fig. 2 (**a**) Clipped view of the meristem after the "Edge Detect" process. The work store is shown in *red* and the main store in *green*. Where there is signal from both stores, the color visible is *yellow*. The *red* should not be visible above the *green* of the main store. (**b**) Surface of the meristem with the bottom and the sides selected, ready for deletion

5. Use the mesh process "Marching Cubes Surface" in the "Creation" folder to extract the surface [8]. The cube size should be around half the size of a cell, so as to capture the shape of the meristem but not surface noise (*see* **Notes 10** and **11**).
6. To reduce the number of vertices required, the bottom of the mesh can be removed. First visualize the triangle mesh by ticking the "Mesh" checkbox and un-ticking the "Surface" checkbox of the stack 1.
7. With the mouse move the camera to view the meristem from the side. Once the view is close to an axis, you can double-click the left button to align it to the axis. This will make it easier to select the bottom row of vertices. Using the "Select points in mesh" tool, select the bottom and the sides of the mesh (Fig. 2b). The vertices should turn red once selected. Press the "Delete" key to remove the selected vertices. Remove all parts of the mesh you do not need/want.
8. Smooth and refine the mesh using the mesh processes "Smooth" and "Subdivide" in the "Structure" folder. Apply them both until you get about 5–6 points across each cell. Be careful not to run this too many times, as the vertex count will roughly quadruple each time. You may want to smooth several times when you first start to remove small bumps in the surface.
9. Un-tick the mesh checkbox and tick the surface and main check boxes in the stack 1 to show both the surface of the mesh and the original, cleaned image.
10. Use the mesh process "Project Signal" in the "Signal" folder. The parameters indicate how deep into the stack the signal is taken from. Be sure that the main store is selected and that the work store is not. Otherwise you will project the uniform signal from the filled-in stack. The minimum distance should be

set large enough to avoid the surface wall, and the maximum distance should be just above the height of the smallest cell on the surface. Always check the clarity of the signal and try different ranges if there are problems.

11. Use the mesh process "Smooth Mesh Signal" in the "Signal" folder. This will reduce noise and remove any small holes in the cell walls.
12. At this stage, check that you can see the contour of all the cells. If some cells are too coarse to be seen correctly, repeat **steps 8–11**. Try not to make the mesh too fine, the goal is to get an initial segmentation with as few vertices as possible.

3.3 Meristem Segmentation

The segmentation is an iterative process that starts with a rough segmentation which is then subsequently refined. The goal is to get the best cell shape possible, while only refining the structure where needed, i.e., where the cell walls are. The method used is a hand-seeded watershed that is implemented on the surface mesh.

1. Tick the "Label" radio button next to the surface check box to enable the viewing of surface labels. Each cell will be given a unique label number and will be colored accordingly.
2. Ensure the transfer function for the surface is gray-scaled (click on the first rainbow button below the "Surface" check box).
3. Select the "Add new seed" tool, and seed the cells by left-clicking while pressing the "Alt" key. Each time you click on a triangle, it will advance to the next label and color the triangles with this until you release the mouse button. Each cell you want to segment should be seeded with at least one triangle.
4. After you have finished seeding all the cells of interest, you should place a line of seeds around the seeded area (Fig. 3a). This will stop the cells from "bleeding" into the surrounding area during the watershed segmentation.
5. Use the mesh process "Segment Mesh" in the "Segmentation" folder. The only parameter is the number of iterations between updates of the view. This will allow you to watch the watershed flood-fill as it proceeds. The next section explains how to correct segmentation errors.
6. At this point you should have an initial segmentation, although the walls may be fairly jagged. This is due to the coarseness of the mesh which will now be refined, near the cell walls where more resolution is needed (Fig. 3h, i).
7. Use the mesh process "Subdivide Adaptive Near Borders" in the "Structure" folder. This process will subdivide all triangles that are larger than the "Max Area" parameter, which are near the cell walls with the distance specified in the "Border Dist" parameter. The "Max Area" parameter should be around half

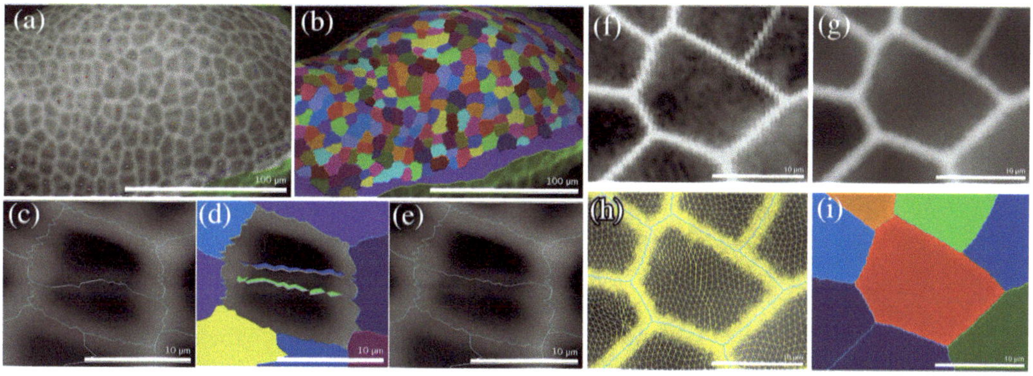

Fig. 3 (**a**) Surface with projected signal and seeds. Note the line of purple seeds all around that will mark the outside of the area of interest. (**b**) Result of the first (coarse) segmentation step. (**c**) Segmentation error. (**d**) Erasing the labels and redrawing close to the cell wall. (**e**) Corrected segmentation. (**f**) Projected image with very small triangles. The voxels (0.5 μm) can be seen in the projection. (**g**) Smoothed projection. (**h**) Different sizes of triangles (in *yellow*) depending on their position with respect to the cells. The boundary between cells is shown as a *blue line*. (**i**) High-quality segmentation

the area of a voxel, and the "Border Dist" should be between the max size of a triangle and under half the size of the smallest cell. If a triangle is not entirely in the border, it won't get subdivided (*see* **Note 12**); however, if the "Border Dist" value is too large, it will clear the small cells entirely. When experimenting with parameters here, be sure to save your initial segmentation to a file so that it can be restored if labels get accidentally cleared.

8. Use the mesh process "Project Signal" in the "Signal" folder, followed by "Smooth Mesh Signal." This will re-project and smooth the signal, which should make the walls look smoother with each iteration (Fig. 3g). Next, run the "Segment Mesh" process to re-segment the newly subdivided surface.
9. Repeat **steps 6** and 7 until the cells walls are sufficiently smooth or the number of vertices in the mesh stops increasing (visible in the status bar at the bottom of the MorphoGraphX window). When the triangles get smaller than the voxels, you will see squarish blocks of signal on the surface when you zoom in (Fig. 3f). At this point further subdivision will not increase the quality of the segmentation.
10. Use the mesh processes "Fix Corners," followed by "Segment Mesh" in the "Segmentation" folder. This will subdivide the corners between cells that were not labelled by the segmentation process and relabel them. Iterate this step until the "Fix Corners" process reports that no vertex has been deleted and that no triangle has been subdivided (visible in the status bar at the bottom).

3.4 Correcting Segmentation Errors

Segmentation errors fall into three categories: (1) over-segmentation when a cell has been divided into many cells, (2) under-segmentation when two or more cells have been merged into one, and (3) incorrect segmentation when the boundaries detected are inconsistent with the data.

1. To correct an over-segmentation error, select the "Pick label" tool (i.e., the pipette in the mesh toolbar). Press the "Alt" key and left-click on one of the segmented cells. Then select the "Fill label" tool (i.e., the bucket in the mesh toolbar), press the "Alt" key, and left-click on all the labels that need to be merged. You will see them being merged after each click.
2. To correct an under-segmentation or incorrect segmentation error, click on the "Label color" button to clear the current label. Then select the "Fill label" tool (i.e., the bucket in the mesh toolbar), press the "Alt" key, and left-click on the cells to clear them. Select the "Add new seed" tool and reseed the cells. In the case of incorrect segmentation, the watershed can be guided by seeding the triangles close to the walls to constrain the segmentation (Fig. 3c–e). Use the mesh process "Segment Mesh" in the "Segmentation" folder to re-segment the affected cells.

3.5 Co-segmentation

To perform time-course studies of the meristem, we need to make sure that each cell has the same label in all time points. After segmentation of the first time point, the segmentation protocol is adapted to reuse the cell labels from the first time point.

1. Load the data into stack 2 and extract the surface of the second meristem as before (*see* **Note 13**).
2. Load the segmented mesh of the previous time step into stack 1.
3. On stack 1, tick the "Mesh" check box and select "Cells" in the drop-down menu next to the check box in the stack 1. Un-tick the "Surface" check box. You should now see a wire-frame outline of only the cells from stack 1.
4. Tick the "Surface" check box and the "Label" radio button in the stack 2.
5. Tick "Stack 1" in the control-key interaction box.
6. Position the camera so as to see both meristems from the side.
7. Holding the control key will allow you to move stack 1 independently from stack 2. While holding the control key, move the stack 1 upward (Fig. 4a) and move the camera to get a view from the top.
8. Align some cells of the stack 1 with some cells of the stack 2. It can be convenient to scale the stack 1 if the cells are significantly smaller (Fig. 4b).

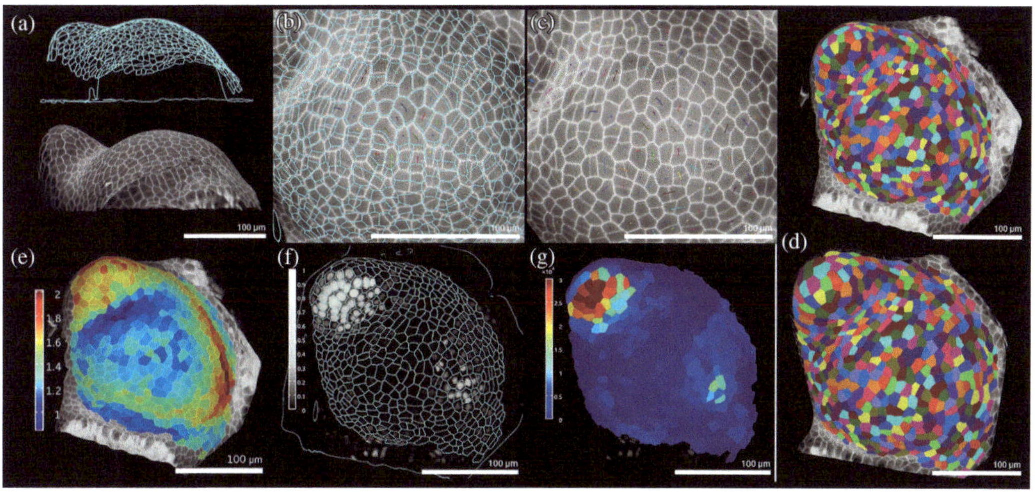

Fig. 4 (**a**) The cell outlines of the previous time point are placed above the surface of the new time point. (**b**) The central cells are aligned to seed the cells using the same labels. (**c**) Result of the co-labelling. Labelling more than one triangle makes it more obvious which cells have been labelled already. (**d**) Result of the co-segmentation: *top*, the previous time point; *bottom*, the new time point. (**e**) Growth heat map based on the co-segmentation of the tissue. (**f**) Projection of DR5::VENUS expression pattern onto the segmented mesh. (**g**) Heat map of DR5::VENUS average intensity per cell

9. Set the stack 2 tab as current, so the stack 2 is the active one.
10. Use the tool "Grab seed from other surface" (i.e., the hand in the mesh toolbar). You are now ready to label stack 2.
11. Press "Alt + Left click" to label the cells. The label used is the one of the cell of the stack 1 through which you click (Fig. 4c). If a cell has divided, make sure all daughter cells get labelled with the same seed. A convenient way to achieve this is to draw a line between the cells.
12. Continue the co-segmentation using the **steps 5–9** of the single tissue segmentation process.

3.6 Computing Geometric Properties of a Tissue

After segmentation, MorphoGraphX offers a set of tools for the analysis of cell shape and the signal intensity of fluorescent markers.

To compute geometric properties (area) of a cell/tissue:

1. Load the segmented mesh, for example, by dropping the mesh file onto the visualization area.
2. Use the mesh process "Heat Map" in the "Heat Map" folder. The "Heat Map Type" should be "Area" and the "Heat Map Visualization" should be set to "Geometry." You can select a file in which the output will be written, which can be read by most spreadsheet and analysis software. You can also specify a range for the color scale, or leave it blank and the color will be auto-scaled to the data.

3. Press "OK" to compute the heat map.
4. You should see a color bar on the top left of the visualization area, and the cells are now colored by area.

3.7 Computing Marker Expression Intensity

To compute the expression intensity of an intracellular marker (e.g., nuclear or cytoplasmic marker), the data must have been acquired at the same time as the cell wall marker in a different channel so that there is no displacement between the two.

1. Load the stack containing the marker of interest and the segmented mesh in the same stack.
2. Use the mesh process "Project Signal" in the "Signal" folder. The distances should be set so as to select most of the cells (*see* **Note 14**, Fig. 4f).
3. Use the mesh process "Heat Map" in the "Heat Map" folder. The heat map type should be set to "Area" and the heat map visualization should be set to "Total Signal" (*see* **Note 15**, Fig. 4g). You can also save the result to a spreadsheet file for further analysis.

3.8 Comparing Data from Two Time Points

Both area and fluorescence expression analysis can be compared over different time points:

1. Load both segmented meshes with the appropriate signal projection, making sure they are both visible.
2. Select the stack tab for the time point that you want to use to visualize the result.
3. Use the mesh process "Heat Map" in the "Heat Map" folder.
4. Select the parameters for the analysis desired (i.e., area, signal).
5. Tick the "Change map" check box and set the other parameters for the change map. If you are analyzing two time points of growth, and are visualizing the result on the first time point, then you would set the change map to "Increasing." This will color the cells with the most expansion in red and those with the smallest in blue. MorphoGraphX will auto-scale the color range to the data. This behavior can be overridden by using the "Use manual range" option and is important when comparing different repeats of an experiment.

4 Notes

1. Most microscopes offer a false coloring of the image using contrasted colors for both ends of the spectrum. Use this mode and ensure you have a few black and a few saturated pixels in the zone of interest. For the purpose of cell segmentation, you

can also use the adaptive gain and threshold to change both continuously, trying to keep the dynamic range constant.

2. If your microscope already creates TIFF images, you should be able to load them into MorphoGraphX (just drag and drop the file into the MorphoGraphX window). In some cases, the voxel size is not read correctly. If this happens, MorphoGraphX includes a stack process called "Change Voxel Size," in the "Canvas" folder. In this process, you specify the size of the voxel. If any dimension is set to 0, it won't be modified.
3. The terminal window of MorphoGraphX is really important. You should never close it. Closing this window will close MorphoGraphX without any hope of saving anything. Also, many algorithms provide more details on the processing or warnings/errors in the terminal window.
4. When acquiring 12 bit images (and sometimes 8 bit), some microscopes store the information in the least significant bits of each voxel, leading to a very dark image. The image needs to be normalized to use the whole range of values. Use the stack process "Autoscale Stack" in the "Filters" folder. You can achieve a similar result without editing the image. First, tick the "16bit" checkbox next to the "Main" checkbox. Then, click on the rainbow button, and click the "Auto Adjust" button. You can then apply the result of this change using the stack process "Apply Transfer Function" in the "Filters" folder.
5. To observe the internal parts of a sample, MorphoGraphX provides clipping planes. There are three clipping planes, each made of two parallel planes. The clipping planes are setup in the "View" tab (Fig. 1i). When enabled, only what is between the two planes is displayed. Clipping planes can be moved by first selecting it in the "Control-Key Interaction" box in the "Main" tab and using the mouse while pressing the "Control" key on the keyboard. The distance between the parallel planes is set by the slider placed on the clipping plane tab.
6. The transfer function is defined per store and can be edited using the rainbow button placed under the checkbox of the corresponding store. The transfer function dialog box shows a histogram of the image, colored with the colors used in the image (Fig. 1g). The simplest way to use the transfer function dialog box is to select a predefined transfer function from the list and use the "Auto Adjust" button to optimize the transfer function to the effective range of values of the image. For a finer control, on the top of the histogram, you can see triangular ticks (on most transfer functions, there are only two: on the extreme left and extreme right). Each tick corresponds to a defined color. In between ticks, the color is linearly interpo-

lated. To edit a tick, simply double-click on the histogram under the tick. A new tick can be added by double-clicking where you want a new one. A color-editing dialog box will appear. Remember than changing the transfer function doesn't change the data, only how it is visualized. If at any point you want to "apply" the current transfer function, you can use the stack process "Apply Transfer Function" in the "Filters" folder. This will change the data to match the transfer function and reload a new transfer function.

7. How to choose the sigma of a Gaussian blur. The Gaussian blur can be used for two purposes: (a) removing noise or (b) removing "flat" areas (i.e., contiguous area where all the voxels have close to the same value) and filling in gaps. (a) A Gaussian function is significant up to a distance of 2–3 times sigma from its center. As such, to remove noise, a sigma about the size of the noise is usually appropriate. (b) To fill gaps and remove flat areas, sigma should be about half of the size of the largest gap or area.

8. The voxel editing tools can be used only on the work store. If you just opened an image you can use the stack process "Copy Main to Work Stack" in the "Multi-stack" folder. The voxel editing tools work by replacing a cylinder in the volume by a given value (by default: 0). However, only the parts visible on the screen will be replaced, and the cylinder is always in the axis of the camera (i.e., it is everything below the circle that becomes visible by pressing the "Alt" key). Orient the view and use the clipping planes to edit only what you need to change and no more. There are three behaviors possible for the editing tools: erase, fill, and seed. The mode depends on the checkboxes "Fill" and "Seed" in the "View" tab. If none is checked, the mode is "erase." If only "Fill" is checked, the behavior depends on the stack mode: if labelled, it will use the current label; otherwise, it will use 65,535. At last, if both "Fill" and "Seed" are checked, and if the stack is labelled, every click will create a new label and use it to fill the selected area.

9. The stack process "Closing" in the "Morphology" folder offers a more general algorithm to fill in (i.e., close) holes. It will work for any shape but is slower and will tend to fuse objects close to each other. The parameter of the process is the size, in voxels, of the largest hole you want to fill in. It is worth noting that if you use increasing values for the radius, the result will be the same as if you had used directly the last value. Therefore, the best strategy to optimize the value is to start with small values, check with clipping planes that all the cells are filled in, and increase the radius if needed. There is no need to go back to the original image in the process.

10. If you want to segment cells in an internal layer, you cannot use this method for extracting the surface. Instead, you should use the cutting surface, either the single plane of the Bezier surface. Move the cutting surface to match the shape of the surface you want to extract, and then use the mesh process "Mesh Cutting Surface" to create the surface. To change the shape of the Bezier surface: (a) draw the grid, (b) select the control points by drawing rectangle around them by pressing the "Alt" key and use the left mouse button, and (c) move them by pressing "Shift" and "Alt," left-click, and drag the control points.
11. If the surface is not visible, make sure the "Surface" check box is ticked. It should appear in white. Show also the original dataset (i.e., the main store) and check that the extracted surface is close to the surface of the meristem on the image.
12. If some cells are too small and the adaptive subdivision erases them, use the mesh process "Subdivide" in the "Structure" folder.
13. You can hide the stack 1, or even reset it temporarily if you don't have enough memory to load both images at the same time. All we need to keep from the stack 1 is the segmented surface.
14. It is possible to project the subepidermal layer if the thickness of the cells is constant enough. Set the minimum distance to a value slightly above the thickness of the epidermal layer and the maximum distance to a value slightly below the distance to the third layer.
15. It is also possible to get the distribution of membrane markers. In the heat map dialog box, you can select "Border signal," which is then the signal from the border to a distance smaller than the "Border Size" parameter of the process. The opposite is the "Interior signal," which is the signal in the cell, but not in the border. At last, you can compute a series of ratios, which have been documented for endocytosis studies [2].

Acknowledgments

We gratefully acknowledge all those involved in the development of the MorphoGraphX—Naomi Nakayama, Thierry Schuepbach, Alain Weber, Anne-Lise Routier-Kierzkowska, and Micha Hersch. We thank Daniel Kierzkowski for time series images and Cris Kuhlemeier for discussions. This work was supported by SystemX.ch, the Swiss National Science Foundation, EMBO Long-Term Fellowship to S.R., and the University of Bern.

References

1. Kierzkowski D, Nakayama N, Routier-Kierzkowska A-L et al (2012) Elastic domains regulate growth and organogenesis in the plant shoot apical meristem. Science 335:1096–1099
2. Nakayama N, Smith RS, Mandel T et al (2012) Mechanical regulation of auxin-mediated growth. Curr Biol 22:1468–1476
3. Santuari L, Scacchi E, Rodriguez-Villalon A et al (2011) Positional information by differential endocytosis splits auxin response to drive arabidopsis root meristem growth. Curr Biol 21:1918–1923
4. Schindelin J, Arganda-Carreras I, Frise E et al (2012) Fiji: an open-source platform for biological-image analysis. Nat Methods 9:676–682
5. Abramoff M, Magalhaes P, Ram S (2004) Image processing with imagej. Biophoton Int 11:36–42
6. Schneider CA, Rasband WS, Eliceiri KW (2012) NIH image to imagej: 25 years of image analysis. Nat Methods 9:671–675
7. Linkert M, Rueden CT, Allan C et al (2010) Metadata matters: access to image data in the real world. J Cell Biol 189:777–782
8. Lorensen WE, Cline HE (1987) Marching cubes: a high resolution 3d surface construction algorithm. Comput Graph 21:163–169

Chapter 11

Mechanical Measurements on Living Plant Cells by Micro-indentation with Cellular Force Microscopy

Anne-Lise Routier-Kierzkowska and Richard S. Smith

Abstract

Indentation methods on the micro- and nanoscale are increasingly used to assess mechanical properties of living plant tissues. These techniques rely on recording the force resulting from indenting the cell surface with a small probe. Depending on the scale of indentation and the indenter shape, force-indentation data will reflect several factors such as cell wall elasticity, turgor pressure, cell and tip geometry, and contact angle. Cellular force microscopy is a micro-indentation method that was designed to precisely measure and apply forces on living plant cells. Here we explain how to use this method to map the apparent stiffness in single cells and tissues.

Key words Plant mechanics, Micro-indentation, Turgor pressure, Primary cell wall, Elastic modulus, Stiffness, Mechanosensing

1 Introduction

The variety of plant cell shape and growth patterns can be attributed to two primary factors, turgor pressure and cell wall mechanical properties. The cell walls, which are only a few hundred nanometers thick, support large tensile stresses (100–1,000 atm) created by the high internal turgor pressure, typically 3–10 atm [1]. Different methods have been used to measure mechanical properties of the primary cell wall and turgor pressure [2, 3]. Indentation methods can be used to measure cellular mechanical properties by deforming the cell surface with an indenter while monitoring both force and tip displacement. These methods are used at various scales, either deforming the whole cell with a large probe or indenting the cell wall very locally with a sharp needle. Compression of whole cells is used to evaluate turgor pressure as well as cell wall elasticity and wall strength, i.e., the force needed to rupture it [4–8]. Nano-indentation experiments, on the other hand, rely on very localized indentation of the cell wall with the sharp tip of an atomic force microscope (AFM) to deduce wall

Viktor Žárský and Fatima Cvrčková (eds.), *Plant Cell Morphogenesis: Methods and Protocols*, Methods in Molecular Biology, vol. 1080, DOI 10.1007/978-1-62703-643-6_11, © Springer Science+Business Media New York 2014

elasticity [9–11]. Micro-indentation techniques are situated in between, measuring both the effects of turgor pressure, cell wall in-plane elasticity, and local compression of the cell wall [12–19]. The cell geometry and tip shape also have an influence on cell deformation and the force measured. Mechanical models are needed in order to interpret the data and untangle the contribution of different factors on the force-indentation curves [13, 15–18].

Cellular force microscopy (CFM) has been designed for force-indentation measurements on living plant tissues [15, 16, 20]. CFM can be used to measure a large range of forces (sub to hundreds of microNewtons), can make large displacements (up to centimeters), and is well adapted to the high forces involved in the measurement of stiffness on turgid plant cells. The CFM can also be used as a plant microsurgery device by applying forces sufficient to puncture the cell wall. The robot consists of a capacitive force sensor mounted on a 3-axis piezo positioner. The system is automated and offers a high resolution in position (nanometer scale) and force (sub microNewton). CFM can be used to scan the sample surface and extract its shape, as well as measure stiffness from the analysis of force-indentation curves (Fig. 1). The setup can easily be integrated with light microscopy for simultaneous imaging during the measurements. Large probe length (1–2 mm) allows measurement in liquids and increases probe access on curved samples. The robot can also be used to mechanically stimulate the cells by applying a controlled force or deformation for a certain period of time. Here we will focus on stiffness measurements in single cells and tissues in liquid.

During a typical measurement with CFM, the sample surface position is first detected, and then a force-indentation curve is acquired for the stiffness measurement. CFM offers two possibilities for detection of sample surface: it may be based on a change in force or stiffness. The force criterion assumes that the force is close to zero before touching the sample and rises only if there is contact with the surface. For measurements in liquid, however, the surface tension of water creates a force on the sensor tip, up to several microNewtons. In this case, a stiffness threshold can be used, with the stiffness coming from surface tension usually in the range of 0.01–0.1 N/m. This mode of surface detection can also be used to avoid issues due to sensor voltage drift caused by light, temperature, and other environmental influences. Once the surface has been detected, the offset on the force before contact (e.g., due to liquid surface tension) can be computed and used to correct the force readout. Force-indentation data is then acquired by slowly indenting the probe into the sample until a predefined force threshold or indentation depth is reached.

The robotic positioner has two modes of actuation, stepping mode and scanning mode. Stepping mode involves a stick-slip movement and is used for large displacements, but it is jittery due to

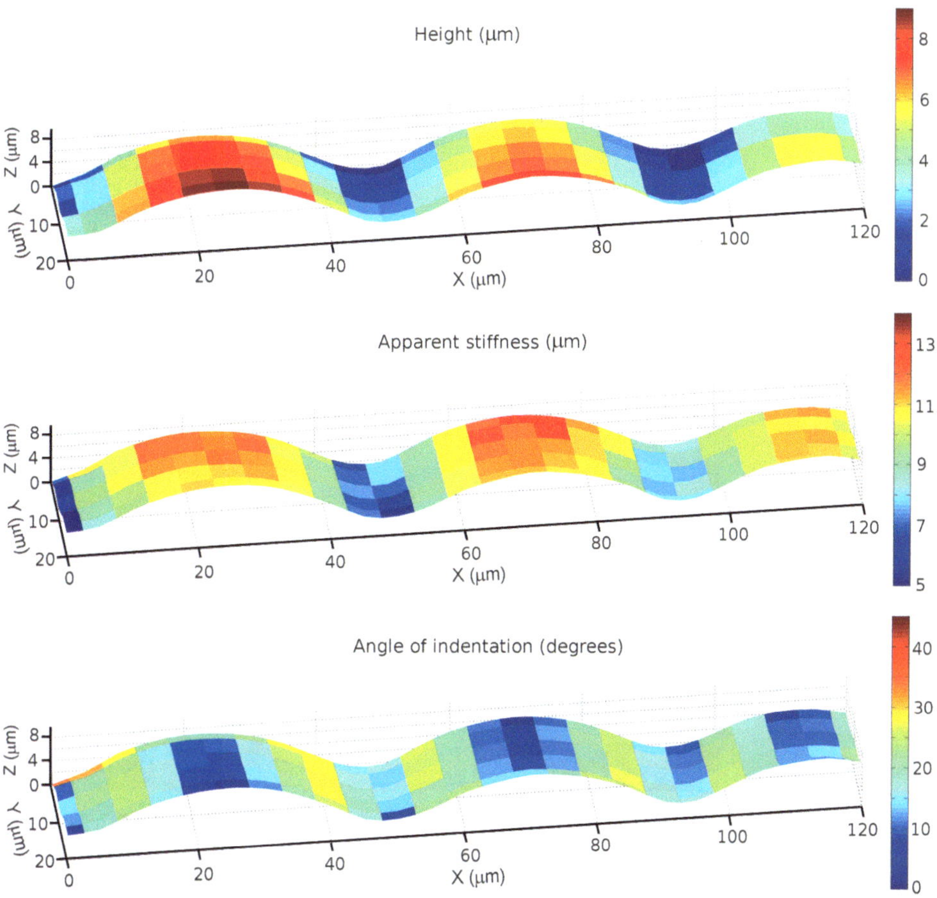

Fig. 1 Examples of data obtained with CFM on onion epidermal peels. The apparent stiffness is dependent on the angle of indentation. Comparison of stiffness values should be done only for data obtained on surfaces almost perpendicular to the sensor axis (e.g., angle <10°)

repositioning of the piezo crystal against the rails. Scanning mode is preferred for the stiffness measurements as the movement is smooth and very precise. The CFM actuator has closed loop positioning, implemented with an internal optical sensor, allowing movements towards a predefined position, with precision in the nm range.

Proper setup preparation and calibration are important factors for data quality. The force sensor is composed of four springs that deflect under load, changing the distance between pairs of combs on the sensor which act as capacitors (Fig. 2). This movement, as well as the slight deflection of other elements of the setup (e.g., connecting parts, probe tip), results in a difference between the actual displacement of the sensor tip and the position reported by the actuator itself. It is therefore necessary to quantify the inherent sensor stiffness by measuring force against displacement while indenting a very stiff sample (e.g., a piece of glass or metal).

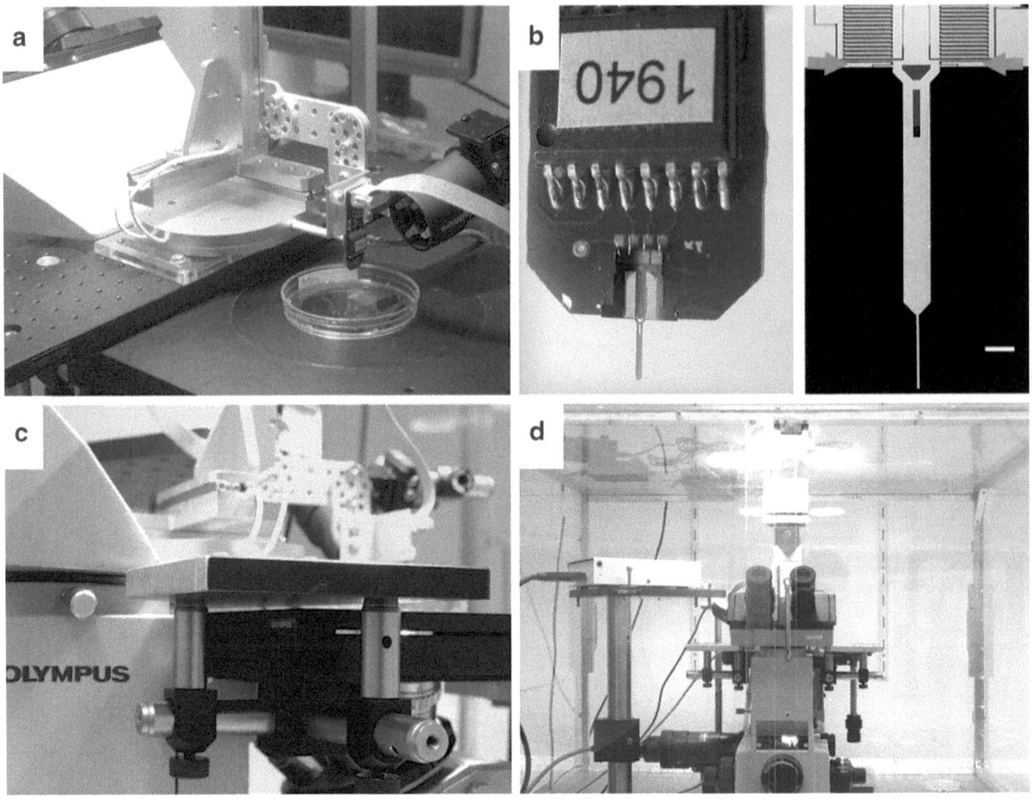

Fig. 2 CFM setup combined with an inverted microscope. (**a**) The force sensor is connected to a 3-axis actuator by aluminum arms. A digital microscope is used to monitor sensor immersion into liquid from the side. (**b**) Close up on the force sensor. The tungsten needle (on the *right*) is glued to the sensor tip. Sensor cantilevers (*red arrows*) deflect when a load is applied to the tip, changing the distance between capacitor plates. Scale bar = 300 μm. (**c**) Side view of the custom stage supporting the CFM setup, above the microscope stage. The posts are directly connected to the nonmoving part of the microscope stage, in order to minimize the total thermal expansion of elements in between the sample and sensor tip. (**d**) The whole setup is placed on an anti-vibration table, isolated by a plexiglass enclosure and illuminated with LED sources outside the box

Sample preparation consists in immobilizing the tissue and providing experimental conditions for the sample that result in a stable turgor pressure within the cells. This will prevent movements during the experiments that would cause aberrant surface reconstruction and biased force measurements (Fig. 3c, d). The manipulations should be as noninvasive as possible and ensure a large contact area between the sample and a stiff substrate. Be aware that samples may react to light or gravity with a fast growth response that could result in non-negligible movements and disturb the measurements.

The choice of parameters for the experiment depends on the sample dimensions, its stiffness, and the time available before the sample starts moving. The force-indentation curves are sensitive to the indentation angle, and any comparison of results should be

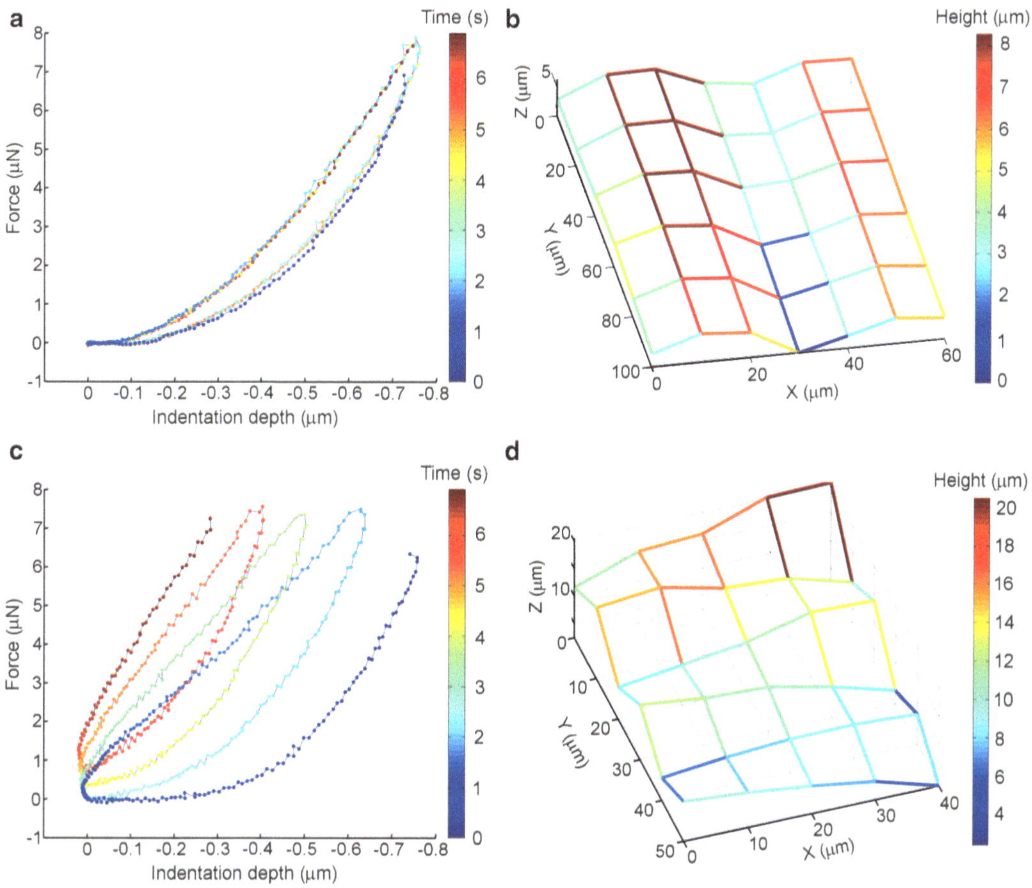

Fig. 3 The quality of experimental data depends on sample movement during the measurements. (**a**, **b**) Shows correct data, for a properly immobilized onion peel. The force-indentation curve (**a**) represents 3 loading/unloading cycles of a total duration of 7 s. The height map (**b**) reflects the proper shape of the sample (here, two adjacent cells). (**c**, **d**) Shows artifacts in data acquired along a tomato stem that was laid horizontal for the experiment. The gravitropic response induced an upward movement of the sample. Force-indentation curve (**c**) shows a drift in the position during the 3 loading-unloading cycles, in contrast with the overlapping curves of the previous example (**a**). The distance between curves at maximal force (7 μN) shows a movement of approximately 50 nm/s (3 μm/min). The height map (**d**) presents a characteristic zigzag shape in the *XZ* plane. This artifact is due to the upward movement of the sample over time while the probe goes back and forth along the *X*-axis to scan the surface

made only for the parts of the sample which are perpendicular to the force sensor axis. This can be achieved by performing a scan grid of the tissue surface and detecting the surface slope (Fig. 1). The grid size and lateral resolution should be sufficient to recognize individual cells and compute the indentation angle accurately at each point. However, too high a resolution slows down the experiment and makes it more invasive. Force-indentation curves can be acquired up to a predefined force threshold or indentation depth. The step size in *Z* direction should be adjusted so that the curves contain enough points for data interpretation, and the maximal deformation should be chosen such that the indentation is not too invasive for the cells, especially if the experiments have to be repeated (time lapse study). Loading and unloading curves are often different and show hysteresis, which can be due to viscoelastic effects and/or adhesion between the tip and the sample. If loading and unloading cycles are repeated, a drift in the force-indentation curves may be due to creep (irreversible, time-dependent deformation), viscous effects, or sample movement (Fig. 3c). Overlapping curves, on the other hand, indicate reversible (elastic) deformations and proper stabilization of the sample (Fig. 3a).

During data analysis, the *Z* position is first corrected to account for sensor deflection under load, using the stiffness value obtained by measuring on glass or metal [15]. Further analysis of the raw force-indentation curves will depend on the experiment purpose, and different methods can be applied. The contact point can be detected based on a force or stiffness criterion or by fitting the entire force-indentation curve. Each method relies on the fact that the force-indentation curve presents two different regimes, before and after the contact point. Typically, values of force and stiffness are low before contact and present a nearly linear increase of force with indentation (in the order of 10^{-2}–10^{-1} N/m) which is due to surface tension of the medium. After contact, the force rises sharply with indentation depth, often in a nonlinear fashion. The contact point can be poorly detected in very soft samples which do not largely differ in stiffness compared to values measured in liquid alone. Once the contact point has been detected, force-indentation curves can be used to extract an apparent stiffness or fit model curves. The apparent stiffness is useful for qualitative comparison between cells. The stiffness can be extracted as the tangent of force-indentation curves at a particular force or indentation depth [15]. Note that in that case no particular assumption is made on the exact shape of the force-indentation curves. In quantitative studies, the experimental data are fitted to simulated force-indentation curves obtained with a mechanical model [4, 5, 7, 8, 16–18]. Unique solutions for the cell wall elastic modulus or the turgor pressure can be obtained only if enough model parameters are known, such as the cell wall thickness, cell geometry, initial stretch ratio, and Poisson's ratio. Depending on the timescale of indentation, it may also be required to consider cell wall permeability to water [4, 7].

2 Materials

2.1 CFM Setup

The actuator robot is a modular system composed of three linear micro-positioners (SmarAct GmbH, SLC-1780) and controller (SmarAct GmbH, MCS-3D). The controller has knobs for the manual positioning prior to experiments and a USB interface that allows automated control during experiments. The capacitive force sensor (FemtoTools GmbH, http://www.femtotools.com) is mounted on a custom aluminum arm which is fixed to the microrobot. We used the FT-S540-type robot, which has a force range of ±180 μN with a gain around 90 μN/V, to perform noninvasive measurements. FemtoTools sensors from the current line are available in ± 100, ± 1,000, and ± 10,000 μN ranges. The sensors can be ordered with a variety of tungsten probes glued to the tip (Picoprobe tungsten probe tips T-4-22/T-4-35/T-4-60, GGB industries Inc., http://www.ggb.com). The probes stick out of the sensor by 1–2 mm. They are thin cylinders (diameter from 22 to 60 μm) with conical tip that is rounded to a diameter of approximately 1–3 μm. The force signal is collected throughout a data acquisition card from National Instruments (http://www.ni.com) that is connected to the computer either by USB (NI USB-6009) or PCI (NI PCIe-6321).

The CFM robot used here was mounted on a custom-made stage (optomechanical components from Thorlabs GmbH, http://www.thorlabs.com) on the top of an inverted microscope (Olympus IMT-2) (Fig. 2). The whole setup sits on an anti-vibration table (TMC 63-530 series, http://www.techmfg.com) and is covered with a custom plexiglass box to minimize disturbance coming from acoustic vibrations and air currents. The table and setup are also electrically grounded. The light source for the microscope was replaced with LED rings (AmScope, http://www.amscope.com) placed on the top of the plexiglass enclosure to avoid sources of heat within the box. It is important to keep in mind that the capacitive force sensors are sensitive to light. When choosing a light source, make sure that there will be no strong gradient in light that could affect force measurements. In case of CFM on inverted microscope, imaging in reflection mode can be an issue because the light beam is concentrated through the objective. Prefer transmission with light sources far away from the sensor. While performing force measurements in liquid, it is crucial to keep the capacitor plates and electronics of the sensor dry. A high magnification digital microscope (DigiMicro 2.0, dnt GmbH, http://www.dnt.de) was used to provide a side view of the sensor tip (Fig. 2) and check the liquid level. Make sure that the camera focal length is long enough for use with the setup (a few cm).

The controlling software comes in different versions, either custom applications implemented in LabVIEW (National Instruments, http://www.ni.com, used in ref. [15]) or a stand-alone C++ library currently under development. The software is freely available from the authors (http://www.MorphoRobotX.org).

2.2 Biological Samples

Flat samples, such as epidermal peels, can be immobilized using laboratory tags (Tough-Tags, Diversified Biotech, http://divbio.com). Otherwise, they can be partially embedded in high-concentration (2 % or more) low-melting-point agarose (Sigma-Aldrich, http://www.sigmaaldrich.com); make sure that the sample surface is not below the agarose surface. Slides coated with a tissue section adhesive (Biobond, BBInternational, http://www.buybbi.com) improve adhesion of single cells, such as tobacco culture cells, on the substrate. Cell viability can be checked with propidium iodide (0.1 %) staining (Sigma-Aldrich, http://www.sigmaaldrich.com) or, if the cells are large and transparent enough, by monitoring cytoplasmic streaming visually.

3 Methods

In this section we describe how to produce stiffness maps on samples immersed in liquid with a CFM setup mounted on an inverted microscope. Other applications, such as cell puncture or samples that require a horizontal setup to avoid gravitropism, are mentioned in the notes.

3.1 Setup Preparation

It is crucial to make sure that the setup and sensor are working correctly before each experiment. A good practice is to position the robot, mount the sensor, arrange the lighting, and make all the necessary connections to the computer, and then let the sensor stiffness calibration run while preparing the biological samples:

1. Let the entire system warm up in order to stabilize the temperature and minimize thermal expansion. Leave doors and windows closed; turn on all lights and equipment that will be used during the experiment several hours before use.
2. Mount the sensor (*see* **Note 1** for choice of sensor and tip size) and make sure that there is as little play as possible in the sensor arm (screws for the sensor are tight).
3. Perform multiple stiffness measurements on a hard surface (clean glass slide) to check the sensor stiffness. Use a relatively large grid (e.g., 10×10 points with lateral step size 10 μm). The reconstructed surface of the glass should be flat and (nearly) aligned with the $[x,y]$ plane of the actuator. Sensor stiffness should not vary more than a few percent between the grid points, and the force-indentation curves should not present drift in successive loading-unloading cycles. The sensor stiffness should not deviate more than a few percent from when it was first used. A large decrease in sensor stiffness could indicate damage, as they are very fragile (*see* **Note 2**).

3.2 Sample Preparation

1. Remove any obstacle that the sensor tip could collide with during the measurements (e.g., older primordia around the meristem, trichomes on the leaf surface). The sample surface should be clean.

2. Prevent sample movements in all directions (*see* Subheading 2.2). Let the sample recover from mechanical stress inflicted during the preparation. In the case of epidermal peels, this can take several hours.
3. Let the sample adapt to osmotic conditions. Changes in osmolarity result in modification of the turgor pressure and cell deformation. For example, the shoot apex needs at least 30 min to reach a stable size after a change of osmolarity. Long relaxation times can be caused by viscous behavior of the cell wall or osmoregulation. Check if the sample is stable with time lapse microscopy both before and during the experiments (e.g., using a confocal microscope for opaque samples, or inverted microscope for transparent ones).
4. Check that the sample was not damaged during preparation, for example, with PI staining, and make sure the sample and the liquid is dust-free. Dust particles on the sensor tip can disturb the measurements, and removing them is time consuming.

3.3 Poking in Liquid Medium

1. When using a tungsten tip, only the tip itself should be immersed in water (*see* **Note 3**). The sensor probe should not touch the water during the measurements. As a rule of thumb, the tungsten tip should be only half immersed, meaning there should be around 1 mm of water above the sample. Make sure that the liquid is not too deep by using the image from the side digital microscope.
2. The liquid surface tension will affect each force sensor in a different way, depending on the tip size and roughness. The force or stiffness threshold used for contact detection during the measurements should be higher than values resulting from the liquid alone. The effects of surface tension have to be measured for each tip, by monitoring the force while moving the tip in water without coming into contact with the sample.
3. Turgor pressure has a large influence on the measurements. When measuring in medium, check the solution osmolarity regularly, since it could change due to evaporation during long experiments.

3.4 Choice of Parameters

1. Choose the grid resolution depending on the size of surface details. Typically, it is good to have at least 5–10 points across one cell. Too fine a grid should be avoided, as the measurement time will increase dramatically (*see* **Note 4**). Start by scanning small areas with a coarse grid to check if the sample is properly stabilized.
2. The choice of indentation depth/force threshold for stiffness measurements depends on what is being measured (turgor, cell wall elasticity, cell wall strength). A general rule would be to avoid too shallow indentations, which result in very small

forces and noise from contact. For stiffness mapping, a maximal force of 1–40 μN for an indentation depth of 500 nm–5 μm is a good place to start. The cell stiffness is in the order of 1–50 N/m.

3.5 Data Analysis

1. Correct the position along the vertical axis for sensor deflection, using the stiffness values obtained while indenting a hard surface. For any position, the sensor deflection is equal to the force measured divided by the sensor stiffness.
2. Detect the contact point. The value of force at the contact point can be used as an approximation of the force offset due to surface tension, sensor drift, light, etc.
3. The stiffness can be computed in different ways. It can be extracted as the tangent of the force-indentation curves [12] at a fixed force [15] or predefined indentation depth. If the shape of the nonlinear force-indentation curve is known a priori, the function can be fitted to the data [5, 7].
4. Data interpretation of stiffness measurements is easier and less prone to bias performed on the "flat" parts of the sample, i.e., where the surface is perpendicular to sensor axis. Both the force measured and the indentation depth are influenced by the angle under which the probe touches the surface. When comparing stiffness values, make sure that the indentation angle did not exceed a few degrees (e.g., 10°).
5. Stiffness values do not directly reflect the cell wall elasticity or the cell wall turgor pressure. A mechanical model is needed to interpret the data to extract these quantities.

4 Notes

1. The choice of rounded tip of sizes larger than the cell wall thickness ensures that the cell wall will not be damaged during the experiment, provided that the applied force is not too high. For cell wall puncture, use stiffer sensors (e.g., FT-S1000) and a tungsten tip with a small tip diameter (e.g., T4-22).
2. If possible, use the same sensor (same stiffness, same gain, same tip shape) for the whole set of experiments, in order to compare results between samples. This also allows to check the sensor state by monitoring its stiffness over successive days.
3. Use a side camera to check if the sensor tip is not too much into the water. Dirt floating on the water surface can also influence the results if it adheres to the tip. This can be prevented by checking the surface regularly with the side camera as well.

4. Check the timescale of sample growth compared to the time needed for the measurements. The CFM setup can be adapted to perform measurements horizontally instead of vertically. In case the organ presents a fast gravitropic response when laid horizontally, it can be then kept in an upright position during the experiments with a setup modified to measure horizontally.

Acknowledgments

We gratefully acknowledge all those involved in the development of the CFM, Dimitris Felikis, Simon Muntwyler, Felix Beyeler, Brad Nelson, Cris Kuhlemeier, and Michal Huflejt, as well as those who helped to perfect the protocol in the first experiments, Petra Kochova and Alain Weber. We thank Gabriella Mosca, Pierre Barbier de Reuille, and Sarah Robinson for discussions and Willi Tanner for custom machine work. This work was supported by SystemX.ch and the University of Bern.

References

1. Cosgrove DJ (2005) Growth of the plant cell wall. Nat Rev Mol Cell Biol 6:850–861
2. Burgert I (2006) Exploring the micromechanical design of plant cell walls. Am J Bot 93:1391–1401
3. Geitmann A (2006) Experimental approaches used to quantify physical parameters at cellular and subcellular levels. Am J Bot 93:1380–1390
4. Smith AE, Moxham KE, Middelberg APJ (1998) On uniquely determining cell-wall material properties with the compression experiment. Chem Eng Sci 53:3913–3922
5. Smith AE, Moxham KE, Middelberg APJ (2000) Wall material properties of yeast cells. Part II. Analysis. Chem Eng Sci 55: 2043–2053
6. Blewett J, Burrows K, Thomas C (2000) A micromanipulation method to measure the mechanical properties of single tomato suspension cells. Biotechnol Lett 22:1877–1883
7. Wang CX, Wang L, Thomas CR (2004) Modelling the mechanical properties of single suspension cultured tomato cells. Ann Bot 93:443–453
8. Wang CX, Wang L, McQueen-Mason SJ et al (2008) pH and expansin action on single suspension-cultured tomato (Lycopersicon esculentum) cells. J Plant Res 121:527–534
9. Milani P, Gholamirad M, Traas J et al (2011) In vivo analysis of local wall stiffness at the shoot apical meristem in Arabidopsis using atomic force microscopy. Plant J 67: 1116–1123
10. Fernandes AN, Chen X, Scotchford CA et al (2012) Mechanical properties of epidermal cells of whole living roots of *Arabidopsis thaliana*: an atomic force microscopy study. Phys Rev E Stat Nonlin Soft Matter Phys 85:021916
11. Radotić K, Roduit C, Simonović J et al (2012) Atomic force microscopy stiffness tomography on living *Arabidopsis thaliana* cells reveals the mechanical properties of surface and deep cell-wall layers during growth. Biophys J 103: 386–394
12. Parre E, Geitmann A (2005) Pectin and the role of the physical properties of the cell wall in pollen tube growth of *Solanum chacoense*. Planta 220:582–592
13. Bolduc J-E, Lewis LJ, Aubin C-E et al (2006) Finite-element analysis of geometrical factors in micro-indentation of pollen tubes. Biomech Model Mechanobiol 5:227–236
14. Zerzour R, Kroeger J, Geitmann A (2009) Polar growth in pollen tubes is associated with spatially confined dynamic changes in cell mechanical properties. Dev Biol 334:437–446
15. Routier-Kierzkowska A-L, Weber A, Kochova P et al (2012) Cellular force microscopy for in vivo measurements of plant tissue mechanics. Plant Phys 158:1514–1522

16. Vogler H, Draeger C, Weber A et al (2013) The pollen tube: a soft shell with a hard core. Plant J. 73:617–627
17. Hayot CM, Forouzesh E, Goel A et al (2012) Viscoelastic properties of cell walls of single living plant cells determined by dynamic nanoindentation. J Exp Bot 63:2525–2540
18. Forouzesh E, Goel A, Mackenzie SA et al (2012) In vivo extraction of Arabidopsis cell turgor pressure using nanoindentation in conjunction with finite element modeling. Plant J. doi:10.1111/tpj.12042
19. Peaucelle A, Braybrook SA, Le Guillou L et al (2011) Pectin-induced changes in cell wall mechanics underlie organ initiation in arabidopsis. Curr Biol 21:1720–1726
20. Felekis D, Muntwyler S, Vogler H et al (2011) Quantifying growth mechanics of living, growing plant cells in situ using microbotics. Micro Nano Lett 6:311–316

Chapter 12

High-Pressure Freezing and Low-Temperature Processing of Plant Tissue Samples for Electron Microscopy

Ichirou Karahara and Byung-Ho Kang

Abstract

Use of electron tomography methods improves image resolution of transmission electron microscopy especially in the *z*-direction, enabling determination of complicated 3D structures of organelles and cytoskeleton arrays. The increase in resolution necessitates preservation of cellular structures close to the native states with minimum artifacts. High-pressure freezing (HPF) that immobilizes molecules in the cell instantaneously has been used to avoid damages caused by convention chemical fixation. Despite the advantages of HPF, cells could still be damaged during dissection prior to HPF. Therefore, it is critical to isolate cells/tissues of interest quickly and carefully. The samples frozen by HPF are often processed by freeze substitution (FS), and FS should be carried out under appropriate conditions. Here we describe dissection, HPF, and FS methods that we have utilized to prepare plant samples for electron tomography/immuno-electron microscopy.

Key words High-pressure freezing, Freeze substitution, Microfilament, Microtubule, Coated pit and vesicle, Onion, Maize endosperm

1 Introduction

Slow penetration of chemical fixatives has been shown to cause artifactual morphological changes at the cellular level that are readily detected in electron micrographs [1]. To overcome this problem, cryofixation techniques, such as cold metal block freezing, plunge freezing, spray freezing, propane jet freezing, and high-pressure freezing [2], have been developed to immobilize cellular components at rates much faster than those of chemical fixatives [3]. Among these cryofixation techniques, high-pressure freezing (HPF) is especially effective for thicker samples, such as plant tissues. HPF can freeze samples as thick as 100–300 μm, which is far thicker than the other techniques can freeze [3–5], without the tissue-damaging formation of ice crystals. A combination of cryofixation and freeze substitution (FS) has been used to visualize subcellular structures in plant cells by electron microscopy (EM; 6, 7).

Viktor Žárský and Fatima Cvrčková (eds.), *Plant Cell Morphogenesis: Methods and Protocols*, Methods in Molecular Biology, vol. 1080, DOI 10.1007/978-1-62703-643-6_12, © Springer Science+Business Media New York 2014

These techniques are further combined with electron tomography (ET) to visualize subcellular structures, including microtubule organization [8–10], endomembrane compartments [11–13], and coated vesicles [14, 15], and to analyze these structures in three dimensions at nanometer-level resolutions.

Despite the advantages of HPF and FS, the protocols for individual samples or for particular organelles need to be optimized because there is no single HPF or FS protocol that works for every tissue sample. For example, ultrastructural visualization of microfilaments has been difficult, although they have been reported in some cell types [16]. Although an improved freeze substitution method has been developed, using tannic acid after treatment with a mixture of OsO_4 and uranyl acetate (UA) during the FS process [17], this method appears to be successful only with limited cell or tissue types. Murata et al. optimized the conditions for visualizing microfilaments by treating with 2 % OsO_4, followed by 5 % UA at −40 °C [18]. The improved visualization of microfilaments made it possible to examine microtubules and coated pits in combination with microfilaments in the cortical cytoplasm, facilitating accurate ultrastructural analyses of microtubule ends [18], clathrin-coated pits, clathrin-coated vesicles [15], and multivesicular bodies [19].

Processing samples for TEM by HPF and FS involves a number of steps which require more than several days. There are many small details that researchers have to pay attention to in order to achieve successful results. However, the amount and quality of structural information acquired from well-frozen cell samples is worth the effort and time. For example, budding of COPII vesicles inside a ribosome-excluding scaffold [12], formation of clathrin-coated vesicles from the plasma membrane under the preprophase band [15], and the distinctive morphology of the plus-end of phragmoplast microtubules were all observed in plant cell samples processed by the combined use of HPF and FS [8]. Cryofixation is gaining popularity for TEM analysis, not only of Arabidopsis but also of other plant species as more researchers realize its superiority over chemical fixation in capturing detailed and transient cellular structures [20–23]. Procedures for Arabidopsis samples are available in other reports [10, 13, 24]. Here we describe procedures of HPF and FS for non-Arabidopsis materials, including onion seedlings, BY-2 suspension-cultured cells, and maize endosperm cells.

2 Materials

2.1 High-Pressure Freezing

1. High-pressure freezer, BAL-TEC HPM 010 (Boeckeler, Tucson, AZ) or Leica HPM100 (Leica Microsystems, Buffalo Grove, IL).
2. Specimen carrier: brass type A and type B planchettes (Ted Pella, Redding, CA) or aluminum type A and type B planchettes (Technotrade International, Manchester, NH).

3. Liquid nitrogen.
4. Onion seeds.
5. BY-2 suspension-cultured cells (liquid culture).
6. Fresh ear of corn.
7. Cryoprotectant: 0.1–0.15 M sucrose solution in distilled water or 20 % dextran (molecular weight: ~40,000, Cat no. 31398 from Sigma) in BY-2 cell culture medium.
8. Vacuum filtration funnel and 30 μm pore size nylon filter.
9. Biopsy punch (diameter: 2 mm; Miltex, Plainsboro, NJ).

2.2 Freeze Substitution

1. 2 % OsO_4 in acetone: dissolve 1 g of OsO_4 in 50 ml anhydrous acetone. Aliquot into cryovials (approximately 1 ml per vial). Freeze in liquid nitrogen and store frozen until use.
2. 0.1 % glutaraldehyde (GA) and 0.25 % UA in acetone: prepare 25 % uranyl acetate in anhydrous methanol. Dilute a 10 % GA solution in acetone (Electron Microscopy Sciences, Hatfield, PA) and the 25 % methanolic UA solution with acetone to final concentrations of 0.1 % GA and 0.25 % UA. Aliquot into cryovials, freeze in liquid nitrogen, and store frozen until use.
3. AFS2 automatic freeze substitution machine (Leica Microsystems, Buffalo Grove, IL).

2.3 Embedding

1. Embed-812 kit or low viscosity Spurr's kit (Electron Microscopy Sciences, Hatfield, PA).
2. Lowicryl HM20 resin kit (Electron Microscopy Sciences, Hatfield, PA).
3. Flat-bottomed BEEM capsules (Electron Microscopy Sciences, Hatfield, PA).

2.4 Ultramicrotomy and Post-staining

1. UC-7 (Leica Microsystems, Buffalo Grove, IL) or a similar ultramicrotome.
2. Formvar-coated EM grids (copper or nickel slot grids, 1 mm × 2 mm slot size).
3. Diamond knife.
4. 2 % (w/v) UA in 60 % ethanol or 2 % aqueous UA and Reynolds' lead citrate solution.

3 Methods

3.1 Growing Onion Seedlings

1. Sow onion (*Allium cepa L.* cv. Highgold Nigou) seeds on a piece of filter paper (No. 1) wetted with 0.05 M sucrose in distilled water in a plastic Petri dish (*see* **Note 1**). Use parafilm to seal the lid of the dish to avoid drying of the filter paper. Cover the container with a sheet of aluminum foil to allow the

seeds to germinate in the dark. Place the container at a 45° angle so that the seedlings will grow fairly straight.

2. Leave the plates in the dark for 2 days at 25 °C.
3. Open the container, select seedlings with less than 1 mm long roots, and transfer them onto a piece of filter paper wetted with 0.1 M sucrose in distilled water.
4. Culture the seedlings further for 1 day.

3.2 High-Pressure Freezing (Onion Seedlings)

1. Place a glass plate under a dissecting microscope.
2. Put a few drops of an aqueous solution of 0.1 M sucrose on the glass plate, and make a pool of this solution (*see* **Note 1**). Prepare two brass freezer planchettes. One is for the bottom and the other is for the top.
3. Using forceps, immerse the bottom planchette in the sucrose pool. Be careful not to create air pockets or bubbles inside the bottom planchette, as these would collapse during pressurization and cause damage to the sample (*see* **Note 2**; ref. [3]).
4. Select a seedling in which the length between the bending point of the hook to the base of the cotyledon is <10 mm.
5. Pick the seedling up by gently catching the residual seed in forceps, and immerse the seedling in the sucrose pool. Proceed as quickly as possible from this step to the freezing step to avoid drying of samples [4].
6. Hold the seed part with forceps in one hand, and cut a basal part of the cotyledon 1–2 mm in length with a razor blade using the other hand under the dissecting microscope. Split the cotyledon section into two pieces longitudinally.
7. Put the tissue pieces in the bottom planchette. Wet the bottom of the top planchette with the solution, and put it on top of the bottom planchette. Again try to remove any air bubbles trapped inside the bottom planchette with forceps.
8. Pick up the sample enclosed in the planchettes, and load it into a sample holder immediately.
9. Freeze the sample using a high-pressure freezer.
10. Transfer the high-pressure frozen tissues with the planchettes to liquid nitrogen in cryovials for storage. (Precautions for this process are described in detail in ref. [25]; *see* **Note 3**).

3.3 High-Pressure Freezing (BY-2 Cells)

1. Set up a vacuum filtration funnel with a 30 μm pore size nylon filter.
2. Pour fresh BY-2 culture into the funnel and apply vacuum. Vacuum out the culture medium until cells' concentrate forms a slurry on top of the filter. Never dry the cells.

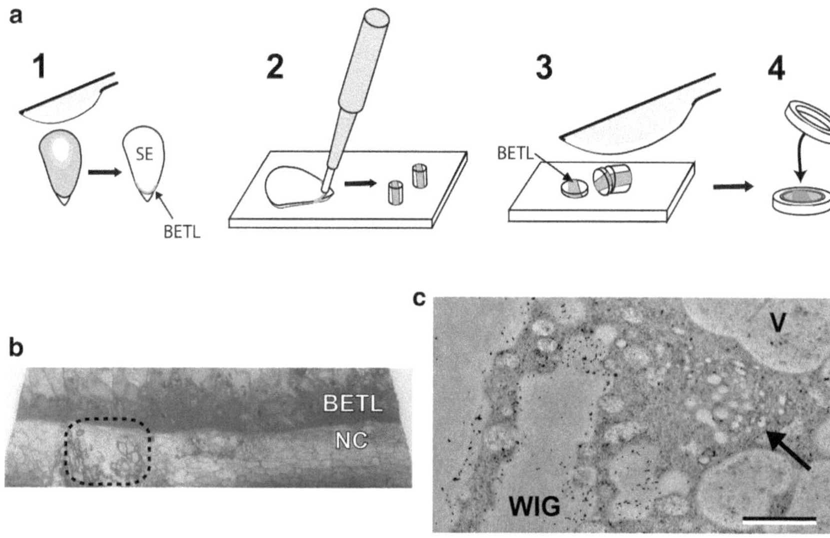

Fig. 1 (**a**) Dissection of maize kernels and loading of basal parts of the endosperm specimens into HPF planchettes. Steps in dissecting and loading are explained in the text. *SE* starchy endosperm, *BETL* basal endosperm transfer cell layer. (**b**) Semi-thin section (500 nm) of a BETL and the nucellar placento-chalazal cell (NC) zone of a maize kernel processed by HPF/FS. NC and the underlying integument were damaged during dissection, as outlined by the *dashed box*. (**c**) TEM micrograph of a basal endosperm transfer cell in which a secretory protein was immunolabeled (gold particles—15 nm). A Golgi/trans-Golgi network complex is marked with an *arrow*. *V* vacuole, *WIG* cell wall ingrowth. Scale bar: 1 μm

3. Add 20 % dextran dissolved in the BY-2 culture medium to the cell slurry and gently mix (*see* **Note 4**).
4. Pipette out the excess dextran solution, and transfer the resulting cell slurry to an A-type HPF planchette. Cover with the flat side of a B-type planchette.
5. Freeze and recover samples as described in Subheading 3.2.

3.4 High-Pressure Freezing (Developing Maize Kernel)

1. Harvest maize ears at the particular stages of interest and bring them to the lab quickly. Have the HPF machine ready before harvesting ears.
2. Remove husks and pluck individual kernels from the ear using a knife tip or other tool, without damaging the kernels (*see* **Note 5**).
3. Cut a kernel into approximately 2–3 mm slices (Fig. 1a-1). Using a 2 mm biopsy punch, extract the endosperm tissues (Fig. 1a-2). Slice the tissue thinly enough for fitting into a B-type planchette (Fig. 1a-3).
4. Transfer the tissue to a B-type planchette and cover with another B-type planchette (Fig. 1a-4).
5. Freeze and recover samples as described in Subheading 3.2.

3.5 Freeze Substitution and Embedding in Spurr's Resin

1. Day 1. Transfer the samples from the liquid nitrogen to 2 % (w/v) OsO_4 in acetone, contained in cryovials, and store at −80 °C for 72 h. Use an aluminum block with holes for holding cryovials. The aluminum block carrying the cryovials can be placed in a Styrofoam box containing dry ice to maintain a −80 °C temperature for 3 days.
2. Day 4. Transfer the samples to −20 °C and incubate for 24 h.
3. Day 5. Move the samples to a 4 °C refrigerator and incubate for an additional 18 h.
4. Day 6. Warm the samples to room temperature for 1 h. Incubate the samples, which are kept in the same 2 % (w/v) OsO_4 solution, at 40 °C for 4 h.
5. Remove the osmium solution and replace with acetone; keep the samples in acetone at 4 °C for 2 h.
6. Cool the samples to −20 °C for 1 h.
7. Remove the acetone and replace with 5 % (w/v) UA in methanol and incubate for 2 h. The UA solution should be cooled to −20 °C before use.
8. Transfer the samples to acetone cooled to −20 °C and store for 2 h. Warm the samples to room temperature. Tissues are teased from the planchettes during this time.
9. Transfer the samples to 1 % (v/v) Spurr's resin in acetone and store for 12 h. Use a resin mix without dimethylaminoethanol (DMAE).
10. Day 7. Infiltrate the samples with a graded series (2, 4, 6, 8, 10 % (v/v)) of Spurr's resin diluted in acetone for 2 h each.
11. Day 8. Replace the solution with 25 % (v/v) Spurr's resin (without DMAE) and infiltrate for 12 h.
12. Replace the solution with 50 % (v/v) Spurr's resin (without DMAE) for 12 h.
13. Day 9. Replace the solution with 75 % (v/v) Spurr's resin (without DMAE) and infiltrate for 24 h.
14. Day 10. Replace the solution with 100 % (v/v) Spurr's resin (without DMAE) and infiltrate for 24 h.
15. Day 11. Replace the solution with fresh 100 % (v/v) Spurr's resin (without DMAE) and infiltrate for 24 h.
16. Day 12. Replace the solution with 100 % (v/v) Spurr's resin (with DMAE) and infiltrate for 24 h.
17. Day 13. Replace the solution with fresh 100 % (v/v) Spurr's resin (with DMAE) and infiltrate for 24 h.
18. Polymerize the resin at 70 °C for 12 h in a vacuum oven.

3.6 Freeze Substitution and Embedding in HM20 Resin

1. Day 1. Transfer the samples to 0.1 % GA and 0.25 % UA in anhydrous acetone and incubate at −80 °C for 72 h in the AFS2.
2. Day 4. Slowly warm the AFS2 chamber to −45 °C at a rate of 1 °C/h.
3. Day 6. Wash the samples with precooled anhydrous acetone three times, and start resin embedding by incubating samples in 33 % resin in acetone. Until UV polymerization is completed, all steps are carried out at −45 °C (from Day 6 to Day 8).
4. Day 7. Increase resin concentration to 66 % and then to 100 %. Wash samples two times with 100 % resin. Separate samples from planchettes if they are still attached to the planchettes. Transfer the samples to flat-bottomed BEEM capsules, and place under UV illumination (*see* **Note 6**).
5. Day 8. Check whether the HM20 resin has been cured. Allow the AFS2 chamber to warm to room temperature once the samples are hardened.

3.7 Sectioning and Staining

1. Wash slot grids (2 × 1 mm, SPI Supplies, PA) with 1,2-dichloroethane using a sonicator.
2. Clean a glass slide with ethanol or lab detergent and wipe clean with tissues. Coat the grids with 0.5–1 % (w/v) formvar solution in 1,2-dichloroethane (*see* **Note 7**).
3. For a preliminary observation, cut semi-thin sections (500 nm) using an ultramicrotome.
4. Stain the sections with 0.1 % (w/v) toluidine blue O in an aqueous solution containing 1 % (w/v) sodium borate. Check the sections under an optical microscope to make sure that the cells are not damaged (Fig. 1b).
5. Prepare ultrathin sections, post-stain with UA and lead citrate solutions. Examine the sections under an electron microscope to confirm that cells are well preserved.
6. For localization of macromolecules, perform immuno-gold labeling (as described in refs. [26, 27]) prior to post-staining (Fig. 1c).

4 Notes

1. We have used an aqueous solution of 0.1 M sucrose for this purpose. Preservation of tissues after high-pressure freezing is better when seedlings have been grown with sucrose added to the culture medium [28]. Because the germination of onion seeds was inhibited in the presence of 0.1 M sucrose, seeds

were germinated in the presence of 0.05 M sucrose for 2 days. Germinated seedlings were then grown in the presence of 0.1 M sucrose for 1 day and were then processed for high-pressure freezing. Tobacco seedlings were grown in the presence of 3.5 % (w/v) sucrose before root tips were removed for HPF [18]. In general, non-penetrating cryoprotectants are preferred because they are considered to disrupt cell structure less than penetrating cryoprotectants [4]. 1-Hexadecene [29] can be used for plant tissue [30]; we have obtained good results with 1-hexadecene for onion cotyledon. However, after treating 1-hexadecene-protected specimens with OsO_4, the FS medium turns black, making it difficult to see specimen pieces. Be careful not to discard specimens when removing the OsO_4 solution and when washing with acetone.

2. Plant cells are surrounded by a dense meshwork of cell wall polysaccharides. Aerial organs of plants are covered with hydrophobic waxes. These protective layers on plant tissues interfere with the penetration of chemical cross-linkers. The large vacuoles that most plant cells contain also pose problems for chemical fixation. Chemical fixatives sometimes break vacuoles, and the leakage of acidic vacuolar contents could disrupt cellular structures and can also dilute the fixative. These shortcomings of conventional chemical fixation can be overcome by the use of HPF. However, plant tissues are often too large to fit into HPF planchettes. When dissecting plant tissues, stress and damage to the tissues prior to freezing should be minimized. Therefore, it is crucial to dissect rapidly and carefully. It is important to practice dissecting material and transferring samples to planchettes before attempting to freeze critical samples.
3. Poor preservation of a high-pressure frozen sample is primarily due to damage to the fine structure of subcellular components caused by the formation of ice crystals in the cytoplasm or nucleus during freezing. Examples of poor freezing are shown in refs. [25, 26, 31]. Not all plant samples are suitable for HPF processing. Waxy and hard tissues, like tree leaves, are difficult to dissect and to submerge in cryoprotectant. For such samples, it is better to resort to conventional fixation [32]. It is also difficult to freeze plant cells in which the volume is occupied almost entirely by vacuoles, because of the high water content in the vacuoles.
4. Dextran is a glucose polymer that binds water molecules with its hydroxyl group. Dextran serves as a cryoprotectant because it suppresses ice crystal growth outside BY-2 cells. Another role of dextran is to hold BY-2 cells together throughout the FS and resin-embedding steps.

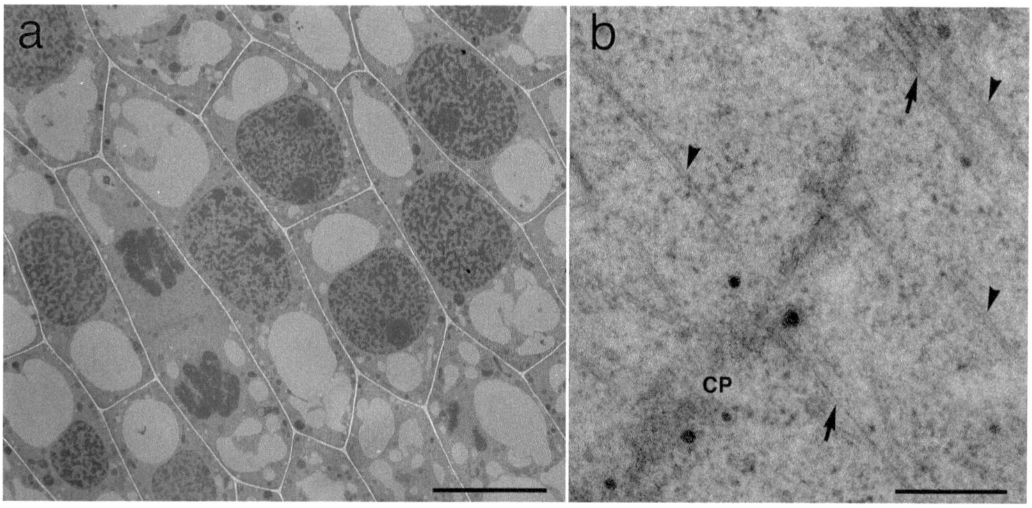

Fig. 2 Tangential longitudinal sections of high-pressure frozen onion epidermal cells embedded in Spurr's resin. (**a**) A lower magnification view. Scale bar: 10 μm. (**b**) A higher magnification view showing a portion of a developing cell plate (CP). Microfilaments (*arrows*) are clearly seen, as are microtubules (*arrowheads*). Scale bar: 0.25 μm

5. Using a steak knife tip or other sharp tool, remove kernels by cutting through the cob under the kernel. Place the cutting tool far enough away from the kernel so that the basal part of the kernel is not damaged.
6. Samples should look yellowish in 100 % HM20 resin because the FS medium contained UA.
7. For conventional electron microscopy, prepare formvar film with a thickness of ~80 nm. The film thickness can be estimated from its interference color; the color of an ~80 nm film is silver. For electron tomography semi-thin sections (e.g., 250 nm or more) are used, and thus the interference color of the formvar film should be dark gold, because the formvar film can be easily broken if thinner. It is convenient to use a glass slide lifter (MA-1, Nisshin EM, Tokyo, Japan) for formvar coating, if available. Electron micrographs of ultrathin sections of typical well-frozen onion samples are shown in Fig. 2.

Acknowledgments

This work was supported by JSPS KAKENHI Grant (No. 24620003) to I. K. and NSF (No. MCB 0958107) and USDA (AFRI 2010-04196) to B.-H. K. We are thankful to Dr. Mineyuki (University of Hyogo) and Donna S. Williams (University of Florida) for their careful reading and comments.

References

1. Gilkey JC, Staehelin LA (1986) Advances in ultrarapid freezing for the preservation of cellular ultrastructure. J Electron Microsc Tech 3:177–210
2. Moor H (1987) Theory and practice of high pressure freezing. In: Steinbrecht RA, Zierold K (eds) Cryotechniques in biological electron microscopy. Springer, Berlin, pp 175–191
3. Dahl R, Staehelin LA (1989) High-pressure freezing for the preservation of biological structure: theory and practice. J Electron Microsc Tech 13:165–174
4. McDonald K (2007) Cryopreparation methods for electron microscopy of selected model systems. Methods Cell Biol 79:23–56
5. Sartori N, Richter K, Dubochet J (1993) Vitrification depth can be increased more than 10-fold by high-pressure freezing. J Microsc 172:55–61
6. Hess MW (2007) Cryopreparation methodology for plant cell biology. Methods Cell Biol 79:57–100
7. Parthasarathy MV (1995) Freeze-substitution. Methods Cell Biol 49:57–69
8. Austin JR II, Seguí-Simarro JM, Staehelin LA (2005) Quantitative analysis of changes in spatial distribution and plus-end geometry of microtubules involved in plant-cell cytokinesis. J Cell Sci 118:3895–3903
9. Mineyuki Y, Karahara I, Suda Y et al (2006) Electron tomographic analysis of cortical microtubule ends in plant interphase and prophase cells preserved by high-pressure freezing. Proceedings of the 16th international microscopy congress, doi: 10.1016/S0304-3991(08)00010-7, p 17
10. Otegui MS, Austin JR (2007) Visualization of membrane-cytoskeletal interactions during plant cytokinesis. Methods Cell Biol 79:221–240
11. Staehelin LA, Kang BH (2008) Nanoscale architecture of endoplasmic reticulum export sites and of Golgi membranes as determined by electron tomography. Plant Physiol 147:1454–1468
12. Kang BH, Staehelin LA (2008) ER-to-Golgi transport by COPII vesicles in Arabidopsis involves a ribosome-excluding scaffold that is transferred with the vesicles to the Golgi matrix. Protoplasma 234:51–64
13. Kang BH, Nielsen E, Preuss ML et al (2011) Electron tomography of RabA4b- and PI-4Kβ1-labeled trans Golgi network compartments in Arabidopsis. Traffic 12:313–329
14. Donohoe BS, Kang BH, Staehelin LA (2007) Identification and characterization of COPIa- and COPIb-type vesicle classes associated with plant and algal Golgi. Proc Natl Acad Sci USA 104:163–168
15. Karahara I, Suda J, Tahara H et al (2009) The preprophase band is a localized center of clathrin-mediated endocytosis in late prophase cells of the onion cotyledon epidermis. Plant J 57:819–831
16. Tiwari SC, Wick SM, Williamson RE et al (1984) Cytoskeleton and integration of cellular function in cells of higher plants. J Cell Biol 99:63s–69s
17. Ding B, Turgeon R, Parthasarathy MV (1991) Microfilaments in the preprophase band of freeze substituted tobacco root cells. Protoplasma 165:209–211
18. Murata T, Karahara I, Kozuka T et al (2002) Improved method for visualizing coated pits, microfilaments, and microtubules in cryofixed and freeze-substituted plant cells. J Electron Microsc (Tokyo) 51:133–136
19. Karahara I, Staehelin LA, Mineyuki Y (2012) The role of endocytosis in the creation of the cortical division zone in plants. In: Ceresa B (ed) Molecular regulation of endocytosis. InTech, Rijeka, pp 41–60
20. Kang BH, Xiong Y, Williams DS et al (2009) Miniature1-encoded cell wall invertase is essential for assembly and function of wall-in-growth in the maize endosperm transfer cell. Plant Physiol 151:1366–1376
21. Toyooka K, Goto Y, Asatsuma S et al (2009) A mobile secretory vesicle cluster involved in mass transport from the Golgi to the plant cell exterior. Plant Cell 21:1212–1229
22. Washida H, Sugino A, Doroshenk KA et al (2012) RNA targeting to a specific ER sub-domain is required for efficient transport and packaging of α-globulins to the protein storage vacuole in developing rice endosperm. Plant J 70:471–479
23. Reyes FC, Chung T, Holding D et al (2011) Delivery of prolamins to the protein storage vacuole in maize aleurone cells. Plant Cell 23:769–784
24. Otegui MS, Staehelin LA (2004) Electron tomographic analysis of post-meiotic cytokinesis during pollen development in *Arabidopsis thaliana*. Planta 218:501–515
25. Meehl JB, Giddings TH Jr, Winey M (2009) High pressure freezing, electron microscopy, and immuno-electron microscopy of *Tetrahymena thermophila* basal bodies. Methods Mol Biol 586:227–241
26. Kang BH (2010) Electron microscopy and high-pressure freezing of Arabidopsis. Methods Cell Biol 96:259–283

27. Takeuchi M, Takabe K, Mineyuki Y (2010) Immunoelectron microscopy of cryofixed and freeze-substituted plant tissues. In: Schwartzbach S, Osafune T (eds) Immunoelectron microscopy: methods and protocols. The Humana Press Inc., Totoa, pp 155–164
28. Mineyuki Y, Karahara I, Murata T et al (2001) High pressure freezing in plant tissues. Denshi Kenbikyo 36:105–107 (in Japanese)
29. Studer D, Michel M, Müller M (1989) High pressure freezing comes of age. Scanning Microsc Suppl 3:253
30. Kaneko Y, Keegstra K (1996) Plastid biogenesis in embryonic pea leaf cells during early germination. Protoplasma 195:59–67
31. Mineyuki Y, Suda J, Karahara I (2004) Electron tomography. Plant Morphol 16: 21–30
32. Koh EJ, Zhou L, Williams DS et al (2012) Callose deposition in the phloem plasmodesmata and inhibition of phloem transport in citrus leaves infected with "Candidatus *Liberibacter asiaticus*". Protoplasma 249: 687–697

Chapter 13

Reconstructing Plant Cells in 3D by Serial Section Electron Tomography

Kiminori Toyooka and Byung-Ho Kang

Abstract

In micrographs acquired with a transmission electron microscope, 3-dimensional (3D) objects are superimposed onto a 2D screen. This reduction in dimension necessarily leads to a degradation of image resolution. To overcome this problem, 3D microscopy techniques, such as tomography and single particle analysis, have been developed. Tomography has been used to visualize cells in 3D, and single particle analysis has been used to investigate macromolecules and viral particles. In this chapter we will describe how we have collected tilting series micrographs from plant cells and how we have reconstructed the cellular volumes using dual axis electron tomography.

Key words Electron tomography, IMOD software package, Serial section, eTomo, 3D reconstruction

1 Introduction

Micrographs acquired by transmission electron microscopy (TEM) are projections of 3D volumes onto a 2D screen. The reduction of dimensionality inherently causes a loss of spatial information because varying densities from cellular components are superimposed along the direction of projection [1]. TEM samples are sliced into 60–80 nm thin sections. By preparing serial thin sections, it is possible to obtain 3D information about cellular structures [2–4]. However, the resolution of the TEM in the direction of the electron beam is, at best, limited to the section thickness. In theory, image resolution could be improved by preparing thinner sections, but reliably acquiring sections thinner than 40 nm from resin embedded samples is challenging, even for skilled ultramicrotomy specialists.

Employing a tomography algorithm can partially retrieve the information by reduction in dimension. Tomography is a technique for generating images representing 3D structures of an object from a set of 2D-projected images of the object taken with penetrating radiation at many angles. Tomography is most commonly used in

Viktor Žárský and Fatima Cvrčková (eds.), *Plant Cell Morphogenesis: Methods and Protocols*, Methods in Molecular Biology, vol. 1080, DOI 10.1007/978-1-62703-643-6_13, © Springer Science+Business Media New York 2014

modern medicine to produce image slices of specific body areas in a patient. Electron tomography (ET) is an extension of TEM in which the resolution of the TEM is enhanced by incorporating tomography methods [5]. Images of a 200–300 nm thick resin section are captured from a wide range of directions in a transmission electron microscope in ET, and the images are subsequently projected back into a virtual 3D volume [6, 7]. Images representing 1–4 nm thin slices, termed tomographic slices, can be extracted from the virtual volume. Because tomographic slices are far thinner than the resin sections, organelles and macromolecular complexes in the slices could be examined at resolutions far better than by conventional TEM especially in the z-direction. 3D views of subcellular components can be constructed from the tomographic slices. Furthermore, by joining tomograms of a cell from serial sections, ET can visualize relative large volumes of a cell in 3D [8].

With its improved resolution and 3D capability, ET has contributed to upgrading our understanding of cellular structures. ET is especially useful for investigating organelles with complicated 3D architectures. Examples of the published electron tomography analyses include studies on the *trans*-Golgi network [9], the cell plate [10–12], thylakoids in the chloroplast [13, 14], endoplasmic reticulum membranes [15], and multivesicular bodies [16]. It has also enabled detailed structural characterizations of entities smaller than the thickness of TEM sections, including transport vesicles and ribosomes [17–19]. To overcome artifacts of conventional chemical fixation, cryofixation methods for TEM, such as high-pressure freezing, have been practiced since the 1980s [20]. Combined with high-pressure freezing, ET can capture short-lived events in the cell in 3D at nanometer-level resolutions [21].

ET was historically a technique utilized by a handful of labs with sophisticated transmission electron microscopes and high-performance computers. With the rapid increase in computer power and easy access to intermediate voltage transmission electron microscopes, ET has become more available to plant biologists, whose research has benefitted from incorporating high-resolution 3D imaging. In this chapter, we will describe how we prepare TEM sections for ET and calculate tomograms from tilt image stacks. We have used SerialEM for acquiring raw data and IMOD for generating tomograms. Both programs are open-source software freely available from The Boulder Laboratory for 3D Electron Microscopy of Cells at the University of Colorado (http://bio3d.colorado.edu/SerialEM/ and http://bio3d.colorado.edu/imod/). We will describe operation of the eTomo module of the IMOD package. An excellent tutorial is included in the IMOD package (version 4.5.7), and we recommend that readers consult the tutorial for detailed information. The Boulder Laboratory for 3D Electron Microscopy of Cells has a YouTube channel containing tutorial videos, including videos for eTomo (http://www.youtube.com/bl3demc). Due to the page limitations of this chapter, the details for modeling organelles and for 3D quantitative analysis of cellular volumes will be discussed elsewhere. There are also many great reviews about 3D analysis using ET [22–25].

2 Materials and Equipment

2.1 Ultramicrotomy and Grid Preparation

1. UC-7 (Leica Microsystems, Buffalo Grove, IL) or a similar ultramicrotome.
2. Formvar-coated EM grids (copper or gold slot grids, 1 mm × 2 mm slot size).
3. Diamond knife.
4. 2 % (w/v) uranyl acetate (UA) in 70 % methanol or 2 % aqueous UA and Reynolds' lead citrate solutions.
5. Unconjugated 15 nm gold particles (Catalog number: 15704-20, Ted Pella, Redding, CA; *see* **Note 1**).

2.2 Acquisition of Tilting Series

1. Intermediate voltage transmission electron microscopes, such as TF20/TF30 (FEI, Hillsboro, OR) or JEM-2200FS/FEM-3100F (JEOL, Peabody, MA), equipped with a motorized eucentric goniometer for collecting tilting series.
2. Grid holder for dual axis tomography, like the Fischione Model 2040 holder (Fischione, Export, PA).

2.3 3D Reconstructions

1. Personal computer running a Linux or Unix operating system or a Windows computer with a Linux emulator such as Cygwin (*see* **Note 2**).
2. Monitor resolution at least 1,600 × 1,200.
3. Mouse with 3 buttons (left, middle, and right).

3 Methods

3.1 Serial Sectioning of Resin Blocks and Preparation of Grids for ET

1. Trim blocks into trapezoids which contain the tissues/cells of interest. Ensure that the top and bottom sides of the trapezoids are parallel to keep the serial section ribbon straight.
2. Collect thin sections (60–70 nm) from the samples to locate the desired regions for further analysis by ET and to assess preservation quality in those regions.
3. Once regions for ET have been chosen, collect ribbons of 200–300 nm thick serial sections on Formvar-coated slot grids (2 mm × 1 mm; *see* **Note 3**).
4. Stain the thick section ribbons with UA and Reynolds' lead citrate solutions. If needed, perform immunogold labeling prior to heavy metal staining.

5. Coat the stained grids with 15 nm unconjugated gold particles. Put 10 μl of the gold colloid solution onto one side of the grids. Wait 10 min, rinse by dipping the grids in distilled water, and dry the grids. Repeat for the other side of the grids.

3.2 Acquisition of Tilting Series

1. Examine the serial sections in a standard transmission electron microscope operated at 100–120 kV to confirm that the regions for ET are free of damage in all sections and to capture low magnification images. Prepare a map to assist in readily locating regions for ET in the intermediate voltage TEM (*see* **Note 4**).
2. Collect tilted images at 1° intervals, from -60° to +60° (or from -70° to +70°) in each section. Transmission electron microscopes for ET are equipped with eucentric tilting stages and with an automatic image acquisition programs such as SerialEM. Record images at 15,500× or 20,000× to accommodate an approximately 3 micrometer × 3 micrometer or 2 micrometer × 2 micrometer field of view, respectively, in the tomograms.
3. After acquiring tilting series from all sections on the grid, rotate the grid 90° and record a tilting series of the same areas in the sections for dual axis ET analysis. Name the pair with a root name followed by a and b, e.g., example_a.st and example_b.st.

3.3 Reconstruction of Tomograms with eTomo

1. Create a data folder and copy the pair of tilting series files (also termed image stack files) from the dual axis recording to this folder. Create a subfolder for backup files. Start the eTomo program and choose "New tomogram" in the "Front Page – eTomo" window to open the "Setup Tomogram – eTomo" window (Fig. 1a). In the dataset box, type the name for the first of the two tilting series (e.g., example_a.st), or press the "open file" button to select a tilting series. Enter the location of the backup file folder in the window.
2. Click the "Scan Header" button to get values for pixel size (nm) and image rotation (degrees). The two boxes should autopopulate after clicking the button. Enter the size of the gold fiducial in the fiducial diameter (nm) box (e.g., 15). To examine the tilting series, click the "View Raw Image Stack" button. To go to the next step, click on "Create Com Scripts."
3. The main window has several process control buttons (Fig. 1b). "Pre-processing" is for erasing bad pixels resulting from X-rays and/or removing bad CCD rows from the image stacks. Press the "Pre-Processing" button and choose "Find X-rays (Trial Mode)." Wait several seconds for processing, and click the "Create Fixed Stack" button to write a new image stack file. Press the "Use Fixed Stack" to replace the original

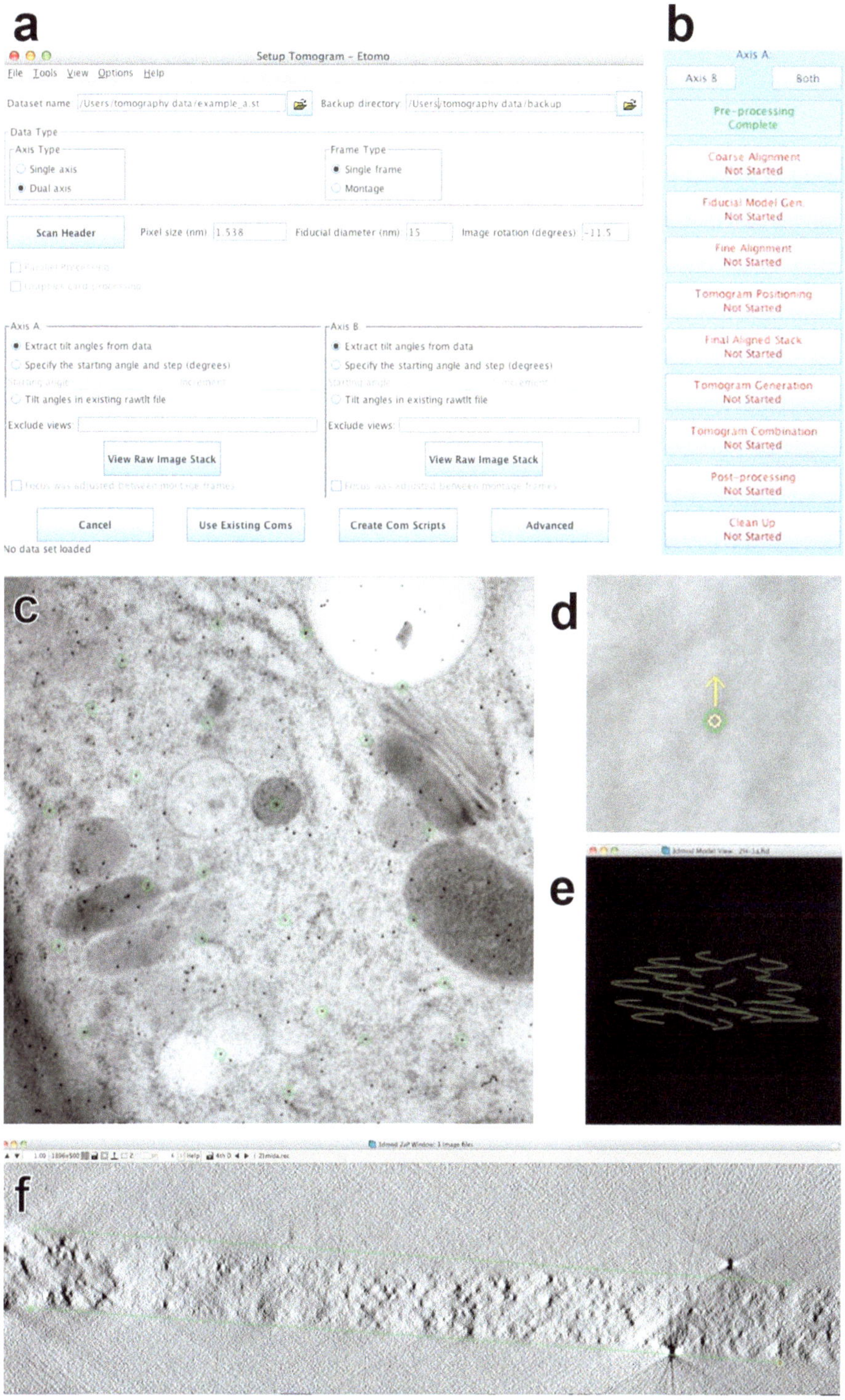

Fig. 1 (**a**) "Set up tomogram" window for generating command scripts. (**b**) "Process Control" window that opens after scripts are ready. The buttons are arranged in the order of use for calculating a dual axis tomogram. (**c**) Seed model showing 25 fiducials (marked with *green circles*) on the zero tilt angle image. (**d**) In the "Bead Fixer" window, eTomo suggests amounts to shift a fiducial with an *arrow*. (**e**) Fiducial model showing the trajectories of the 25 fiducials in the tilt image series. (**f**) 3dmod zap window showing a sample tomogram for creating a boundary model. Two *lines* were drawn by placing two *points* on both the *top* and *bottom* surfaces. In the panel, the two *green lines* encompassing the section are shown

file with the newly prepared fixed stack file. Click "Done" to go to the next step.

4. The "Coarse Alignment" step is for adjusting rotations and scales of images in the stack, using a cross-correlation method, to prepare a roughly aligned image stack. Click the "Calculate Cross-Correlation" button to determine adjustments between sequential images. Once the calculation is completed, press "Generate Coarse Aligned Stack" to create a coarsely aligned image stack. The pre-aligned stack (example_a.preali) can be viewed in the 3dmod interface by pressing "View Aligned stack in 3dmod." If cross-correlation cannot fix large shifts in images, the images should be aligned in the Midas program. Press "Fix Alignment With Midas" to open the program. Click "Done" to go to the Fiducial Model Generation step.
5. "Fiducial Model Generation" is where fiducial models are created using the gold particles deposited on the top and bottom surfaces of the sections. After checking the "Make seed and track box," press the "Seed Fiducial Model" button. Two tool dialog boxes "3dmod: example_a.preali" and "Bead Fixer" will open. Choose 20–40 gold particles in the image at the center of the tilting series (Fig. 1c, *see* **Note 5**). Each point should belong to one contour in a single model. After saving the model, click the "Track Seed Model" button. This will produce a fiducial model from the seed model by tracking the gold particles in all images in the stack. Once the tracking is done, click the "Fix Fiducial Model" button. Several gaps will appear on the "Bead Fixer" window. These gaps represent images in which the tracking program failed to find the positions of fiducial particles. Connect the trajectories of fiducial particles by adding new point(s) in the image(s). Pressing the "Go to Next Gap" button will move to the next image with a missing point. Repeat until the message, "No more gaps found," shows up in the main window after pressing the "Go to Next Gap" button. Save the file, and press "Done" (*see* **Note 6**).
6. Click "Compute Alignment," with the default parameters. Press the "View Residual Vectors" button to examine residuals, the distances between actual positions of the gold fiducial markers and their predicted positions (Fig. 1d). If the residuals are too large, fix them in the "bead fixer" manually or by using the "Move Point" feature. To view trajectories of fiducials in the tilting series, press "View 3D model" (Fig. 1e).
7. "Tomogram Positioning" is used to locate the section in the reconstructed volume to make the final tomogram flat and to minimize blank slices in it. This can be achieved by determining tilting and shift of the section in 3 sampled regions. First

enter the sample tomogram thickness. Set it large enough so that the sample regions can contain the entire tomogram. For example, for a 300 nm thick section at 15,500× magnification (pixel size = 1.54 nm), enter 300–400, which is 50–100 % larger than the 200 expected from the pixel size (300 nm ≈ 1.54 nm × 200). Click the "Create sample tomogram" button. After the sample tomogram is prepared, click the "Create Boundary model" button. It will open "top.rec" in the Zap window of the 3dmod interface (*see* **Note 7**). Draw a line along the boundary at each edge of the section as shown in Fig. 1f. Repeat in "mid.rec" and "low.rec" tomograms. Use the page up and page down buttons to move between the sample tomograms. Do not put more than two points in the line. After drawing six lines, two per sample tomogram, click "Compute Z shift & Pitch angle." Finish the step by pressing the "Create Final Alignment" button.

8. Open the Final Aligned Stack page, and press "Create Full Aligned Stack" to generate the final image stack for tomogram calculation. The stack can be examined in the 3dmod program by pressing "View Full Aligned Stack."
9. The Tomogram Generation page is used for calculating tomograms from the final aligned stack. Simply press the "Generate Tomogram" button. Once the tomogram calculation is completed, examine the tomogram by pressing the "View Tomogram in 3dmod" button.
10. The next step is to generate the second axis tomogram. In the eTomo window, press axis B (Fig. 1b). The procedure for the second axis (axis B) tomogram is almost the same as for the first axis (axis A) tomogram, except that fiducial selection is transferred over from the first axis image stack instead of being selected de novo. After doing the pre-processing and coarse alignment, open the "Fiducial Model Generation" page and press "Transfer Fiducials From Other Axis" button. eTomo displays the result of the fiducial transfer from the first axis after looking for corresponding fiducials in the two image series. Occasionally eTomo fails to find all matching fiducials in the two orthogonal image stacks, but a failure in transferring a few fiducials does not hinder calculation of the axis B tomogram.
11. Press "Track Fiducial Seed Model" and proceed through the rest of the steps as completed for the first axis tomogram (Fig. 2a).
12. After generating the axis B tomogram, return to the axis A page (Fig. 1b). Press the "Tomogram Combination" button. The Tomogram Combination page consists of three tabs. As the combination proceeds, the page moves from the "Setup" tab to the "Initial Match" tab and to the "Final Match" tab

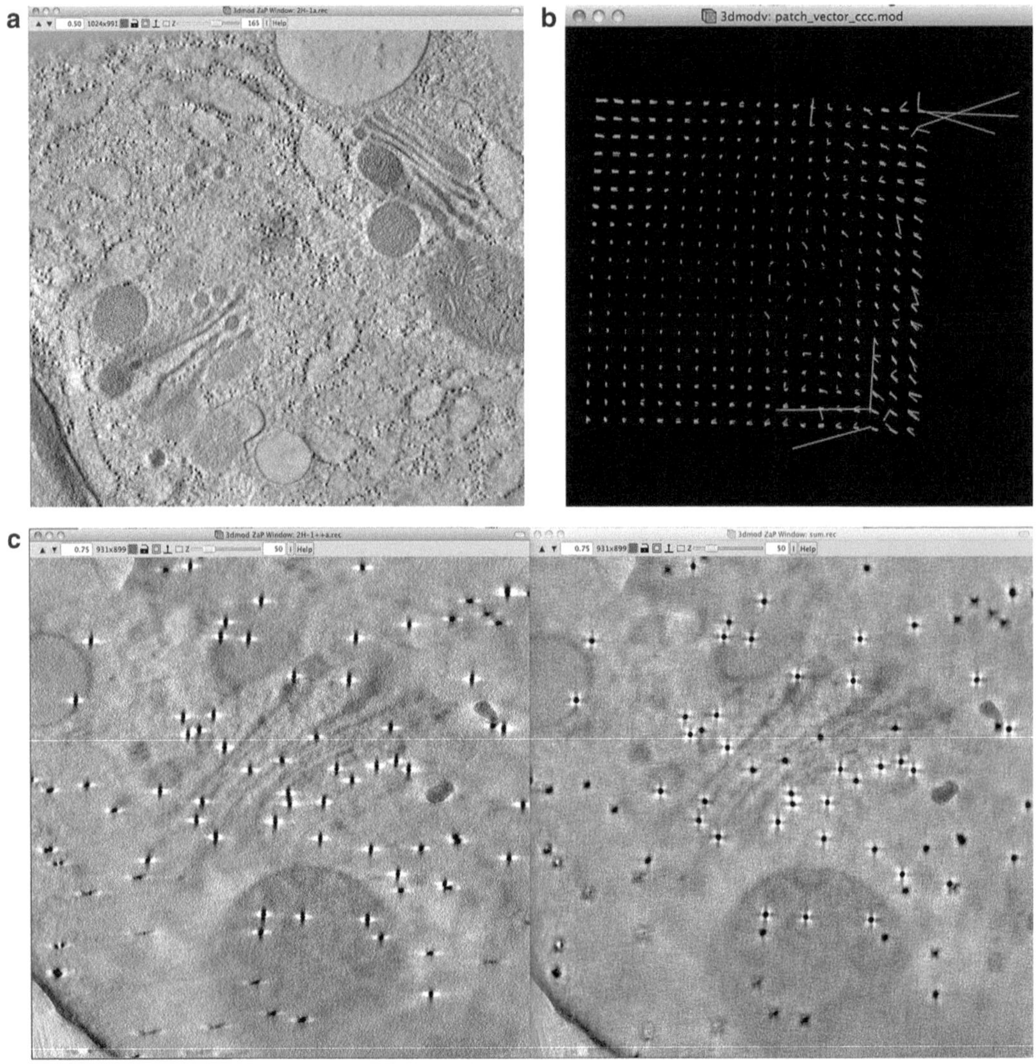

Fig. 2 (**a**) A slice image of the axis A tomogram from the section in Fig. 1. (**b**) Patch vector model generated for combining the axis A and axis B tomograms. There are several abnormally long vectors in the *upper right corner* and in the *lower right corner*. Combination can restart after deleting the abnormal vectors. (**c**) Slice images showing fiducial particles of a single axis tomogram (*left*) and a dual axis tomogram after combination (*right*)

automatically. In the Setup tab, information for starting tomogram combination is given. Usually the axis B tomogram is matched to the axis A tomogram and is chosen accordingly in the "Tomogram Matching Relationship" box. Indicate that fiducials are present on both sides of the section in the "Solvematch Parameters" box. In the "Patch Parameters for Refining Alignment" box, select "Medium Patches." If the tomogram has large areas which contain no structural information for matching the two tomograms, it is advantageous to use the "patch region" model. For example, large vacuoles or thick cell wall areas do not have features that can be utilized for finding corresponding volumes in the two tomograms from each axis. By excluding the empty regions, tomogram combination is performed more efficiently. Leave the "Volcombine Controls" and "Intermediate Data Storage" boxes unchanged. Press the "Create Combine Script" button and then the "Start Combine" button (*see* **Notes 8** and **9**).

13. In the post-processing step, the noncellular volume is trimmed, and the tomogram is reoriented so that images of the XY plane are loaded when 3dmod opens the tomogram. The combined tomogram is always named sum.rec. Open the sum.rec file by pressing "3dmod Full Volume." Scan through the tomogram and determine the X, Y, and Z ranges to be included in the final tomogram. Type in the X, Y, and Z ranges, and choose either "Swap Y and Z dimensions" or "Rotate around X axis." As long as the selection is consistent for all tomograms from serial sections, either selection will suffice. Press "Trim Volume."
14. The "Clean Up" step is used to delete large temporary files. If the final tomogram is satisfactory, highlight all the intermediate files displayed in the window and press "Delete Selected."

3.4 Joining Tomograms from Serial Sections

1. First collect all tomograms and place in a new folder. Determine the order of the serial tomograms, from the bottom to the top. The lowest Z end of the bottom tomogram will be the lowest Z end of the final tomogram, and the top Z end of the top tomogram will be the top Z end of the final tomogram. It is helpful to include a number when naming the tomograms to denote their order, such as example1.rec, example2.rec, and example3.rec. Make a working directory in the folder for temporary files and for files produced during joining.
2. Open eTomo and choose "New Join" in the "Front Page – eTomo" window. The setup panel will open in the "New Join – eTomo" window. Give the path and name of the working directory and the root name for the output file. Press the "Add" button in the section table to choose tomograms. Start from the bottom tomogram (Z Order 1) to the top tomogram. In the "Final" column, enter the ranges of tomograms to be included in the joined tomogram by typing the range (start and end slice numbers). To prepare sample slices for aligning tomograms, enter 5–10 extra slices on both sides of each tomogram. Press the "Make Samples" button to prepare an image stack showing boundaries between tomograms. The image stack will be used for aligning tomograms. When the process is done, click the "Align" tab to open the Align panel.
3. Click "Midas" to open the sample slices in the Midas program (Fig. 3). This program is useful for adjusting alignment between pairs of images by manipulating translation, rotation, magnification, and stretching of the images. For more detail, see the Midas page in the 3dmod package manual. After completing alignment at each boundary, save the transformation file.

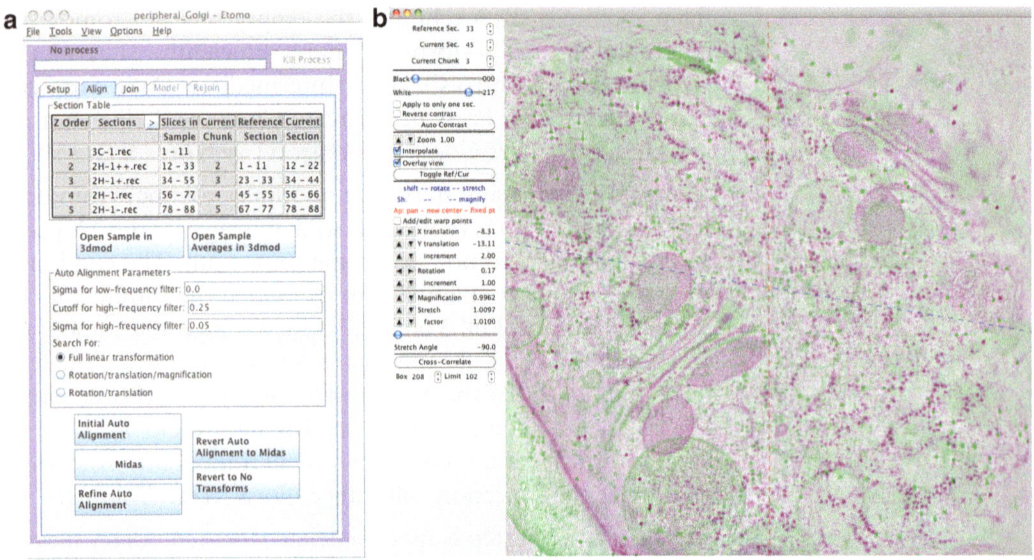

Fig. 3 (**a**) "Midas" window. (**b**) Two images are overlaid to align them. *Top* and *bottom* images are shown in two colors (*magenta* and *green*). Using 4 operations, translation, rotation, magnification, and stretching, the flanking sections can be transformed to maximize overlap between them

4. Go to the "Join" panel and make a trial tomogram to assess the range and alignment of the joined volume. If the trial tomogram is satisfactory, click "Finish Join" button.

4 Notes

1. The gold particles serve as markers, termed fiducials, for accurate image alignment. Positions of each particle in an image series are automatically tracked to generate its trajectory. In our experience, 15 nm particles are more easily tracked than smaller gold particles.
2. The IMOD software package has eTomo, a graphic user interface, for generating tomograms from tilt image series, which runs on Windows, Mac OS X, and Linux computers. The program runs more smoothly on computers with large memory (8–32 GB) capacity, to which entire tilting series or tomograms can be uploaded.
3. For a detailed explanation about sample preparation and serial sectioning, *see* ref. [26].
4. Image contrast is reduced in the intermediate voltage TEM, and locating specific areas in the ribbon of serial sections can be difficult. The standard TEM previewing also "bakes" the sections, preventing drift, when tilted images are collected in the intermediate voltage TEM.

5. Choose approximately 20–30 fiducial particles from both the top and bottom surfaces in the zero tilt image. The selected fiducial particles should be spread evenly over the image. Mark the center of each particle with a point. Each point should belong to a different contour.
6. To view the trajectories of fiducials in the image tilt series, open the "model view" window. The trajectories should be smooth. After shifting deviant points, press the "Track with Fiducial Model as Seed" button to generate a new model.
7. If the section does not fit into the sample tomogram, increase the thickness of the sample tomogram and start over.
8. If combination stops at the Matchorwarp program, examine the patch vector model. If the model has abnormally long vectors, simply delete them, and save the model (Fig. 2b). Click "Replace Patch Vectors" and "Restart at Matchorwarp." If the cellular volume in the tomogram has large empty spaces, such as large vacuoles, the eTomo program cannot find accurate matches, and the patch vector model will have a lot of large false vectors. In such cases, create a patch region model to exclude the empty spaces when the axis A and B tomograms are compared for combination. If the residual error is slightly larger than the threshold values, simply increase the threshold values, and restart Matchorwarp.
9. Because TEM sections cannot be imaged at high angles near 90°, image resolution is not uniform from all directions. This artifact can be observed in the reconstruction of fiducials. Fiducials have haloes that are uneven in the x–y plane (Fig. 2c, left panel), and they are elongated in the z direction (Fig. 1f). By combining the two orthogonal tomograms, the artifact can be alleviated (Fig. 2c, right panel; ref. [7]).

Acknowledgments

This work was supported by NSF (No. MCB-0958107) and USDA (No. AFRI 2010-04196) to B. -H. K, and the JSPS Institutional Program for Young Researcher Overseas Visits to K. T.

References

1. Hoenger A, McIntosh JR (2009) Probing the macromolecular organization of cells by electron tomography. Curr Opin Cell Biol 21:89–96
2. Toyooka K, Goto Y, Asatsuma S et al (2009) A mobile secretory vesicle cluster involved in mass transport from the Golgi to the plant cell exterior. Plant Cell 21:1212–1229
3. Breckenridge DG, Kang BH, Xue D (2009) Bcl-2 proteins EGL-1 and CED-9 do not regulate mitochondrial fission or fusion in *Caenorhabditis elegans*. Curr Biol 19:768–773
4. Koh EJ, Zhou L, Williams DS et al (2011) Callose deposition in the phloem plasmodesmata and inhibition of phloem transport in citrus

leaves infected with "Candidatus *Liberibacter asiaticus*". Protoplasma 249:687–697
5. McIntosh R, Nicastro D, Mastronarde D (2005) New views of cells in 3D: an introduction to electron tomography. Trends Cell Biol 15:43–51
6. Frank J (2010) Electron tomography: methods for three-dimensional visualization of structures in the cell. Springer, New York
7. Mastronarde DN (1997) Dual-axis tomography: an approach with alignment methods that preserve resolution. J Struct Biol 120:343–352
8. Donohoe BS, Mogelsvang S, Staehelin LA (2006) Electron tomography of ER, Golgi and related membrane systems. Methods 39: 154–162
9. Kang BH, Nielsen E, Preuss ML et al (2011) Electron tomography of RabA4b- and PI-4Kbeta1-labeled trans Golgi Network compartments in Arabidopsis. Traffic 12:313–329
10. Otegui MS, Staehelin LA (2004) Electron tomographic analysis of post-meiotic cytokinesis during pollen development in *Arabidopsis thaliana*. Planta 218:501–515
11. Otegui MS, Mastronarde DN, Kang BH et al (2001) Three-dimensional analysis of syncytial-type cell plates during endosperm cellularization visualized by high resolution electron tomography. Plant Cell 13:2033–2051
12. Segui-Simarro JM, Austin JR 2nd, White EA et al (2004) Electron tomographic analysis of somatic cell plate formation in meristematic cells of Arabidopsis preserved by high-pressure freezing. Plant Cell 16:836–856
13. Shimoni E, Rav-Hon O, Ohad I et al (2005) Three-dimensional organization of higher-plant chloroplast thylakoid membranes revealed by electron tomography. Plant Cell 17:2580–2586
14. Austin JR 2nd, Frost E, Vidi PA et al (2006) Plastoglobules are lipoprotein subcompartments of the chloroplast that are permanently coupled to thylakoid membranes and contain biosynthetic enzymes. Plant Cell 18:1693–1703
15. Leitz G, Kang BH, Schoenwaelder ME et al (2009) Statolith sedimentation kinetics and force transduction to the cortical endoplasmic reticulum in gravity-sensing Arabidopsis columella cells. Plant Cell 21:843–860
16. Otegui MS, Herder R, Schulze J et al (2006) The proteolytic processing of seed storage proteins in Arabidopsis embryo cells starts in the multivesicular bodies. Plant Cell 18: 2567–2581
17. Kang BH, Staehelin LA (2008) ER-to-Golgi transport by COPII vesicles in Arabidopsis involves a ribosome-excluding scaffold that is transferred with the vesicles to the Golgi matrix. Protoplasma 234:51–64
18. Donohoe BS, Kang BH, Staehelin LA (2007) Identification and characterization of COPIa- and COPIb-type vesicle classes associated with plant and algal Golgi. Proc Natl Acad Sci USA 104:163–168
19. Lee KH, Park J, Williams DS et al. (2013) Defective chloroplast development inhibits maintenance of normal levels of abscisic acid in a mutant of the Arabidopsis RH3 DEAD-box protein during early post-germination growth. Plant J 73:720–732
20. Gilkey JC, Staehelin LA (1986) Advances in ultrarapid freezing for the preservation of cellular ultrastructure. J Electron Microsc Tech 3:177–210
21. Staehelin LA, Kang BH (2008) Nanoscale architecture of endoplasmic reticulum export sites and of Golgi membranes as determined by electron tomography. Plant Physiol 147:1454–1468
22. Hoenger A, Bouchet-Marquis C (2011) Cellular tomography. Adv Protein Chem Struct Biol 82:67–90
23. Marsh BJ (2007) Reconstructing mammalian membrane architecture by large area cellular tomography. Methods Cell Biol 79:193–220
24. Otegui MS (2011) Electron tomography and immunogold labelling as tools to analyse de novo assembly of plant cell walls. Methods Mol Biol 715:123–140
25. Barcena M, Koster AJ (2009) Electron tomography in life science. Semin Cell Dev Biol 20:920–930
26. Kang BH (2010) Electron microscopy and high-pressure freezing of Arabidopsis. Methods Cell Biol 96:259–283

Chapter 14

Imaging Plant Nuclei and Membrane-Associated Cytoskeleton by Field Emission Scanning Electron Microscopy

Jindřiška Fišerová and Martin W. Goldberg

Abstract

Scanning electron microscopy (SEM) is a powerful technique that can image exposed surfaces in 3D. Modern scanning electron microscopes, with field emission electron sources and in-lens specimen chambers, achieve resolutions of better than 0.5 nm and thus offer views of ultrastructural details of subcellular structures or even macromolecular complexes. Obtaining a reliable image is, however, dependent on sample preparation methods that robustly but accurately preserve biological structures. In plants, exposing the object of interest may be difficult due to the existence of a cell wall. This protocol shows how to isolate plant nuclei for SEM imaging of the nuclear envelope and associated structures from both sides of the nuclear envelope in cultured cells as well as in leaf or root cells. Further, it provides a method for uncovering membrane-associated cytoskeletal structures.

Key words Scanning electron microscopy, Plant, SEM, Tobacco, Critical point drying, Coating, Wet fracture, Nucleus, Nuclear envelope, Cytoskeleton

1 Introduction

Scanning electron microscopy (SEM) provides useful 3D information about structural details of surfaces of tissues, cells, organelles, or even single proteins. Unlike light microscopy, which can take advantage of in vivo detection of cells, cellular movements, or movements of individual proteins, the sample to be observed under a conventional SEM must be fixed, dehydrated, and metal coated first. Thus, the sample is subjected to significant manipulation, and damage to the structures of interest may occur. Therefore, it is

Viktor Žárský and Fatima Cvrčková (eds.), *Plant Cell Morphogenesis: Methods and Protocols*, Methods in Molecular Biology, vol. 1080, DOI 10.1007/978-1-62703-643-6_14, © Springer Science+Business Media New York 2014

most important to choose an appropriate and gentle fixation technique to preserve the cellular content in as close to in vivo state as possible. A variety of fixation methods can be tested for the desired structural preservation [1–3]. Because field emission in-lens SEM (feiSEM) achieves better than 0.5 nm resolution, it is a powerful tool to provide structural information complementary to light microscopy. Labelling of proteins and other macromolecules with antibodies conjugated to gold particles helps with interpretation of the data.

For feiSEM the surface of interest must be exposed. If the cell wall of plant culture cells is of interest, simply wash off the medium and replace it with a suitable fix. If the surface is buried deep in the cell or integrated into a tissue, it may be necessary to test a variety of options.

Some cellular components can be uncovered by fracturing techniques such as freeze fracture or dry fracturing [4], but these methods usually preclude immunolabelling because the sample is either frozen or fixed and dried, respectively, before the labelling can be done. For many intracellular components it may be necessary to isolate them. In this way membranes can be preserved, but morphological changes are possible. Unlike purification of organelles for biochemical analysis, however, it is unnecessary to aim for highly purified preparations where damage is likely to occur. We have developed methods for isolating nuclei from plants suitable for SEM and feiSEM and immunogold labelling [5]. The procedure involves removal of the cell wall and then bursting the resulting protoplasts either osmotically or using mechanical pressure, e.g., pushing through a syringe needle or wet fracturing as described in the method of Collings et al. [6]. This technique can be similarly adapted when other plant cell types such as leaf or root cells and their organelles are of interest; protocols for making protoplasts from *Arabidopsis* or tobacco leaf or root cells can be found in [7, 8]. Handling protoplasts requires some experience, though. It is important to choose the appropriate enzyme cocktail for cell wall removal and to ensure the enzymes would not cause any unwanted damage to the cellular content by washing off thoroughly before bursting the protoplasts.

Prior to imaging by SEM, the procedure further involves post-fixation of membranes with osmium tetroxide, dehydration of the sample, and metal coating, details of which can be found in [9]. In the protocol presented here, we describe how to isolate and fix nuclei of plant culture cells for feiSEM (Fig. 1). The method can be easily adapted for observing the membrane-associated cytoskeletal structures (also detailed in the protocol; Fig. 2) or nuclei of leaf or root cells (by adapting the protoplasting procedure accordingly).

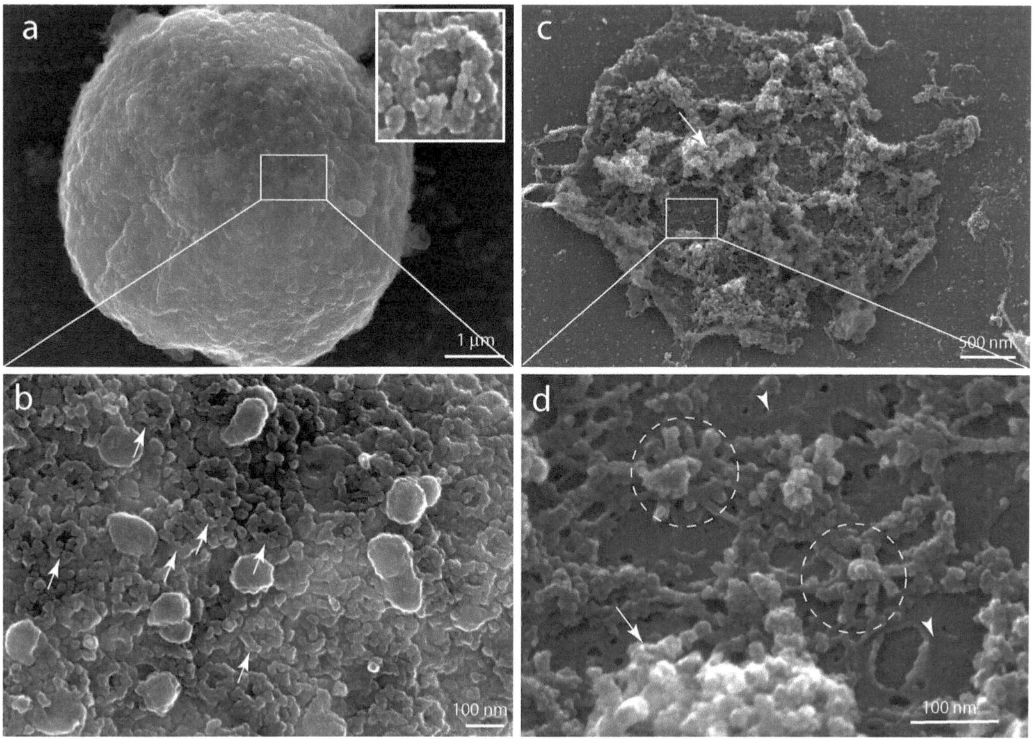

Fig. 1 Tobacco BY-2 cell nucleus and details of a nuclear envelope viewed by feiSEM. (**a**) Nucleus. (**b**) Detail of the nuclear envelope viewed from a cytoplasmic side. *Arrows* point the nuclear pore complexes. (**c**) Nuclear envelope viewed from a nucleoplasmic side after wet fracturing of the nucleus. Remnants of chromatin are marked by *arrow*. (**d**) Detail of the nuclear envelope and nuclear pore complex baskets (*encircled*) from a nucleoplasmic side. Remnants of chromatin are marked by *arrow*. Remnants of the inner nuclear membrane are depicted by *arrowhead*

2 Materials

2.1 Preparing the Nuclei and a Disclosure of the Membrane-Associated Cytoskeleton

1. *Nicotiana tabacum* L. cv. Bright Yellow 2 cells.
2. MS medium: 1x MS salts, 1 mg/l thiamine, 200 mg/l KH_2PO_4, 100 mg/l inositol, 30 g/l sucrose, and 0.2 mg/l 2,4-dichlorophenoxyacetic acid (2,4-D), pH 5.8 (all chemicals obtained from Sigma-Aldrich, St. Louis, USA).
3. 100 ml Erlenmeyer flasks.
4. Horizontal shaker with orbital diameter 30 mm (IKA KS501, IKA Labortechnik, Germany).
5. 50 ml Falcon tubes.
6. Centrifuge (Sigma 3 K18, Rotor Nr. 11133, 47/100, Sigma Laboratory Centrifuges, Osterode am Harz, Germany).

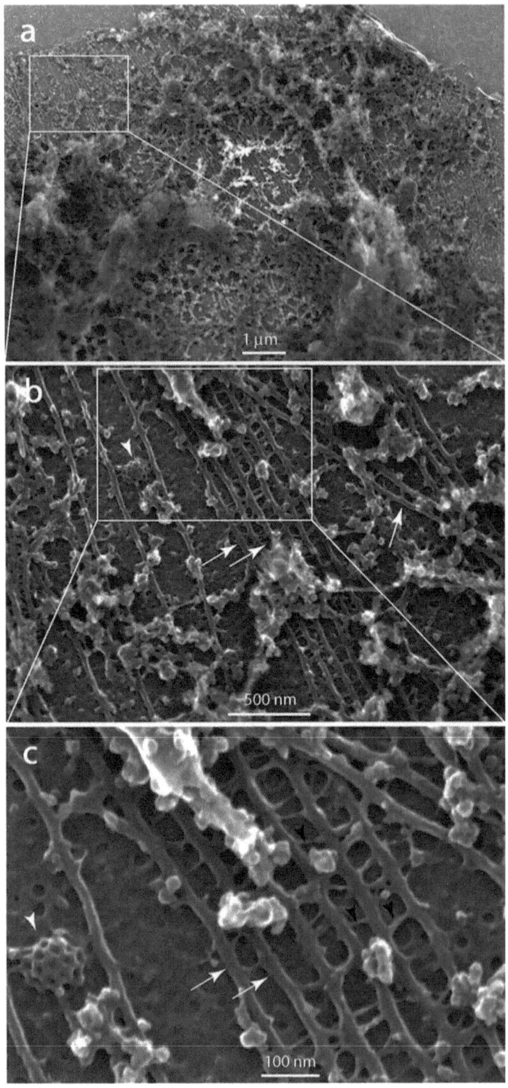

Fig. 2 Tobacco BY-2 cytoplasmic membrane exposed from cytoplasmic side and viewed by feiSEM. (**a**) Low magnification image. (**b**) and (**c**) Details of the microtubule array attached to the plasma membrane. *Arrows* indicate the microtubules, *black arrowheads* show microtubule associated proteins, and *white arrowhead* marks detail of a clathrin-coated pit

7. Enzyme solution: 1 % (w/v) cellulase (Serva), 1 % (w/v) pectolyase Y-23 (Kyowa Chemical Products Co. LTD, Tokyo, Japan), and 0.45 M mannitol (Sigma-Aldrich; *see* **Note 1**).
8. Wash buffer: 10 mM PIPES (Sigma-Aldrich), pH 6.4, 100 mM KCl (VWR International Ltd., Lutterworth, UK), and 0.285 mannitol (Sigma-Aldrich; *see* **Note 2**).
9. Lysis buffer: 10 mM Tris–HCl (Melford Laboratories Ltd., Ipswich, UK), pH 7.2, 5 mM $MgCl_2$ (Sigma-Aldrich), 2 mM

DTT (dithiothreitol, Sigma-Aldrich), and 0.2 mM PMSF (phenylmethylsulfonyl fluoride, Sigma-Aldrich). Make fresh as required (*see* **Note 3**).

10. 0.5 × 16 mm syringe needle (Terumo Europe, Leuven, Belgium).
11. 5 ml syringe (VWR International, Lutterworth, UK).
12. Membrane filter, pore size 0.2 μm (VWR).

2.2 Processing Samples for feiSEM

1. 5 mm × 5 mm silicon chips (Agar Scientific Ltd., Stansted, UK).
2. Acetone.
3. Poly-L-lysine: 1 mg/ml in water (Sigma-Aldrich) stored at −20 °C.
4. Distilled water.
5. Microtubes for making microtube chambers.
6. Microtube cutter (Science Services, München, Germany).
7. Scissors.
8. Diamond scribe for labelling silicon mounts (Agar Scientific Ltd., Stansted, UK).
9. PME buffer: 100 mM PIPES, pH 6.4, 5 mM $MgCl_2$, 10 mM EGTA (Sigma-Aldrich), and 0.2 M sucrose (VWR International; *see* **Note 4**).
10. Fixative 1: 3 % [w/v] paraformaldehyde (Agar Scientific Ltd.) and 0.2 % [w/v] glutaraldehyde (Sigma-Aldrich) in PME buffer. Make fresh as required (*see* **Notes 5** and **6**).
11. Fixative 1 without sucrose. Make fresh as required (*see* **Note 5**).
12. Fixative 2: 2 % [w/v] glutaraldehyde and 0.2 % [w/v] tannic acid in PME buffer (all chemicals obtained from Sigma-Aldrich). Make fresh as required.
13. Cover slides.
14. Dumont tweezers 5/45 (Agar Scientific Ltd.).
15. 35 mm Petri dishes (Jencons-PLS, East Grinstead, UK).
16. 0.1 M sodium cacodylate pH 7.4 (*see* **Note 7**). Store at room temperature up to 1 month.
17. 0.1 % OsO_4 (Agar Scientific Ltd.) in 0.1 M sodium cacodylate pH 7.4. Store at room temperature up to 1 month (*see* **Note 8**).
18. 50, 70, 95, 100 % ethanol (*see* **Note 9**).

2.3 Critical Point Drying

1. Bal-Tec CPD030 (now Leica Microsystems GmbH, Wetzlar, Germany) (other makes are also suitable) with exit gas flow meter and connected to high purity CO_2 cylinder with liquid withdrawal.
2. 100 % ethanol (*see* **Note 9**).

2.4 Chromium Coating

1. Cressington 308UHR or 328UHR coating unit with sputter head, chromium target, swinging shutter, "cryopump," and film thickness monitor (Cressington Scientific Instruments Ltd., Watford, UK).
2. High purity research grade argon gas (BOC, Guilford, UK).
3. Liquid nitrogen.
4. Clean glass slide (VWR).

2.5 Observation by feiSEM

Scanning electron microscope, preferably with field emission source (e.g., Hitachi S-5200 in-lens feiSEM, Hitachi High-Tech, Tokyo), should be used.

3 Methods

3.1 Preparing the Nuclei and Exposing the Membrane-Bound Cytoskeleton

1. Culture cells in dark at 25 °C on a horizontal shaker at 120 rpm.
2. Collect 5 ml of cells by spinning in Falcon tube at 500×*g* for 3 min (*see* **Note 10**).
3. Remove excess medium.
4. Resuspend cells in enzyme solution (*see* **Note 10**).
5. Shake on a horizontal shaker at low rpm (roughly 20–40 rpm) for 3.5–4.5 h until protoplasts are made (*see* **Note 11**).
6. Pellet protoplasts by spinning at 500×*g* for 3 min (*see* **Note 10**).
7. Replace the excess enzyme solution with wash buffer.
8. Place on ice (*see* **Note 12**).
9. Mix the protoplasts 1:1 with ice-cold lysis buffer.
10. For nuclear isolation, pass the resulting suspension through the syringe needle 2–4 times (*see* **Note 13**).

3.2 Processing Samples for feiSEM

1. Prepare microtube chambers as shown in Fig. 3.
2. Scratch numbers on silicon chips using diamond scribe and place in glass dish containing acetone.
3. Pick up chips with fine tweezers and dry with a tissue to clean. Chips must be perfectly clean with no sign of smearing on the surface.
4. Pipette 40 μl of poly-L-lysine onto each chip and allow to adhere for 30 min at 25 °C.
5. Rinse the chips in distilled water and allow to dry (*see* **Note 14**).
6. Place the chips at the bottom of the microtube chambers and layer with 200 μl of ice-cold fixative 1 with sucrose. Keep at 4 °C.

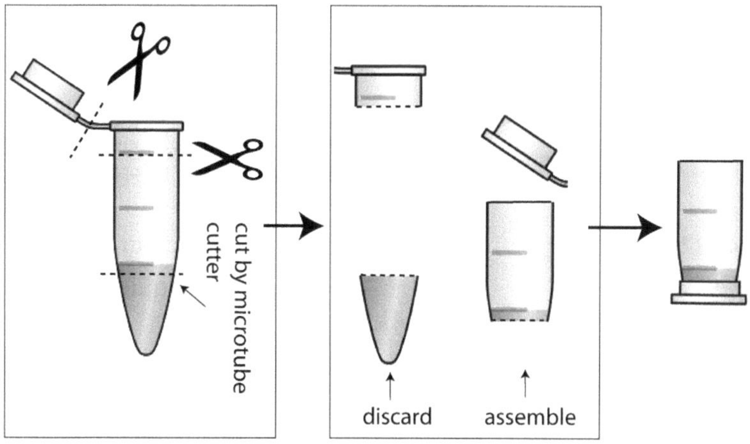

Fig. 3 Diagram of microtube chamber preparation

7. For nuclear isolation, load 10–40 μl of sample on the top of the fixative 1 in microtube chambers (*see* **Note 15**). Spin the sample onto silicon chips at 2,000×*g* for 10 min at 4 °C. Continue with **step 11**.
8. For membrane-bound cytoskeleton, pipette 10–40 μl of the protoplasts from **step 9** of Subheading 3.1 onto a silicon chip (*see* **Note 15**).
9. Allow 30 min to settle at 25 °C.
10. Fracture the protoplasts attached to the chip by gently touching a glass coverslip on top of the silicon chip (wet) fracturing (*see* **Note 16**).
11. Transfer the chips into fixative 1 without sucrose and incubate for 10 min at 4 °C.
12. Transfer the chips into fixative 2 for 10 min at 25 °C (*see* **Note 17**).
13. Rinse the chips in sodium cacodylate for 1 min twice (*see* **Note 18**).
14. Transfer chips into a dish containing osmium tetroxide for 10 min (*see* **Note 19**).
15. Set out six Petri dishes. Fill one with water and the remainder with 50, 70, 95, 100, 100 % ethanol, respectively.
16. Transfer chips to each of these for 2 min each (*see* **Note 20**).

3.3 Critical Point Drying

1. Transfer chips to CPD carrier under 100 % dry ethanol.
2. Fill CPD chamber with 100 % ethanol and place specimen carrier in chamber.
3. Close lid and start cooling.
4. When cooled to ~10 °C, exchange ethanol for liquid CO_2 with at least ten changes until all ethanol is replaced.

5. Leave to stand for 30 min (*see* **Note 21**).
6. Exchange another ten times.
7. Warm chamber to 40 °C.
8. Release CO_2 gas slowly over about 10 min, using gas flow meter to monitor rate of release.

3.4 Chromium Coating

1. Place samples in the vacuum chamber and pump to ~10^{-6} mBar.
2. Cool cryopump with liquid nitrogen and, if appropriate, open isolation valve.
3. Wait for a vacuum of at least 5×10^{-7} mBar.
4. Ensure shutter is in place between the sample and sputter head.
5. Introduce argon atmosphere to a pressure of around 10^{-3} mBar.
6. Sputter for 30 s onto shutter. The plasma should be a sky-blue color (*see* **Note 22**).
7. Open shutter until 1–1.5 nm chromium is deposited on the sample.
8. Turn off current.
9. Close valve on cryopump, if appropriate.
10. Remove sample from chamber.
11. It is useful to simultaneously coat a glass coverslip which should be placed on a sheet of white paper after coating. The color of the metal coat on the glass coverslip should be grey (*see* **Note 23**).
12. Image samples in SEM as soon as possible (*see* **Note 24**).

3.5 Imaging

1. Insert specimen into SEM.
2. Select 10 kV accelerating voltage (*see* **Note 25**).
3. Set emission current to as high as possible (*see* **Note 26**).
4. Use the secondary electron detector to acquire a high-resolution surface image. This detects low-energy electrons ejected from the sample surface (particularly the chromium coat), giving an image of the sample surface.
5. Capture the image and analyze it using appropriate imaging software (*see* **Note 27**).

4 Notes

1. Filter-sterilize before use or storage. Enzyme solution can be stored at −20 °C up to 1 month.
2. Wash buffer can be stored at −20 °C up to 1 month.
3. Make 1 M DTT stock in water and keep at −20 °C up to 6 months. Prepare 200 mM PMSF in isopropanol and keep at

−20 °C up to 6 months. PMSF is toxic and should be handled very carefully in a fume hood.

4. PME buffer can be stored at −20 °C up to 1 month.
5. Caution when working with paraformaldehyde. Paraformaldehyde is toxic by ingestion or inhalation. It can cause skin corrosion and serious eye damage. Handle as a carcinogen, work in a fume hood, and wear gloves.
6. Dissolve PFA by heating the solution in a water bath at 60 °C maximally. It is important to keep the temperature stable as dissolving at higher temperatures the fixative loses its fixing properties whereas at lower temperatures it would take too long to dissolve.
7. Caution when working with sodium cacodylate. May be fatal if swallowed. Known carcinogen in humans. Harmful if inhaled, may be harmful by skin contact. Long-term exposure may lead to kidney and liver damage. Eye and skin irritant. Use gloves, safety glasses, and good ventilation. Handle as a carcinogen.
8. Caution: osmium tetroxide is volatile, toxic by inhalation and by skin contact. Work in a fume hood and wear gloves.
9. 100 % ethanol should be stored with Molecular Sieve to ensure that it is dry.
10. Be extremely cautious and gentle when working with cells and protoplast, use pipette tips with cut tip, and resuspend cells by gently tapping the Falcon tube.
11. In the meantime you can prepare microtube chambers and silicon chips (Subheading 3.2, **Steps 1–6**).
12. All subsequent steps are performed on ice.
13. Use gentle pressure when passing the suspension through the syringe needle.
14. Silicon chips can be stored at room temperature up to 1 month.
15. Experience is needed to estimate the amount. The volume of the loaded sample can be adjusted according to the cellular density. As a rough guide, if the suspension has a nearly milky color, use 10 μl or less; if lightly cloudy, use 20–40 μl.
16. No pressure is required; just place the cover slide on the chip and remove immediately.
17. This step can be performed also at 4 °C for several hours or overnight; thus, the timing can be adapted as needed.
18. We recommend transferring chips between dishes containing the required solution rather than changing solution in one dish.
19. Ten minutes is sufficient for thin samples such as nuclei and subcellular components, but longer times (30–60 min) are required for bulkier samples such as pieces of tissue.

20. Ensure chips are transferred between liquids quickly and there is never any risk of air drying.
21. This allows any ethanol trapped in the sample time to diffuse out.
22. This step allows the oxidized layer of chromium on the target to be removed by pre-sputtering onto the shutter.
23. If there is any hint of brown, it suggests that the coating is contaminated with chromium oxide. Reasons for this include insufficiently initial vacuum prior to introducing argon, vacuum leaks, leaks in the argon pipes, air in the argon pipes after changing the cylinder, and insufficient pre-sputtering onto the shutter. If the coater has not been used for some time, chromium oxide can build up on the target which will then require longer pre-sputtering on the shutter. This can happen even when the unit is kept under vacuum. It is good practice to keep a chromium coating unit pumping continuously to keep the chamber and vacuum meticulously clean.
24. Chromium films on samples oxidize rapidly and samples should be examined ideally on the same day they are coated. There is a noticeable loss of signal over a few hours and a significant loss over a few days. The deterioration of the coat can be slowed by storing samples in a vacuum.
25. 10 kV has been found to be optimal in our Hitachi S5200 SEM but 30 kV was better in the Topcon DS130F; therefore, it is necessary to experiment with different voltages and other conditions to optimize for the sample, gold size, and specific microscope.
26. If beam damage is observed, the current may have to be reduced. Some analytical SEMs (such as the Hitachi SU70) are capable of very high beam currents which would be inappropriate.
27. It requires some experience to search and identify the structure of interest as the sample has many remnants of membranes, organelles, and protein complexes that are not necessarily of interest.

Acknowledgments

This work was supported by grants from the Biotechnology and Biological Sciences Research Council, UK, grant number BB/E015735/1 and BB/G011818/1. Thanks to Christine Richardson and Helen Grindley for technical assistance.

References

1. Bomblies K, Shukla V, Graham C (2008) Scanning electron microscopy (SEM) of plant tissues. CSH Protoc 2008:4933. doi:10.1101/pdb.prot4933
2. Schroeder-Reiter E, Sanei M, Houben A et al (2012) Current SEM techniques for de- and re-construction of centromeres to determine 3D CENH3 distribution in barley mitotic chromosomes. J Microsc 246: 96–106
3. Schwab B, Hulskamp M (2010) Quick and easy fixation of plant tissues for scanning electron microscopy (SEM). CSH Protoc 2010:4934. doi:10.1101/pdb.prot4934
4. Allen TD, Rutherford SA, Murray S et al (2007) Visualization of the nucleus and nuclear envelope in situ by SEM in tissue culture cells. Nat Protoc 2:1180–1184
5. Fiserova J, Kiseleva E, Goldberg MW (2009) Nuclear envelope and nuclear pore complex structure and organization in tobacco BY-2 cells. Plant J 59:243–255
6. Collings DA, Asada T, Allen NS et al (1998) Plasma membrane-associated actin in Bright yellow 2 tobacco cells: evidence for interaction with microtubules. Plant Physiol 118:917–928
7. Abel S, Theologis A (1994) Transient transformation of Arabidopsis leaf protoplasts: a versatile experimental system to study gene expression. Plant J 5:421–427
8. Mathur J, Koncz C (1998) Protoplast isolation, culture, and regeneration. Methods Mol Biol 82:35–42
9. Goldberg MW, Fiserova J (2010) Immunogold labelling for scanning electron microscopy. Methods Mol Biol 657:297–313

Chapter 15

Immunogold Labeling of Resin-Embedded Electron Microscopical Sections

Ilse Foissner and Margit Hoeftberger

Abstract

Knowledge about the spatio-temporal distribution patterns of proteins and other molecules of the cell is essential for understanding their function. A widely used technique is immunolabeling which uses specific antibodies to reveal the distribution of molecular components at various structural levels. Electron microscopy offers the highest resolution of morphological techniques and is thus an indispensable tool for the analysis of molecule distribution patterns at the subcellular level. In this chapter we describe a routine method for labeling ultrathin sections of resin-embedded material with antibodies conjugated to colloidal gold.

Key words Colloidal gold, Immunolabeling, Transmission electron microscopy, Resin-embedded sections, Antibodies

1 Introduction

1.1 General

Immuno techniques use antibodies to detect, characterize, and localize specific antigens in cell extracts (e.g., dot and Western blots, immunoprecipitation) and in situ inside tissues and cells (immunolabeling or immunocytochemistry). To study their distribution antibodies must be conjugated to markers which differ according to the method used for observation [1]. Fluorescent dyes and enzymes which produce colored products with chromogenic substances are suitable for fluorescence and bright field microscopy. For electron microscopy (EM) electron dense gold particles are commonly used [2]. These gold particles can be adsorbed to macromolecules by complex electrochemical interactions and are available in distinct sizes. The colloidal gold method is applicable to most of the electron microscopical systems and techniques (transmission EM, scanning EM; ultrathin sections and freeze fracture). After increasing the size of gold particles by silver enhancement, the distribution of gold-conjugated probes can also be visualized in the bright field microscope.

Viktor Žárský and Fatima Cvrčková (eds.), *Plant Cell Morphogenesis: Methods and Protocols*, Methods in Molecular Biology, vol. 1080, DOI 10.1007/978-1-62703-643-6_15, © Springer Science+Business Media New York 2014

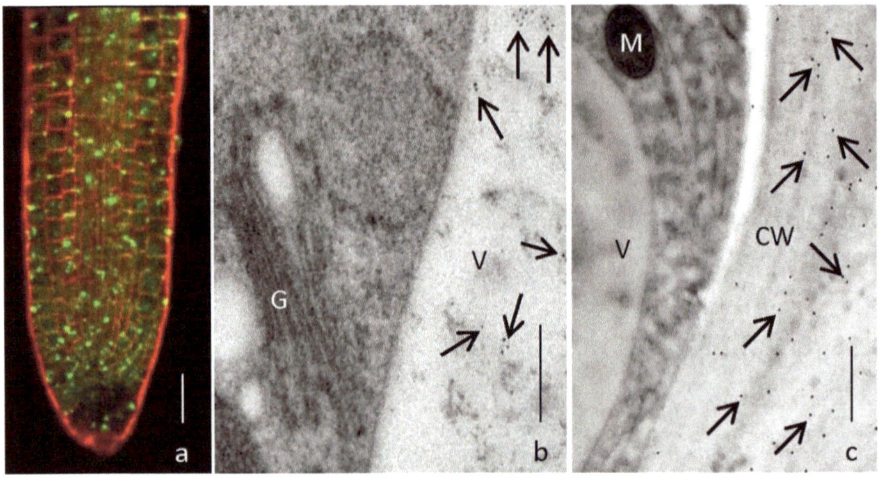

Fig. 1 Immunofluorescence of GFP-labeled protein (**a**) and immunogold-labeled EM sections (**b**, **c**). (**a**) Confocal laser scanning image of a root of *Arabidopsis thaliana* expressing a vacuolar protein fused to GFP. The sample was counterstained with the red fluorescent endocytic marker FM4-64. (**b**) Similar root tip as in (**a**) which was high pressure frozen, cryosubstituted, and embedded in LRGold. The EM image shows a section through an epidermal cell after immunolabeling with anti-GFP. 10 nm gold particles coupled to the second antibody are visible in the vacuole (V; *arrows*). *G* = Golgi stack. (**c**) EM section of high-pressure-frozen, cryosubstituted, and LRGold-embedded internodal cell of the green alga *Chara braunii* after immunolabeling with a primary antibody against a cell wall (CW) polysaccharide and a secondary antibody conjugated to 10 nm gold particles (*arrows*). *M* = mitochondrion. Scale bars are 25 μm (**a**) and 500 nm (**b**, **c**)

EM is complemented by genetic and molecular biological techniques. Proteins can be visualized in the confocal laser scanning microscope by expressing them together with the amino acid sequence of a fluorescent marker, e.g., green fluorescent protein (GFP). Co-expression with known organelle markers tagged with a different fluorescent protein gives a first hint for the localization of the protein under study. After fixation the distribution of the fluorescent markers can be studied at much higher, even "suborganellar" resolution in the EM using specific, commercially available gold-conjugated antibodies (Fig. 1a, b). Immunogold labeling thus bridges the gap between molecular biology and cellular architecture at the fine structural level [3, 4].

1.2 Direct and Indirect Immunolabeling

Markers for immunolabeling can be conjugated either to the primary antibody which binds to the epitope of a protein or other molecule under study or to a secondary antibody which specifically recognizes the primary unlabeled antibody. Accordingly we distinguish between direct and indirect immunolabeling. Direct immunolabeling is used, e.g., for lectins which recognize specific sugar residues and for enzymes to localize their substrates [5]. However, the most widely applied method is indirect immunolabeling. Although it requires additional incubation and washing steps, indirect immunolabeling is, in general, time and cost saving because there is no need to label every primary antibody and one secondary antibody can be used for a variety of primary antibodies. Even more importantly, indirect immunolabeling offers signal amplification (*see* Subheading 1.4).

1.3 Primary Antibodies Used for Immunolabeling

Primary antibodies are either polyclonal or monoclonal [6]. Polyclonal antibodies are collected from the serum of an immunized animal (usually a herbivore) and recognize different epitopes, thus increasing the likelihood of antibody binding. Unfortunately, the serum will contain not only antibodies against the protein, peptide, or carbohydrate used for immunization but also antibodies against contaminant molecules and native antibodies of the immune system, e.g., antibodies raised against the food which is especially disturbing when working with plant cells. Therefore, pre-immune serum should be collected prior to immunization and should be used for control experiments. Monoclonal antibodies are produced by cell cultures originating from fusion of an antibody-producing cell with a myeloma cell and are specific for one epitope, thus reducing unspecific binding. However, there is no guarantee that this epitope is exposed and available for antibody binding.

Different classes of immunoglobulins are present in the serum of immunized animals. Most antibodies used for immunolabeling are IgGs which are purified in order to reduce background staining. The antigen binding sites are localized at the light chains located at the ends of the arms of the Y-shaped immunoglobulin.

1.4 Secondary Antibodies and Gold Markers

The secondary antibody must specifically recognize the first antibody; i.e., it must be directed against the host species of the primary antibody. For example, if the primary antibody is mouse IgG, an anti-mouse immunoglobulin raised in another animal (e.g., goat) is required. The most frequently used secondary antibodies are affinity-purified polyclonal IgGs (whole molecules) generated by immunizing an animal with whole IgG. These secondary antibodies recognize different epitopes so that several molecules with their attached markers will bind to one primary antibody which results in an amplification of the signal as compared with direct immunolabeling. Monoclonal secondary antibodies, fragments of secondary antibodies, and antibodies raised against fragments are likewise commercially available but not so often used for immunolabeling of plant cells.

Gold particles suitable for conjugation with antibodies are available at sizes between 1.4 nm (nanoparticles) and 30 nm. The larger gold particles can be seen at low magnifications of the EM. They have the disadvantage, however, that because of steric hindrance, a lower number of secondary antibodies can bind to the primary antibody which decreases the extent of signal amplification. Therefore, it is better to use antibodies conjugated to small particles and, if necessary, to increase their size by silver amplification. Secondary antibodies with differently sized gold particles can be used for double immunostaining. This is usually done by using primary antibodies raised in different animals (e.g., in mouse and in goat). Instead of secondary antibodies, Protein A-gold or Protein G-gold can be used. Both bind to a single site at the Fc region of antibodies of various host species (but not to all!) which means that signal amplification is not possible. Streptavidin-gold can be used in combination with a biotinylated antibody.

An alternative to conventional colloidal gold is fluorescent nanogold particles [7] and quantum dots [8]. These probes allow studying the same specimen by fluorescence microscopy and by EM (visualization of quantum dots requires the use of an energy-filtered transmission electron microscope).

1.5 Postembedding Immunogold Labeling

The most frequently used method for EM cytochemistry is indirect immunogold labeling of ultrathin sections of resin-embedded material (Fig. 1b, c). A prerequisite for successful immunolabeling is the preservation of the antigen, especially on EM sections where only few binding sites are available for recognition by the antibody. For postembedding immunogold labeling, high-pressure frozen material and cryosubstituted material are frequently used because ultrarapid freeze fixation is considered to preserve structure and antigen binding sites more efficiently than chemical fixation [9]. But often a compromise between preservation of cytoarchitecture and preservation of antigen has to be made. Before sectioning, cells or tissues are embedded in polar acrylic resins [1, 4, 5, 9–12]. Epoxy resins because of their hydrophobicity and the high temperature necessary for polymerization are usually less suited for immunogold labeling.

In the following we describe a protocol for immunogold labeling of resin-embedded sections. Further information can be found in refs. [1–16] as well as in protocols provided on the homepages of EM laboratories or antibody manufacturers. Protocols for pre-embedding immunogold labeling are to be found elsewhere [17] as well as immunolabeling of cryosections [18–20]. The latter method offers the advantage that epitope antigens are much better preserved and accessible for antibody binding but requires the use of a cryostat microtome. Methods to quantify immunogold localization are described in ref. [21].

2 Materials

All solutions should be prepared with double distilled water if available (*see* **Note 1**). Reagents should be of analytical grade and can be stored at room temperature, except bovine serum albumin (BSA, 4 °C) and antibodies (4 °C or frozen, *see* Subheading 2.3). Buffers can be prepared as 10× stock solutions and supplemented by 0.02 % NaN_3 for longer storage. All other solutions should be freshly prepared.

2.1 pH Buffers

We use one of the following pH buffers as a component of the blocking solution. Buffer components are sequentially dissolved in 90 ml water, adjusted to pH 7.4 with HCl or NaOH, and finally made up to 100 ml with water.

1. Tris–HCl 50 mM: 0.605 g Tris (tris(hydroxymethyl) aminomethane) base or 0.758 g Trizma pre-set crystals (Sigma, T7693) (*see* **Note 2**).
2. TBS (Tris-buffered saline): 0.121 g Tris base (final concentration 10 mM) or 0.152 g Trizma pre-set crystals, 0.877 g NaCl (150 mM).
3. PBS (phosphate-buffered saline): 0.818 g NaCl (140 mM), 0.022 g KCl (2.95 mM), 0.032 g KH_2PO_4 (2.38 mM), and 0.108 g Na_2HPO_4 (76.1 mM).

2.2 Blocking Buffer

The blocking solution is made of buffer to provide the necessary pH for antibody binding and of various agents which help to reduce unspecific staining (*see* **Note 3**).

Use a magnetic stirrer at moderate speed (*see* **Note 4**) to mix: 10 ml Tris–HCl, TBS or PBS (*see* **Note 5**), 1 % BSA (100 mg; *see* **Note 6**), 0.1 % Tween 20 (polyoxyethylene-sorbitan monolaurate; 10 mg; *see* **Note** 7), and 50 mM glycine (37.6 mg; for aldehyde fixed cells only; *see* **Note 8**).

2.3 Antibody Solutions

Antibodies are diluted in blocking buffer. The working solutions should be prepared immediately before use and centrifuged (maximum speed of microfuge or up to 10,000 × *g* for 5 min) in order to remove larger aggregates.

Stock solutions of most primary antibodies should be stored at ≤−20 °C and frozen in aliquots in order to prevent repeated freezing and thawing; some primary antibodies require storage at 4 °C or at room temperature. For each primary antibody the optimum concentration must be tested (*see* **Note 9**). Stock solutions of secondary antibodies are usually stored at 4 °C and often can be used for months and even years. For secondary antibodies start with dilutions given in the manufacturer's data sheets (*see* **Note 10**).

2.4 Miscellaneous

1. Equipment: magnetic stirrer, pH meter, microfuge, shaker, and pipettes with adjustable volumes between 0.1 and 10 μl, 10 and 50 μl, and 100 and 1,000 μl.
2. Petri dish with a diameter of 15 cm or larger. Alternatively, several small Petri dishes can be used.
3. Wet chamber (plastic box with rack to support Petri dish or plastic box lined with wet filter paper).
4. Parafilm.
5. Fine, antimagnetic tweezers (e.g., Dumont no 7; *see* **Note 11**).
6. Disposable Pasteur pipettes.
7. Lint-free filter paper.

8. Scissors.
9. Grid mat or grid holder pad.
10. Ultrathin sections collected on nickel or gold grids (*see* **Note 12**).

3 Methods

Work in a dust-free clean environment and avoid contamination of solutions, sections, and material. Wash forceps in distilled water between each step. Cover Petri dish between handling of grids in order to reduce evaporation. Never let grids completely dry out before the end of the incubation procedure! It is advisable to run the necessary negative controls concurrently in the same or another Petri dish (*see* **Note 13**). Before immunolabeling a number of positive controls should be made (*see* **Note 14**).

1. Place a sheet of Parafilm on the bottom of a Petri dish (*see* **Note 15**; Fig. 2a). Be careful not to touch the surface of the Parafilm with ungloved hands.

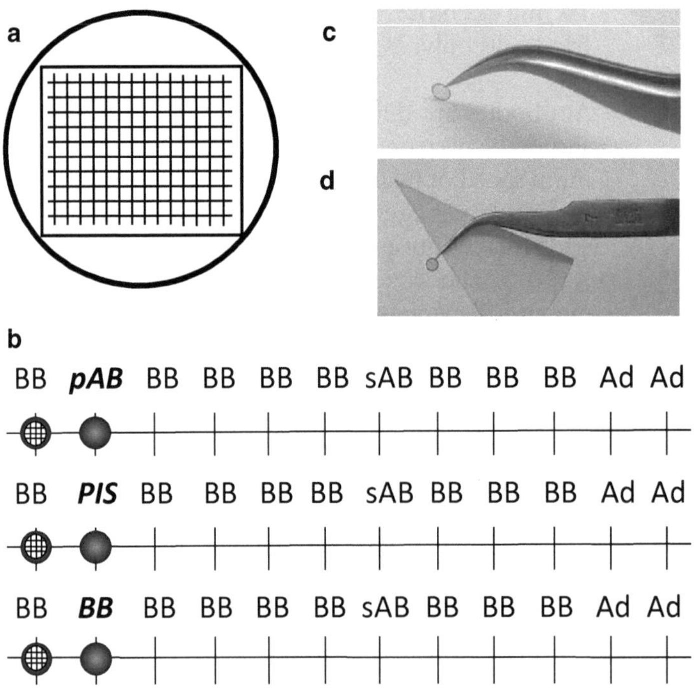

Fig. 2 Immunogold labeling of EM sections. (**a**) Petri dish with Parafilm onto which a mesh-like pattern was engraved using a blunt needle. (**b**) Series of incubation steps for immunogold labeling and two important negative controls (*see* **Note 13**). The sections on the grid in the upper row (now in banding buffer BB) will be incubated in primary antibody (pAB). The sections on the grid in the middle row will be incubated in pre-immune serum (PIS) to check specificity of primary, polyclonal antibody. The sections on the grid in the lower row will be incubated in blocking buffer to check specific binding of the secondary antibody (sAB). All other steps are the same. Ad = distilled water. (**c**) Removal of excess liquid by vertically touching filter paper with grid. (**d**) A wedge of filter paper is used to remove liquid between prongs of tweezers

2. Place 20–50 μl droplets of blocking buffer onto the Parafilm (*see* **Note 16**; *see* Fig. 2b for this and following steps).
3. Put grids on top of droplets with section side facing downwards and incubate for 30 min at room temperature (*see* **Note 17**).
4. Place droplets of the primary antibody solution onto the Parafilm.
5. Remove grids from blocking buffer; remove excess liquid by touching the edge of the grid with filter paper (Fig. 2c) and place on top of droplets containing the primary antibody. Incubation time should be 1–2 h at room temperature or up to 24 h in fridge (*see* **Note 18**).
6. Remove grids from droplet with primary antibody solution, remove excess liquid, and pass grids through a series of at least four droplets (50 μl) of blocking buffer with at least 10 min incubation each (*see* **Note 19**).
7. After removing excess liquid, place grids on top of droplets containing secondary, gold-conjugated antibody diluted in blocking buffer and incubate for at least 1 h at room temperature (or up to 12 h at about 4 °C).
8. Remove excess liquid and pass grids through a series of at least three droplets of blocking buffer (10 min each) and two droplets of distilled water.
9. Remove grids from droplets and rinse both sides by drop-like addition of about 5 × 1 ml of distilled water from a Pasteur pipette. Remove excess liquid as described above and use a small wedge of filter paper to absorb liquid trapped between prongs of forceps that hold grid (Fig. 2d).
10. Allow grids to dry on a gridpad with sections facing upwards.

Option 1: After **step 9** use silver enhancement to enlarge the size of small (2–5 nm) gold particles (*see* **Note 20**).

Option 2: After **step 9** stain sections by placing grids on droplets of 2 % aqueous uranyl acetate for 10 min (*see* **Note 21**).

4 Notes

1. The conductivity of the water should be ≤1 μS m^{-1}. It is advisable to use freshly distilled water not stored for a longer time in plastic containers.
2. We prefer Trizma pre-set crystals because they need no pH adjustment. Notice that certain silver chloride pH electrodes do not give accurate results with Tris buffers. In that case better use pH sticks.

3. The aim of all immunolabeling procedures is to obtain a maximum specific signal (reflecting the distribution of the antigen) with a minimum of nonspecific background. In case of a low signal to noise ratio, first try to find the optimum concentrations of primary and secondary antibodies (compare **Notes 9** and **10**). Then the concentration and composition of blocking buffer should be varied in order to reduce nonspecific binding of antibodies or colloidal gold to the sample (compare **Notes 5–8**). Nonspecific labeling is mostly a problem with primary antibodies. They often stick to cell walls, chloroplasts, and starch granules even in well-fixed and embedded cells or tissues as well as to the Formvar film.
4. Fast stirring will produce excessive BSA foam.
5. If sample contains positively charged components which unspecifically attract negatively charged gold particles, the ionic concentration of the Tris–HCl buffer might be too low. In that case the use of TBS or PBS is suggested.
6. BSA saturates nonspecific binding sites and neutralizes positively charged samples. The BSA should be free of IgGs. Acetylated BSA (available from Aurion) has a much higher binding affinity than non-acetylated BSA and more effectively blocks polycationic sites. The concentration of BSA should vary between 0.1 and 2.5 %. Instead of or additionally to BSA, pre- or nonimmune serum from the host species of the secondary antibody can be used up to 5 %.
7. Tween facilitates wetting of the sections and prevents hydrophobic interactions between gold particles and sample; the concentration should vary between 0.1 and 1 %.
8. Glycine masks free aldehyde groups introduced by chemical fixation; the concentration should vary between 10 and 50 mM. Free aldehyde groups can also be quenched by washing with 10–100 mM NH_4Cl, pH 7, or by 0.1 % $NaBH_4$.
9. The appropriate concentration of the antibodies is crucial for successful immunolabeling. When a new primary antibody is applied, a dilution series between 1:1 and 1:5,000 should be tested. In general, EM immunolabeling requires an up to 10× higher concentration of primary antibody than used for Western blotting because of the lower number of binding sites. Ideally, the primary antibody should be diluted so that nonspecific binding is reduced below the level of detection.
10. At the optimum concentration no gold conjugates should be present on sections in the absence of the primary antibody (compare **Note 13**).
11. Tweezer points should always be protected with a cap or fixed by wire wrapped around the levers. Damaged tweezers can be deburred by pulling a fingernail file or a double-sided sandpa-

per between the closed forceps prongs but often they are no longer suited for picking up EM grids.

12. Copper grids may form precipitates when in contact with buffers.

13. Several negative controls are needed in order to confirm the specificity of staining, especially when working with polyclonal primary antibodies. But also monoclonal and secondary antibodies may bind nonspecifically to the sample. The most important negative controls are as follows (Fig. 2b): (a) In case of polyclonal antibody, replace primary antibody with pre-immune serum collected from the animal before immunization (*see* Subheading 1.3). If pre-immune serum is not available, use at least nonimmune (normal) serum of the same species but this is considered not to be an adequate substitute. (b) Replace the primary antibody with a control antibody from the same species raised against an antigen known to be absent in the tissue. (c) If available, use mutants lacking the antigen or, when the distribution of transgenic proteins is studied, use non-transformed cells or plants as controls. (d) Absorb the primary antibody with its antigen by addition of antigen which should abolish binding. (e) Use blocking buffer instead of primary antibody to get information about the background signal caused by unspecific binding of the secondary antibody.

14. Dot blots can be used to prove the presence of the antigen in cell extracts and to detect inhibitory effects of fixatives (e.g., glutaraldehyde or OsO_4) on antibody binding. Western blots are required to show that the antibody recognizes a (single) protein with the appropriate mass. It is also a good idea to first test a new antibody for immunofluorescence where more antigens are available for binding. When immunofluorescence fails, labeling of EM sections is unlikely to be successful. It must also be kept in mind that antibodies suited for Western blots are not necessarily suited for immunostaining. In Western blots antibodies recognize the denatured antigen whereas in samples for immunolabeling antigens are supposed to be in a more natural conformation. But it is also possible that the antigen has become inaccessible, destroyed, or extracted during preparation.

15. A small droplet of water between Parafilm and Petri dish helps to fix the sheet if required.

16. Droplets placed on the intersections of a mesh-like pattern scratched into the Parafilm with the tip of blunt tweezers or needles are more likely to remain in place. But be careful not to perforate the Parafilm!

17. Do not immerse grids into droplet. If that happens accidentally, either thoroughly wash the grid in distilled water and proceed again or continue by immersing the grid with section side facing upwards into the droplets until the end of the

procedure. If sections are collected on uncoated nickel or gold grids, both surfaces can be stained by fully submerging grids in reagents. Such grids can also be used for double labeling when each side is exposed to different primary and secondary antibodies (but not immersed!).

18. For prolonged incubation place covered Petri dish in a wet chamber. Vibration of the samples during incubation helps to shorten incubation time and may also reduce nonspecific background labeling [1]. The dish of shakers may become hot after prolonged use. A thin sheet of polystyrene beneath the Petri dish or the moist chamber prevents heating of samples.
19. All washing steps given in this protocol are minimums! In case of high background, staining or dirt on sections increase number of washing steps and incubation times.
20. For easier detection at low magnification, the size of the gold particles can be increased with silver enhancement kits available from different companies.
21. This step is only required for sections with very low contrast and when using conventional EMs without energy filtering [22]. Too much contrast makes recognition of small gold particles difficult. Thoroughly wash grids with water after staining.

Acknowledgement

This work was supported by the Austrian Science Fund (project no. P 22957-B20 to IF).

References

1. Hawes CR, Satiat-Jeunemaitre B (eds) (2001) Plant cell biology: a practical approach. Oxford University Press, Oxford
2. Hayat MA (ed) (1991) Colloidal gold. Principles, methods and applications, vol 1–3. Academic, London
3. Koster AJ, Klumperman J (2003) Electron microscopy in cell biology: integrating structure and function. Nat Rev Mol Cell Biol 2003:SS6–SS10
4. Schwartzbach SD, Osafune T (eds) (2010) Immunoelectron microscopy: methods and protocols. Springer Protocols. Humana Press Inc., Totowa, NJ
5. Hall JL, Hawes CR (eds) (1991) Electron microscopy of plant cells. Academic, London
6. Javois LC (ed) (1991) Immunocytochemical methods and protocols. Methods in molecular biology, vol 115. Humana Press Inc., Totowa, NJ
7. Takizawa T, Robinson JM (2000) FluoroNanogold is a bifunctional immunoprobe for correlative fluorescence and electron microscopy. J Histochem Cytochem 48: 481–486
8. Nisman R, Dellaire G, Ren Y et al (2004) Application of quantum dots as probes for correlative fluorescence, conventional, and energy-filtered transmission electron microscopy. J Histochem Cytochem 52:13–18
9. Verkleij AJ, Leunissen JL (eds) (1989) Immuno-gold-labeling in cell biology. CRC, Boca Raton
10. Griffiths G (1993) Fine structure immunocytochemistry. Springer, Berlin, Heidelberg, New York
11. Bozzola JJ, Russell LD (1999) Electron microscopy: principles and techniques for biologists. Jones and Bartlett Publishers International, London

12. Newman GR, Hobot JA (2001) Resin microscopy and on-section immunocytochemistry. Springer Lab Manual, Springer, Berlin
13. VandenBosch K (1992) Localization of proteins and carbohydrates using immunogold labeling in light and electron microscopy. In: Gurr SJ, McPherson MJ, Bowles DJ (eds) Molecular plant pathology—a practical approach, vol 2. IRL, Oxford, New York, Tokyo, pp 31–44
14. Hayat MA (1995) Immunogold-silver staining: principles, methods and applications. CRC, Boca Raton
15. Maunsbach AB (1998) Immunolabeling and staining of ultrathin sections in biological electron microscopy. In: Celis JE (ed) Cell biology: a laboratory handbook, vol 3. Academic, New York, pp 268–276
16. Sarraf CE (2000) Immunolabeling for electron microscopy. Methods Mol Med 40: 439–452
17. Humbel BM, de Jong MD, Muller WH et al (1998) Pre-embedding immunolabeling for electron microscopy: an evaluation of permeabilization methods and markers. Microsc Res Tech 42:43–58
18. Tokuyasu KT (1997) Immunocytochemistry on ultrathin cryosections. In: Spector DL, Goldman RD, Leonwand LA (eds) Cells: a laboratory manual. Cold Spring Harbor Laboratory Press, Cold Spring Harbor, pp 131.1–131.27
19. Van Donselaar E, Posthuma G, Zeuschner D et al (2007) Immunogold labeling of cryosections from high-pressure frozen cells. Traffic 8:471–485
20. Ripper D, Schwarz H, Stierhof Y-D (2008) Cryo-section immunolabeling of difficult to preserve specimens: advantages of cryofixation, freeze-substitution and rehydration. Biol Cell 100:109–123
21. Mayhew TM (2011) Quantifying immunogold localization on electron microscopic thin sections: a compendium of new approaches for plant cell biologists. J Exp Bot 62: 4101–4113
22. Lütz-Meindl U, Aichinger N (2004) Use of energy-filtering transmission electron microscopy for routine ultrastructural analysis of high-pressure-frozen or chemically fixed plant cells. Protoplasma 223:155–162

Chapter 16

Live Cell Imaging of Arabidopsis Root Hairs

Tijs Ketelaar

Abstract

Root hairs are tubular extensions from the root surface that expand by tip growth. This highly focused type of cell expansion, combined with position of root hairs on the surface of the root, makes them ideal cells for microscopic observation. This chapter describes the method that is routinely used in our laboratory for live cell imaging of Arabidopsis root hairs.

Key words Root hair, Microscopy, Cell growth, Tip growth, Live cell imaging, Biofoil sandwich

1 Introduction

Root hairs are tubular extensions from trichoblasts, root epidermal cells that are positioned on the anticlinal walls of underlying cortical cells (for review *see* e.g., ref. [1]). Root hair growth is initiated with the formation of a bulge, followed by a rapid elongation phase, during which the root hair expands by tip growth [2, 3]. Tip growth is the highly polarized secretion of cell wall matrix containing exocytotic vesicles in the apex of the root hair. Root hairs expand conveniently from the outer surface of the root, which makes them suitable for microscopic imaging. In addition, due to their rapid, highly polarized growth, root hairs have become a model system to study cell expansion (reviewed in ref. [4]). However, it is technically challenging to reproducibly observe expanding root hairs under a microscope, since their manipulation often causes rapid growth arrest.

The following method that has evolved over the last decade [5–12] is suitable for live cell imaging of root hairs without growth disruption and has been successfully used for drug treatments and stainings (Zhang and Ketelaar, unpublished results).

Viktor Žárský and Fatima Cvrčková (eds.), *Plant Cell Morphogenesis: Methods and Protocols*, Methods in Molecular Biology, vol. 1080, DOI 10.1007/978-1-62703-643-6_16, © Springer Science+Business Media New York 2014

2 Materials

1. Biofoil, cut in rectangles of 24 mm × 30 mm and autoclaved in a dry beaker covered with aluminum foil. Biofoil is a gas permeable foil. Currently it is available from Sarstedt (Lumox foil 25 305 mm × 5 m, product number 94.607317). Besides gas permeability, Biofoil has several properties that make it well suitable for microscopy; it is transparent for visible light, and it is non-reflective and not fluorescent. As an alternative to Biofoil, we have experimented with standard household foil (cling film), which has been working well in our trails.
2. Hoagland's medium (Hoagland modified basal salt mixture; PhytoTechnology Laboratories, KS, USA; http://www.phytotechlab.com) supplemented with 1 % v/w sucrose, pH 5.7, and the same medium complemented with 0.7 % w/v agarose; both autoclaved, the liquid medium at room temperature and the medium with agarose at approximately 60 °C.
3. 70 % v/v ethanol in water.
4. 10–15 % household bleach (4 % sodium hypochlorite) in water with 0.01 % detergent (both Triton X-100 and Tween 20 work fine), freshly prepared.
5. Sterile water.
6. Sterilized (autoclaved if possible) and air-dried (under sterile conditions) consumables: coverslips (50 mm × 24 mm), Petri dishes (110 mm diameter) or 50 ml Falcon tubes, sterile dissecting knife, blue (1.00 ml) pipette tips, Pasteur pipettes.

3 Methods

3.1 Seed Sterilization

The method described below should be used for small amounts of seeds only. The seed volume will increase during the procedure and should not exceed one third of the total volume at any time. The method can easily be scaled up using a larger container for the seeds, for example, a 50 ml Falcon tube.

1. Place a small amount of seeds in an Eppendorf tube and fill the tube with 70 % ethanol. Mix thoroughly but carefully; no air bubbles should form. Allow the seeds to sediment and remove the ethanol by pipetting within 1 min after application.
2. Immediately fill the tube with diluted bleach solution and mix carefully. Incubate for maximally 5 min and rotate the Eppendorf tube frequently. Ensure that there are no seeds sticking to the tube surface above the bleach solution and avoid bubble formation as much as possible. Allow the seeds to sediment and remove the bleach solution by pipetting.

3. Immediately add sterile water and wash the seeds 3 times as described above (*see* **Note 1**). After the last wash, the seeds are stored in sterile water.
4. If your seed stock requires so, seeds can be stratified in the dark at 4 °C before use. Sterilized seeds can be stored for at least 2 weeks under these conditions.

3.2 Biofoil Sandwich Preparation

Work under sterile conditions in a laminar air flow hood.

1. Place coverslips on the lid of a Petri dish and distribute approximately 0.7 ml warm Hoagland's medium containing agarose over the coverslip such that an area of at least 35 mm × 24 mm is completely covered by the medium.
2. Once the medium has gellified, place a sheet of Biofoil (24 mm × 30 mm) on top of the medium so that the short side aligns with the short side (24 mm) of the coverslip that has been covered with medium.
3. With a sterile dissecting knife, cut off the medium that is not covered by the Biofoil straight along the edge of the Biofoil and remove the cutoff part. The Biofoil slide is now ready for application of sterilized seeds.

3.3 Fixing the Biofoil Sandwich in a Petri Dish or Falcon Tube

Biofoil sandwiches can be placed either in a 110 mm Petri dish or in a 50 ml Falcon tube.

1. In a Petri dish, make a support by sliding two 1 ml pipette tips into each other. By cutting the pointed tip, the length of the support is adjusted such that it is as wide as the Petri dish. The Biofoil sandwich is then placed with the end not covered by medium on the support so that its angle is approximately 20° from the horizontal axis, and fixed with an excess of warm, still liquid Hoagland's medium with agarose. This prevents the Biofoil sandwich from drying out and fixes its position.
2. In a 50 ml Falcon tube, the Biofoil sandwich is simply inserted with the end not covered with medium towards the top of the tube. Due to the tube diameter, the angle of the Biofoil sandwich is approximately 70° from the horizontal axis. The bottom of the tube is filled with liquid Hoagland's medium so that the bottom of the Biofoil sandwich is just covered with the medium when the tube is standing up straight.

3.4 Seed Application

Sterilized seeds are carefully placed on the cutoff edge of the medium using a sterile Pasteur pipette. We use approximately ten seeds per slide since in our hands root hair growth is better when the seed density is not too low. Excess water is removed using a Pasteur pipette.

3.5 Plant Growth and Microscopy

1. Petri dishes with Biofoil sandwiches are closed with Parafilm and Falcon tubes are closed by screwing on their lid before transferring the material to an incubator. We incubate the sandwiches at 25 °C under long-day conditions (16 h light, 8 h darkness) for 4–6 days (*see* **Note 2**).
2. When seeds germinate, roots grow into the agar against the coverslip (*see* **Note 3**). As soon as the roots are at least 5 mm in length, they can be used for imaging (*see* **Note 4**). We image the Biofoil sandwiches on an inverted microscope by clamping them into a stage insert with adjustable sliders. Growing root hairs can be imaged for prolonged time (at least 4 h).

3.6 Drug Treatments and Stainings

Drugs and dyes can be applied to roots grown in Biofoil sandwiches by submerging the sandwich in liquid Hoagland's medium containing the chemical of interest. It takes some time for chemicals to diffuse into the sandwiches, but we have been able to successfully depolymerize microtubules within 20 min after application of 10 μM oryzalin and we have observed FM4-64 internalization within 15 min (Zhang and Ketelaar, unpublished results).

4 Notes

1. If seeds on your Biofoil sandwich germinate, but do not form roots, the seed sterilization may have damaged the root meristem. Try reducing sterilization times or increase the number of washes after sterilization.
2. If a white cloud appears after several days, seed sterilization has not been sufficient or contamination has occurred during the procedure.
3. If roots do not grow into the medium, the Biofoil sandwich may have dried too much. Try applying colder medium, cover with Biofoil immediately after solidification of the medium, and do not leave the Biofoil sandwiches uncovered in the flow hood for prolonged periods.
4. If roots grow nicely into the agar, but do not form root hairs, the sandwich may be too thick or thin, or the number of seeds is not high enough.

References

1. Schiefelbein J (2000) Specification of root hair cells. In: Ridge RW, Emons AMC (eds) Root hairs. Cell and molecular biology. Springer, Tokyo, Berlin, Heidelberg, New York, pp 197–210
2. Miller DD, De Ruijter NCA, Bisseling T et al (1999) The role of actin in root hair morphogenesis: Studies with lipochitooligosaccharide as a growth stimulator and cytochalasin as an actin-perturbing drug. Plant J 17:141–154
3. Miller DD, Leferink-ten Klooster HB, Emons AM (2000) Lipochito-oligosaccharide nodulation factors stimulate cytoplasmic polarity with longitudinal endoplasmic reticulum and vesicles at the tip in vetch root hairs. Mol Plant Microbe Interact 13:1385–1390

4. Emons AMC, Ketelaar T (eds) (2008) Root hairs. Plant cell monographs, vol 12. Springer, Berlin, Heidelberg
5. Ketelaar T, Faivre-Moskalenko C, Esseling JJ et al (2002) Positioning of nuclei in Arabidopsis root hairs: an actin-regulated process of tip growth. Plant Cell 14:2941–2955
6. Ketelaar T, De Ruijter NCA, Emons AMC (2003) Unstable F-actin specifies area and microtubules direction of cell expansion in Arabidopsis root hairs. Plant Cell 15:285–292
7. Ketelaar T, Allwood EG, Anthony R et al (2004) The actin interacting protein AIP1 is essential for actin organization and plant development. Curr Biol 14:145–149
8. Ketelaar T, Anthony R, Hussey PJ (2004) GFP-mTalin causes defects in actin organisation and cell expansion in Arabidopsis and inhibits ADFs actin depolymerising activity in vitro. Plant Physiol 136:3990–3998
9. Ketelaar T, Allwood EG, Hussey PJ (2007) Actin organisation and root hair development are disrupted by ethanol induced overexpression of Arabidopsis actin interacting protein 1 (AIP1). New Phytol 174:57–62
10. Deeks MJ, Rodrigues C, Dimmock S et al (2007) Arabidopsis CAP1: a key regulator of actin organisation and development. J Cell Sci 120:2609–2618
11. Van der Honing HS, van Bezouwen LS, Emons AM et al (2011) High expression of Lifeact in *Arabidopsis thaliana* reduces dynamic reorganization of actin filaments but does not affect plant development. Cytoskeleton 68: 578–587
12. Van der Honing HS, Kieft H, Emons AM (2012) Arabidopsis VILLIN2 and VILLIN3 are required for the generation of thick actin filament bundles and for directional organ growth. Plant Physiol 158:1426–1438

Chapter 17

Morphological Analysis of Cell Growth Mutants in *Physcomitrella*

Jeffrey P. Bibeau and Luis Vidali

Abstract

This protocol describes a quantitative analysis of the morphology of small plants from the moss *Physcomitrella patens*. The protocol can be used for the analysis of growth phenotypes produced by transient RNA interference or for the analysis of stable mutant plants. Information is presented to guide the investigator in the choice of vectors and basic conditions to perform transient RNA interference in moss. Detailed directions and examples for fluorescence image acquisition of small regenerating moss plants are provided. Instructions for the use of an ImageJ-based macro for quantitative morphological analysis of these plants are also provided.

Key words Transient transformation, RNAi, *Physcomitrella patens*, Morphological analysis, Solidity, Convex hull, Tip growth, Cell polarization, ImageJ, Plant cell growth, Cytoskeleton, Protonema

1 Introduction

Plants develop the great diversity of structures necessary for their adaption and survival by the coordination of cell expansion and cell division. Hence, in order to understand plant development, we need to investigate the molecular mechanisms that control cell expansion and division, as well as how they are integrated. One of the most useful tools to investigate these mechanisms is the analysis of phenotypic changes caused by mutations. This is a particularly powerful technique when a quantitative method can be used to estimate the changes in phenotype. The simple development program of the moss *Physcomitrella patens* and its ease for genetic manipulation provide an ideal system to investigate the control of plant cell expansion and division [1, 2]. We have implemented a robust and simple method to quantitatively evaluate changes in the growing pattern of moss plants undergoing early development. This method can be combined with transient transformation of RNA interference (RNAi) constructs to investigate the participation of single genes or gene families in plant cell growth, or

Viktor Žárský and Fatima Cvrčková (eds.), *Plant Cell Morphogenesis: Methods and Protocols*, Methods in Molecular Biology, vol. 1080, DOI 10.1007/978-1-62703-643-6_17, © Springer Science+Business Media New York 2014

combined with transient complementation to evaluate the effect of point mutations or gene chimeras [3–7]. The morphological analysis described here can also be used to evaluate the growth characteristics of small moss plants derived from protoplasting and regenerating viable mutants [8, 9], which greatly simplifies the protocol and allows the analysis of a large number of plants without the need for transient transformation.

This methodology takes advantage of the chlorophyll autofluorescence generated by the abundant chloroplasts present in moss protonemal cells; this fluorescence allows determining the plant's shape. Independent images are acquired with a simple fluorescence stereo microscope from regenerating plants. The images are analyzed by an ImageJ macro that provides morphological information. Several hundred plants can be rapidly measured and averaged, and effects can be evaluated by robust statistical analysis. The described protocol can provide evidence for the participation of a gene product in polarized cell growth, protonemal branching, or cell division. In addition, transient complementation analysis allows for the evaluation of fluorescent protein fusion functionality and detailed structure-function analysis using site-directed mutagenesis.

2 Materials

2.1 Moss Cultures

The protocol used here is optimized for polyethylene glycol (PEG)-mediated transient transformation of moss protoplasts, a simple and highly reproducible procedure ([10, 11], *see* **Note 1**). A detailed description of the materials required for moss culture and transformation has been recently compiled and can be easily accessed (together with the relevant experimental protocols) from our recent open-access publication in the Journal of Visual Experiments [10]. This publication includes a detailed video to help those that have no previous experience working with *Physcomitrella*, which can be accessed at http://www.jove.com/video/2560/efficient-polyethylene-glycol-peg-mediated-transformation-moss (*see* also **Note 2**).

2.2 Vectors and Cell Lines

The biological materials listed below are available upon request from the authors or from the Bezanilla Laboratory (http://bezanillalab.com/mossmethods.html).

1. pUGGi-RNAi silencing Gateway-based vector designed by Bezanilla et al. [12]; via a simple LR reaction, this vector generates constructs that silence the gene(s) of interest plus the internal nuclear reporter NLS-GFP-GUS.
2. pUGi-control RNAi plasmid to silence only the NLS-GFP-GUS construct.

3. pTH-Ubi-gate. Vector containing the Pp108 locus targeting site, hygromycin resistance, and the maize ubiquitin promoter for constitutive expression, followed by a Gateway cassette.
4. NLS4 *Physcomitrella* cell line [13]. This line expresses a nuclear GFP-GUS reporter that is silenced by expression of RNAi constructs designed in the pUGGi vector or the control pUGi plasmid.
5. NLS4-LA [14]. This line contains a Lifeact-mEGFP construct in the NLS4 background and can be used to evaluate changes in F-actin resulting from specifically silencing a gene. The Lifeact-mEGFP is not silenced by the RNAi constructs because it does not have the GUS sequences that cause the nuclear GFP reporter to be silenced.

2.3 DNA Purification Kits

Commercial maxi-prep and midi-prep kits can be used for purification of plasmid DNA as well as classic methods for large-scale plasmid preparations. Because of their simplicity and the high purity produced by the existing commercial kits, they should be used when starting (*see* **Note 3**). Once a consistent routine for moss transformation is established and a high number of transformants are routinely attained, the classic "homemade" protocols that present a considerably lower cost should be attempted. When an initial phenotype is investigated, it may be advisable to use a midi-prep size of purification, because if after a few attempts a construct does not produce a phenotype, the DNA obtained from a midi-prep will be all that is needed. Once a phenotype is observed, it is advisable to prepare large quantities of plasmid using a maxi-prep. This becomes critical for the RNAi construct when performing transient complementation assays.

2.4 Software

1. ImageJ. Can be freely downloaded from the NIH: (http://rsb.info.nih.gov/ij/). Additional plug-ins to read a variety of image formats may be necessary; they are available from the Laboratory for Optical and Computational Instrumentation (LOCI) at the University of Wisconsin-Madison (http://loci.wisc.edu/bio-formats/downloads). Alternatively, a distribution version containing the plug-ins and additional materials are available from the Fiji site: (http://fiji.sc/). We provide an optimized ImageJ macro for morphological analysis of small moss plants called MorphologyMacroPublicV2.6.txt that can be downloaded from http://users.wpi.edu/~lvidali/MMB_moss/.
2. Statistical analysis software. We routinely use Origin Pro 8.1(Origin Lab Corp.), but any other statistical software can be used as long as it computes ANOVA and Tukey post hoc tests.

2.5 Microscopy and Photography Equipment

Microscopy and the acquired images are the most critical aspects for success in obtaining quantitative morphological data using the current protocol. It is essential to have the capacity to accurately identify plants that are actively being silenced.

1. This method does not rely on expensive microscopy setups, but can use high-end fluorescence stereo microscopes, which are available in many laboratories to observe small transgenic animals and plants expressing GFP constructs. We have successfully used two types of microscopes, a Leica MZ16FA with a 1x lens and a Zeiss Discovery with a 1× lens. For routine analysis, use a zoom that provides a magnification where one pixel is equivalent to 1 μm. A lower level of magnification is not recommended, but higher magnifications may be used; for this purpose we provide a calibration option in the accompanying ImageJ macro.
2. A filter cube with a 480/40 bandpass excitation, a 505 longpass dichroic mirror, and a 510 longpass emission filter should be used to allow for the visualization of GFP and chlorophyll auto-fluorescence simultaneously.
3. A color digital camera is recommended with a format size of at least 1,000 × 1,000 pixels. A black and white digital camera can also be used, but the active silencing status of the plant will have to be separately recorded during analysis (see below).

2.6 Images

The ImageJ macro provided is designed to work with a variety of images. It is recommended to collect 12 bit/channel color images and saving TIF stacks, with each image in the stack comprised of one plant. The complete stack should correspond to a single "treatment" with one class of RNAi construct. The macro will also work with 8 bit/channel images (RGB) and 8–32 bit black and white images (*see* **Note 4**).

3 Methods

3.1 RNAi Vector Design

1. For the building of RNAi constructs, it is recommended to use the vector pUGGi [12], which contains two inverted Gateway cassettes. This allows for the easy design of plasmids for the expression of hairpin-based RNAi constructs with a single LR reaction. Sequences between 200 and 400 bp are routinely used to silence a gene. When the degree of conservation is high between homologous genes, it is possible to use a single construct directed to the coding sequence of one of the genes to silence two or more genes [5]. For this type of multigene silencing, the approximate cutoff in identity at the nucleotide level is 90 % between genes. When viable, you should use this initial strategy for testing the participation of a gene

product in plant cell growth or proliferation. When a single construct is not anticipated to silence many genes, sequences corresponding to all the genes to be investigated should be strung together [5–7]. This can be easily achieved by the incorporation of restriction sites into the primers and re-amplification of ligated constructs (*see* **Note 5**). Details for the construction of these types of constructs are available from the Bezanilla laboratory website (http://bezanillalab.com/moss-methods.html) and the publications indicated above.

2. For transient complementation coupled with RNAi, it is necessary to use the untranslated regions (UTRs) of the gene for silencing. Sometimes it is difficult to determine the precise location of the 5′ UTR due to lack of conservation between species and between homologous genes, because the gene models corresponding to 5’ UTR regions are less accurate than the coding sequence regions (*see* **Note 6**). From the limited number of constructs we have designed [3, 5, 7], we found that the 5′ UTR regions work well for longer genes, while the 3′ UTR regions are also appropriate for short genes.
3. Once a phenotype is obtained using coding sequence regions, the phenotype should be replicated by the UTR constructs. Using the quantitative approach presented here to measure phenotypic changes (see below), it should be simple to establish the degree to which both constructs cause a similar phenotype. Furthermore, the capacity to silence highly homologous genes separately allows one to determine if genes can perform redundant functions.

3.2 Transient RNAi

1. Transform 30 μg (*see* **Note 7**) of RNAi construct using published protocols for transformation and plate protoplasts on cellophane-covered protoplast regeneration medium-bottom (PRM-B) plates using liquid plating medium (*see* **Note 8**; ref. [10]).
2. Four days after transformation, transfer plants to regular growth medium (PPNH4) plates containing 15 μg/ml hygromycin (*see* **Note 9**).
3. Grow for 3 more days.
4. Proceed to Image Acquisition step.

3.3 Transient RNAi and Complementation

1. In the same transformation tube, add 30 μg of RNAi construct and as a first attempt 30 μg of complementing construct based on the pTH-Ubi backbone.
2. The optimal level of plasmid may need to be determined by titrating the amount of plasmid added and can vary from a few μg to 60 μg (*see* **Note 10**).

3. Plate cells on cellophane-covered PRM-B plates using liquid plating medium.
4. Four days after transformation, transfer plants to regular growth medium (PPNH4) plates containing 15 μg/ml hygromycin.
5. Grow for 3 more days.

3.4 Image Acquisition

1. Using the fluorescence channel that allows the visualization of both the nuclear GFP and the chloroplast auto-fluorescence, identify actively silencing plants. Living plants can be easily distinguished from dead plants by the level of chlorophyll fluorescence; dead plants have very weak or no fluorescence, while live plants show intense red fluorescence. At this stage, there will be several dead plants that were not transformed and were killed by the antibiotic. There will also be two populations of living plants, some containing nuclear GFP and some without nuclear GFP (*see* **Note 11**).
2. The presence of the internal silencing reporter allows one to visually identify the plants that are actively silencing. Under optimal conditions, it should be possible to find 50–100 of these plants per Petri dish of transformed plants.
3. At this point you can remove the lid of the Petri dish under observation to improve the quality of the images. Try to minimize the time that the dishes are open because the level of moisture will rapidly diminish resulting in plant damage. It is also important to avoid constant illumination of a single plant with the blue light; this can result in "crumbling" of plants with extensive exposure to the air.
4. We routinely capture images of individual plants with a color digital camera using a filter that allows for both red and green channels to be captured simultaneously; an example of the types of images acquired is presented in Fig. 1. Alternatively a black and white camera may be used to capture images; under this condition it is important to avoid multiple plants per image; when this is unavoidable the plants that are actively silencing should be identified in a separate record for later morphological analysis.
5. Sometimes two or three actively silencing plants may be present in one field; in those cases we recommend acquiring two or three pictures of the same field to simplify downstream analysis (*see* **Note 12**).

3.5 Examples

Example images are provided to test the ImageJ macro and to become familiar with the types of data sets generated by the morphological analysis. The data set is composed of two separate TIF stacks of images. They can be downloaded from http://users.wpi.edu/~lvidali/MMB_moss/. One data set has 30 images of the

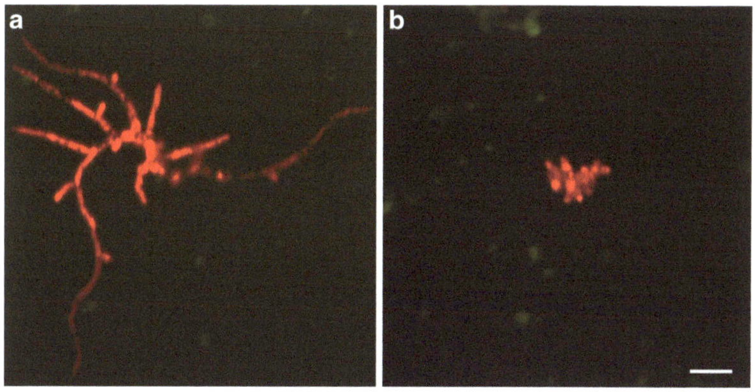

Fig. 1 Representative images of a GUS-RNAi control plant (**a**) and a plant undergoing myosin XI-RNAi (**b**). The images were acquired with a fluorescence stereo microscope and 1× lens. This imaging configuration allows to visualize both the chlorophyll auto-fluorescence and the GFP signal, which in this case is absent from the cell nucleus. Scale bar 100 μm

GUS-RNAi control plants (121031GUS-RNAi.tif). The second data set has 30 images from a myosin XI-RNAi transformation (121031MyosinXI-5UTRi.tif). The myosin XI-RNAi plants are smaller and do not show polarized growth. These plants represent an extreme example of polarized growth inhibition, but they also remain alive showing strong chlorophyll fluorescence that is helpful when performing morphological analysis.

3.6 Morphology Analysis Macro

To quantify the morphology of the regenerated plants, an ImageJ macro is provided. In brief, the macro takes a stack of images corresponding to one RNAi condition and quantifies each individual plant to yield data including area, perimeter, solidity, circularity, convex hull area, and convex hull perimeter. Although the macro is mostly automated, there are some important manual functions that the user will need to input in order for the data processing to be completed.

1. Make image stack. The user must first have the images in an ordered stack. This can be done by dragging a folder containing individual files from a window into image J and creating an image stack. The macro works on all image types including RGBs, RGB stacks, and 8–32 bit grey-scaled images.
2. Name your image stack. Since some of the macro's code is name sensitive, it is important to name your file in compliance with the following format. Include the date as the first six characters of the file name. The remaining characters can be whatever that suits the user and will be displayed on the results table. Although this format is recognized by the code, it is not essential for data processing.

3. Run macro. Drag and drop txt file of the macro into ImageJ. Then select "run macro" from the "macro" menu in the new ImageJ macro window.
4. Input you image scale. Once the macro begins it will first ask the user "What is the image scale in pixels per micrometer?" In the text box the user must input a numeric value to specify the number of pixels it takes to yield one micrometer on the image. This will ensure that the calculations output is in the correct units in the results data file.
5. Input a value for saturation tolerance (percent of pixels). In order for the macro to execute its computations on the entire plant, the plant must be thresholded from the rest of the image. The macro uses the maximum entropy algorithm to threshold the plant of interest. Unfortunately, there may be some cases where dim images are not thresholded properly. For this reason the macro automatically enhances the contrast for each image from the stack using a histogram stretching procedure. The value entered determines the percent of pixels that are allowed to be saturated in the whole image; the default value of 0 % should be tested first, but values between 0.1 and 1 % may improve image thresholding.
6. Enter the type of image. Select the option if your image is color (RGB) or black and white (*see* **Note 13**).
7. Shave images. At this point the macro will produce a window that states, "Finish shaving and press OK to continue." This indicates that the program is waiting for the user to determine which plant on each image the program should perform its calculations. This is done primarily because there are cases when multiple plants appear on the screen. To select the plant of interest, the user must completely encase the plant within the selection bounds. This is done using the mouse to select an area around the plant. Once the plant is selected, the user can press OK to continue to the next image. This process will continue for all the images in the stack.
8. Saving file path. After the shaving process is completed, two windows will appear: a window called Results and a window called Phenotype Results. The Results window displays the calculations of area and perimeter for each plant as they are computed, while the Phenotype Results window is a table where all the data is stored. Once the calculations are complete, another saving message will appear. This one asks if you want to save the files in the same place. (Depending on the processing power of your computer and the size of your data set, it may take a moment for the second saving message to appear.) If "yes" is selected, all of the files selected in the next step will be saved to the same file path. If not, the user will

have to select a unique file path for each file they select. *Important:* clicking cancel while selecting the file path will abort the macro.

9. Selecting the files to save. Once the saving details have been determined, a dialogue box will appear asking the user, "What files would you like to save?" These files include Binary Montage (8 bit Tiff file that displays each thresholded plant sorted by area in a montage), Normal Montage (8 bit Tiff file that displays plants prior to thresholding sorted by area in montage), Phenotype Results Table (text file that contains all the image computations including circularity, solidity, threshold, perimeter, convexity, convex hull area, convex hull perimeter; *see* **Note 14**), Sorted Image Stack (8 bit Tiff file that contains the images sorted by area), and Sorted Binary Image Stack (8 bit Tiff file that contains the binary images sorted by area). Following the completion of the macro, all of these files will be closed so it is important to save all of those the user finds relevant.
10. Macro confirmation. After the files have been selected, Image J will display a message saying "Macro is complete." This statement appears once the files have been saved and the macro is complete. Be sure to review the data and image files to ensure that the binary images are representative of the plant's area and make sure the macro has run properly.

3.7 Statistics

1. For routine analysis, three independent transformations for each construct or complementation set are suggested. This should provide measurements for 100–200 plants.
2. In our experience the variance between experiments does not change significantly and the values for the individual plants can be averaged for their comparisons [3, 5].
3. When more than two conditions are compared, we use an ANOVA test using a Tukey post hoc test for the comparison between means. For the comparison of only two conditions, a *t*-test can be used.
4. For the comparison of area values, we divide the total area of each plant by the average area of the RNAi control plants for each experiment. This provides a consistent value that is not altered by changes in fluorescence intensity resulting from day-to-day fluctuations.
5. The area and circularity values follow a log normal distribution, so the standard statistical analyses such as the *t*-test and ANOVA should be performed on the natural logarithm of the data which results in a normal distribution.
6. The solidity values follow more closely a normal distribution and can be analyzed without conversion.

7. To help appreciate the statistical significance of the differences, plot the mean normalized area and circularity values with the standard error of the mean using a logarithmic scale. The mean solidity values should be plotted with the standard error of the mean on a linear scale.

3.8 Verification of RNAi Activity and Expression of Protein

Sometimes it may be necessary to verify that RNAi constructs are active by a more direct indicator than only the internal nuclear GFP-GUS reporter. This is particularly important in situations where a phenotype is not observed. Two approaches can be used to address the level of silencing attained: reverse transcription followed by quantitative PCR (qRT-PCR), or immunofluorescence. We have used both methods successfully, but they both require a considerable amount of effort, in particular, in order to isolate the actively silencing plants from the background (*see* **Note 15**). The details for this process are beyond the scope of this chapter, but the details to perform these procedures are detailed in the following publications, for immunostaining [3, 5] and for qRT-PCR [6, 7].

4 Notes

1. Moss transformation can also be achieved by other means such as particle bombardment and *Agrobacterium* [15–18], and combined with protoplasting and regeneration, it may be possible to adapt these modes of transformation to this protocol.
2. Another very detailed protocol for culture and transformation has also been recently published in this series [19]; this protocol can be used for transient transformation and should yield very similar results.
3. Sometimes the maxi-preps will have very reduced yield. We believe this is due to the use of ampicillin for selection, which can result in the loss of plasmid in some bacterial cultures. We routinely test an aliquot of the maxi-prep culture with a quick mini-prep to determine that the plasmid concentration is within the expected range for the kit.
4. The nomenclature in ImageJ can be confusing because multichannel color images are referred as RGB stacks, to differentiate them from the conventional RGB images. A TIF stack refers to a series of images in the TIF format and can be 8, 16, or 32 bit. It is possible to have a TIF stack made of RGB stacks, which is the case for TIF stacks of 12 bit/channel color images. Further confusing the nomenclature, most cameras have a 12 bit output which is interpreted by ImageJ as 16 bit image; this is because the 12 bit data is contained within the 16 bit file size.

5. When possible we use exon sequences that can be amplified directly from genomic DNA, but on occasion it may be necessary to use cDNA when two or more exons form part of the construct.
6. Genomic sequence information and the most up-to-date gene model annotations can be retrieved from http://www.cosmoss.org. In addition, Phytozome (http://www.phytozome.net) provides useful comparative analysis that can help refine the automatic models generated by the cosmoss.org site.
7. Sometimes 15 μg is sufficient for a transformation. We find that using 30 μg gives consistent results across several different DNA preparations.
8. This medium is a modification of the original protocols involving top agar for plating. The medium reduces the survival rate of the protoplasts, but they regenerate and grow faster. The other advantage is that the 1-week old plants can be recovered for immunocytochemistry or quantitative PCR.
9. These plates can be prepared in advance; we usually prepare enough for 1–2 months and keep them at 4 °C. It is important to avoid contamination of these plates since they sometimes form condensation.
10. The concentration of complementing plasmid required can vary considerably. Highly abundant proteins such as ADF and profilin complement well at high concentrations of plasmid [3, 5]. Formin, which is present at low concentrations, requires much less plasmid for complementation [7] and, at high concentrations, is toxic. In contrast, myosin XI, which is also expressed at low concentrations, can tolerate high levels of plasmid [6].
11. The precise reason for the presence of these two populations is not completely clear and it varies from construct to construct. In general, constructs with a very weak or no phenotype tend to have fewer of the no-silencing plants; for example, the GUS-RNAi control shows very few plants with nuclear GFP, while constructs with severe growth phenotypes accumulate more plants with nuclear GFP that do not show the growth phenotype.
12. The ImageJ macro requires that images from plants are extracted or "shaved" to isolate them from debris present in the background or from other plants; these isolated images are transferred to a new image stack that automatically matches the same number of frames as the original stack.
13. When processing an RGB stack, the macro assumes that the red color channel, channel one, contains the chloroplast autofluorescence information. If this is not the case, the information must be put in the first RGB channel. This can be done by splitting the RGB stack and rearranging it such that the proper information is in channel one.

14. The macro output file, Phenotype Results, is a .txt file that contains a table of all the data gathered by the ImageJ macro. Within this table are several columns that read Area, Circularity, Solidity, Threshold, Perimeter, Convexity, CHArea, and CHPerim. The area column represents the area occupied by the thresholded plant. The circularity column computes a unitless number ranging from zero to one that describes how well the plant can be defined by a circle with a value of one indicating a perfect circle. Circularity is computed by the following equation: circularity = 4π(area/perimeter^2). Solidity is the ratio between an object's area and the area of the object's convex hull. Solidity is computed by the following equation: solidity = area/convex hull area. The threshold column indicates what intensity was used to threshold each plant. Perimeter is the true perimeter of each plant. Convex hull area is the minimum area needed to encase the entire plant by line segments. Convex hull perimeter is the sum of the length of all the line segments that encase the plant. Convexity = convex hull perimeter/perimeter. We recommend doing statistics on solidity, which is normally distributed, when trying to resolve the difference between stunted and branched plants.
15. An alternative approach to verify gene silencing is to generate a GFP fusion of the gene of interest and quantify the reduction in its level of expression by the reduction in fluorescence.

Acknowledgments

This work was supported by NSF (IOS 1002837 and MCB 1253444 to L.V.). We thank Erin Agar, Jennifer Garbarino, and Graham Burkart for comments on this manuscript and to Luisanna Paulino for sample images. We also thank other members of the Vidali and Bezanilla laboratories for their support.

References

1. Cove D, Bezanilla M, Harries P et al (2006) Mosses as model systems for the study of metabolism and development. Annu Rev Plant Biol 57:497–520
2. Vidali L, Bezanilla M (2012) *Physcomitrella patens*: a model for tip cell growth and differentiation. Curr Opin Plant Biol. 15: 625–631
3. Augustine RC, Vidali L, Kleinman KP et al (2008) Actin depolymerizing factor is essential for viability in plants, and its phosphoregulation is important for tip growth. Plant J 54:863–875
4. van Gisbergen PA, Li M, Wu SZ et al (2012) Class II formin targeting to the cell cortex by binding PI(3,5)P(2) is essential for polarized growth. J Cell Biol 198:235–250
5. Vidali L, Augustine RC, Kleinman KP et al (2007) Profilin is essential for tip growth in the moss *Physcomitrella patens*. Plant Cell 19:3705–3722
6. Vidali L, Burkart GM, Augustine RC et al (2010) Myosin XI is essential for tip growth in *Physcomitrella patens*. Plant Cell 22:1868–1882
7. Vidali L, van Gisbergen PAC, Guerin C et al (2009) Rapid formin-mediated actin-filament

elongation is essential for polarized plant cell growth. Proc Natl Acad Sci USA 106: 13341–13346
8. Augustine RC, Pattavina KA, Tuzel E et al (2011) Actin interacting protein1 and actin depolymerizing factor drive rapid actin dynamics in *Physcomitrella patens*. Plant Cell 23:3696–3710
9. Wu SZ, Ritchie JA, Pan AH et al (2011) Myosin VIII regulates protonemal patterning and developmental timing in the moss *Physcomitrella patens*. Mol Plant 4:909–921
10. Liu YC, Vidali L (2011) Efficient polyethylene glycol (PEG) mediated transformation of the moss *Physcomitrella patens*. J Vis Exp 50:e2560
11. Schaefer D, Zryd JP, Knight CD et al (1991) Stable transformation of the moss *Physcomitrella patens*. Mol Gen Genet 226:418–424
12. Bezanilla M, Perroud PF, Pan A et al (2005) An RNAi system in *Physcomitrella patens* with an internal marker for silencing allows for rapid identification of loss of function phenotypes. Plant Biol 7:251–257
13. Bezanilla M, Pan A, Quatrano RS (2003) RNA interference in the moss *Physcomitrella patens*. Plant Physiol 133:470–474
14. Vidali L, Rounds CM, Hepler PK et al (2009) Lifeact-mEGFP reveals a dynamic apical F-Actin network in tip growing plant cells. PLoS One 4:e5744
15. Cho SH, Chung YS, Cho SK et al (1999) Particle bombardment mediated transformation and GFP expression in the moss *Physcomitrella patens*. Mol Cells 9:14–19
16. Li LH, Yang J, Qiu HL et al (2010) Genetic transformation of *Physcomitrella patens* mediated by *Agrobacterium tumefaciens*. Afr J Biotechnol 9:3719–3725
17. Sawahel W, Onde S, Knight C et al (1992) Transfer of foreign DNA into *Physcomitrella patens* protonemal tissue by using the gene gun. Plant Mol Biol Rep 10:314–315
18. Smidkova M, Hola M, Angelis KJ (2010) Efficient biolistic transformation of the moss *Physcomitrella patens*. Biol Plant 54: 777–780
19. Roberts AW, Dimos CS, Budziszek MJ Jr et al (2011) Knocking out the wall: protocols for gene targeting in *Physcomitrella patens*. Methods Mol Biol 715:273–290

Chapter 18

Plant Cell Lines in Cell Morphogenesis Research

Daniela Seifertová, Petr Klíma, Markéta Pařezová, Jan Petrášek, Eva Zažímalová, and Zdeněk Opatrný

Abstract

Plant organs and tissues consist of many various cell types, often in different phases of their development. Such complex structures do not allow direct studies on behavior of individual cells. In contrast, populations of in vitro-cultured plant cells represent valuable tool for studying processes on a single-cell level, including cell morphogenesis. Here we describe characteristics of well-established model tobacco and Arabidopsis cell lines and provide detailed protocol on their cultivation, characterization, and genetic transformation.

Key words BY-2, VBI-0, Suspension-cultured cells, Cell phenotyping, Cell viability, Cell density, Culture friability, Micromorphology, Subculture interval (SBI)

1 Introduction

The idea of in vitro cultivation of somatic plant cells in the liquid media as the "new type of microorganisms" emerged more than half a century ago [1, 2] in connection with their potential industrial use. However, based on natural cohesiveness of plant tissues, plant cell calli are able to form only more or less viable primary suspension cultures, composed predominantly of cell clumps/aggregates. Therefore, most of the present plant suspension cultures can hardly be denominated "*cell* suspension cultures." Their heterogeneous cell mass is convenient for common metabolic or molecular analyses, but they either hamper or exclude proper simultaneous cytological characterization and further single-cell-based studies.

Nevertheless, there are few exceptions of plant cell lines that are sufficiently and permanently "friable" (i.e., consisting mostly of free cells and small cell aggregates). They are also showing high viability, physiologically and morphologically homogeneous cell population, and, consequently, a stable phenotype. One of the first real plant "cell cultures," suitable for at least general studies on cell morphogenesis and differentiation under the effects of various

Viktor Žárský and Fatima Cvrčková (eds.), *Plant Cell Morphogenesis: Methods and Protocols*, Methods in Molecular Biology, vol. 1080, DOI 10.1007/978-1-62703-643-6_18,

external factors (morphoregulatory cues, xenobiotics, etc.), was the suspension culture of sycamore (*Acer pseudoplatanus* L.; ref. [3]). This culture was still mostly composed of aggregates of various size (containing 10 to more than 100 cells), but these cells exhibited high viability and uniformity of both their size and shape. It also demonstrated general stability of its phenotype ("micromorphology") during long-term (more than 5 years) cultivation.

The first genuine "plant cell line," i.e., the clone of somatic plant cells, growing both on agar and in suspension culture, and composed of only free cells and small cell files or aggregates containing max. of 10–20 cells, was derived in 1967 from the stem pith of adult plants of *Nicotiana tabacum* L. Having originated from callus of cv. Virginia Bright Italia, it was named VBI-0 accordingly [4, 5]. This cell line has been unique not only owing to its permanent and spontaneous high friability but also due to its strictly polar character of both cell division and elongation. Consequently, its cell aggregates are in fact cell files (filaments) of various lengths. The phenotype of the cell population is typical and distinct for individual phases of the subculture interval (SBI). Inocula taken from the stationary phase are composed of elongated free cells or cell couples. Their stepwise division during exponential phase results in formation of cell files, in which the shape of individual cells is cylindrical with roughly the same diameter and length. Beginning of the stationary phase is indicated by the end of cell division, gradual cell elongation, and stepwise disintegration of the long files into free interphasic cells and cell couples. However, the routine use of VBI-0 cell line is to some extent handicapped by its strict dependence on the precise (standard) cultivation regime.

Roughly 15 years later, the first plant "HeLa-like" cell line was reported [6], derived from the callus culture of another *Nicotiana tabacum* L. cultivar, Bright Yellow 2, and again named accordingly, i.e., BY-2. Its phenotype almost copies the abovementioned properties of VBI-0 cell line. However, in contrast to this line and probably as a consequence of its previously intended use as the "industrial cell line" for the production of tobacco biomass, it exhibits enormous growth (i.e., cell multiplication) rate, resistant to various forms of the cultivation stress. Due to this property, BY-2 cells can be also transformed efficiently, and this is another characteristic making them a very useful model for plant research. Consequently, tobacco BY-2 cell line represents the most widely used single-cell-based plant experimental model, the usefulness of which is, among other things, documented also by the two BY-2-related monographies [7, 8]. Nevertheless, some properties of the VBI-0 cell line surpass those of BY-2, and that is why in some types of research, either VBI-0 cell line is used preferentially or both these model cell lines are used simultaneously [9–15].

The advantageous cytological characteristics of these tobacco cell lines are in contrast to the handicap arising from their "pedigree." The complicated genome of *Nicotiana tabacum* L. has not been fully sequenced yet, and concerning these cell lines themselves, detailed gene or even karyological analysis is missing—with one recent exception documenting pronounced genotype heterogeneity of BY-2 cell population, including massive aneuploidy, chromosomal translocation, and variability in satellite DNA [16]. As a logical consequence, there is a continuous effort to derive cell lines or at least cell suspension cultures of acceptable quality from the favorite plant experimental model *Arabidopsis thaliana* Heynh. According to our knowledge and experience, none of the existing Arabidopsis cell suspension cultures reach the quality of cytological characteristics of the tobacco cell lines mentioned above. Nevertheless, at least some of these cell cultures (e.g., *A. thaliana* Heynh., ecotype Landsberg erecta; ref. [17]) are applicable for the experiments aimed for the studies on plant morphogenesis.

Taken together, in populations of in vitro-cultured plant cells, various morphological parameters, including cell shape, cell size, as well as orientation of both cell division and elongation, can be observed directly on a single-cell level. The cell line models allow also to study dynamics of changes in their intracellular structures, in parallel with production of material for simultaneous biochemical analyses. Thus, plant suspension-cultured cell lines represent advantageous experimental models for studies focused on various mechanisms affecting cell morphology and the action of relevant internal and external morphoregulatory factor(s). Their use is complementary to the use of plant tissues, organs, and intact plants.

Here we demonstrate the procedures for long-term cultivation of both cell suspensions of tobacco cell lines BY-2 [6] and VBI-0 [5] and of *A. thaliana*, ecotype Landsberg erecta (LE) cell line [17], together with detailed protocols for their morphological characterization and transformation. These are the techniques that can be used not only in basic research but also in various biotechnological applications.

2 Materials

2.1 Plant Material

1. 7-day-old tobacco *Nicotiana tabacum* L., cv. Bright Yellow 2 (BY-2) cell suspension culture.
2. 14-day-old tobacco *Nicotiana tabacum* L., cv. Virginia Bright Italia 0 (VBI-0) cell suspension culture.
3. 7-day-old *Arabidopsis thaliana*, ecotype Landsberg erecta (LE) cell suspension culture.
4. 30-day-old BY-2, VBI-0, or LE calli.

2.2 Common Solutions

Prepare all cultivation media using deionized water.

1. Stock solutions of plant hormones, 2,4-dichlorophenoxyacetic acid (2,4-D) and naphthalene-1-acetic acid (NAA), both in concentration 0.1 g/l: dissolve 10 mg of 2,4-D or NAA in 1–1.5 ml of 96 % ethanol and refill with hot deionized water to 100 ml. Stir at constant temperature (80 °C) for at least 4 h until the chemicals are fully dissolved.
2. Stock solution of thiamin: dissolve 100 mg thiamin in 100 ml deionized water. Store 2 ml aliquots in small tubes (epi-tubes) at −20 °C.
3. Modified Murashige-Skoog (MS) medium for suspension-cultured BY-2 and LE cells: 30 g/l sucrose, 4.3 g/l Murashige and Skoog salts (Sigma M5524), 100 mg/l inositol, thiamin (1 ml of stock solution per l), 2,4-D (2 ml of stock solution per l), 200 mg/l KH_2PO_4, pH 5.8 (adjust with 3 M KOH). Autoclave at 121 °C for 20 min under 0.1 MPa (*see* **Note 1**).
4. V4 medium for suspension-cultured VBI-0 cells: prepare component solutions: (a) final volume 500 ml: 7.5 g/l KCl, 6 g/l $NaNO_3$, 2.5 g/l $MgSO_4 \times 7H_2O$, 1.25 g/l $NaH_2PO_4 \times 2H_2O$; (b) final volume 500 ml: 7.5 g/l $CaCl_2 \times 2H_2O$; (c) final volume 100 ml: 1 g/l $ZnSO_4 \times 7H_2O$, 0.1 g/l $MnSO_4 \times 4H_2O$, 0.03 g/l $CuSO_4 \times 5H_2O$, 1 g/l H_3BO_3, 0.01 g/l KI, 0.03 g/l $AlCl_3 \times 6H_2O$, 0.03 g/l $NiCl_2 \times 6H_2O$; (d) final volume 500 ml: 3.72 g/l Na_2EDTA, 2.78 g/l $FeSO_4 \times 7H_2O$; (e) final volume 100 ml: 0.01 g/l B1 (thiamin), 0.01 g/l B6 (pyridoxine), 0.05 g/l B3 (nicotinic acid), 0.3 g/l glycine. Mix the following components for final volume 1,000 ml: component A (100 ml), component B (10 ml), component C (1 ml), component D (10 ml), inositol (100 mg/l), casein hydrolysate (1 g/l), sucrose (30 g/l), NAA (10 ml of stock solution; 1 mg/l), 2,4-D (10 ml of stock solution; 1 mg/l), component E (10 ml). Adjust pH to 5.7. Refill with deionized water to final volume. Autoclave at 121 °C for 20 min under 0.1 MPa.
5. Solid (agar) medium for suspension-cultured cells: before autoclaving, add agar (6 g/l) to the particular liquid medium described above and autoclave at 121 °C for 20 min under 0.1 MPa. Pour the medium under sterile conditions into sterile Petri dishes (about 20 or 40 ml to 60 or 90 mm diameter plates, respectively) and leave to solidify.

2.3 Cultivation of Suspension (Liquid) and Callus (Agar) Cultures

1. Sterilized by autoclaving at 120 °C for 20 min: Erlenmeyer flasks (100 and 250 ml) and cylinder, covered with double aluminium foil, regular pipette tips, and pipette tips cut off by ca. 1–1.5 cm, suitable for 1, 5, or 10 ml pipettes (*see* **Note 2**).
2. Metal spatula, parafilm strips, ethanol in a beaker, and gas burner.
3. Sterile plastic Petri dishes (60 or 90 mm diameter).

2.4 Transformation of Cell Cultures

1. Sterile cell filtration set (Nalgene "Filter Holder with Receiver" or similar) with 20 μm nylon mesh filter, sterile Petri dishes (60 and 90 mm diameter), sterile regular pipette tips, and pipette tips cut off by ca. 1–1.5 cm (suitable for 1, 5, or 10 ml pipettes).
2. Liquid YEP medium for cultivation of *Agrobacterium tumefaciens*: 10 g/l select yeast extract, 10 g/l peptone, 5 g/l NaCl, adjust pH to 7.0–7.5 with 3 M KOH.
3. Acetosyringone stock solution (20 mM in ethanol): weigh 39.2 mg of acetosyringone (Fluka-Sigma) in a Falcon tube. Add 10 ml of 96 % ethanol and let dissolve. Filter through Millipore MF membrane filter for sterilization of aqueous solutions (0.22 μm) and pipette 1 ml aliquots into sterile epi-tubes under sterile conditions. Store at −20 °C.
4. Overnight culture (approximately 16 h) of the *A. tumefaciens* strain carrying binary vector with your gene construct (GV2260, C58C1, LBA1115, LBA1100 or others).
5. Stock solution of cefotaxime (Claforan, 100 mg/ml): under sterile conditions add sterilized distilled water into the original Claforan bottle (i.e., 10 ml water to 1 g of cefotaxime) and pipette 1 ml aliquots into sterile epi-tubes. Store at −20 °C.
6. Plates with solid modified MS or V4 medium (*see* Subheading 2.2) containing 100 μg/ml cefotaxime (Claforan) and appropriate selection antibiotics (*see* **Note 3**).
7. Sterile stock solutions of appropriate antibiotics (depending on the construct, i.e., mostly hygromycin or kanamycin; *see* **Note 4**), sterile 3 % sucrose solution (0.5 l for each construct).

2.5 Description, Phenotyping, and Cytological Analysis of Suspension Cultures

1. Pasteur pipette, rack for test tubes, and glass or plastic test tubes.
2. Microscope slides and cover glasses.
3. Fuchs-Rosenthal counting chamber (size of the chamber $4 \times 4 \times 0.2$ mm, total volume 3.2 mm^3).
4. Fluorescein diacetate (FDA) solution. Stock solution: dissolve 20 mg of FDA in 10 ml of acetone, prepare aliquots and store at −20 °C. Working solution to be prepared freshly before each experiment: mix 40 μl of FDA acetone stock solution with 15 ml of culture medium.
5. Trypan blue solution (0.4 %).
6. Upright and inverted light microscope equipped with epifluorescence and camera, image analysis software (e. g., ImageJ, NIS-Elements).
7. Hoechst fluorescent dye for nuclei staining (stock solution 1 mg/ml dissolved in H_2O, aliquots should be stored at −20 °C).
8. 10 % Triton X-100 solution (dissolved in H_2O).

3 Methods

3.1 Cultivation of Cell Suspensions

Perform all steps under sterile conditions in laminar flow box.

1. Day 0: inoculate 1 ml of BY-2 or LE cell suspension (stationary phase) into 30 ml of modified MS medium in sterile 100 ml Erlenmeyer flask. Alternatively, inoculate 2 ml of BY-2 suspension into 100 ml of modified MS medium in 250 ml Erlenmeyer flask. Inoculate 16 ml of VBI-0 cell suspension (stationary phase) into 100 ml of V4 medium in 250 ml Erlenmeyer flask. For cell suspension transfer, use cut off pipette tips. Close the flasks with aluminium foil (Fig. 1a, *see* **Notes 5** and **6**).
2. Cultivate the cells under continuous shaking on an orbital incubator (orbital diameter 30 mm) in darkness at 27 °C and 150 rpm (BY-2 and VBI-0) or 25 °C and 130 rpm (LE).
3. Subculture into the fresh medium every 7 days (BY-2 and LE) or every 14 days (VBI-0). During the SBI (subculture interval), you may check the condition of cultured cells with inverted microscope without opening the Erlenmeyer flask (Fig. 1b).

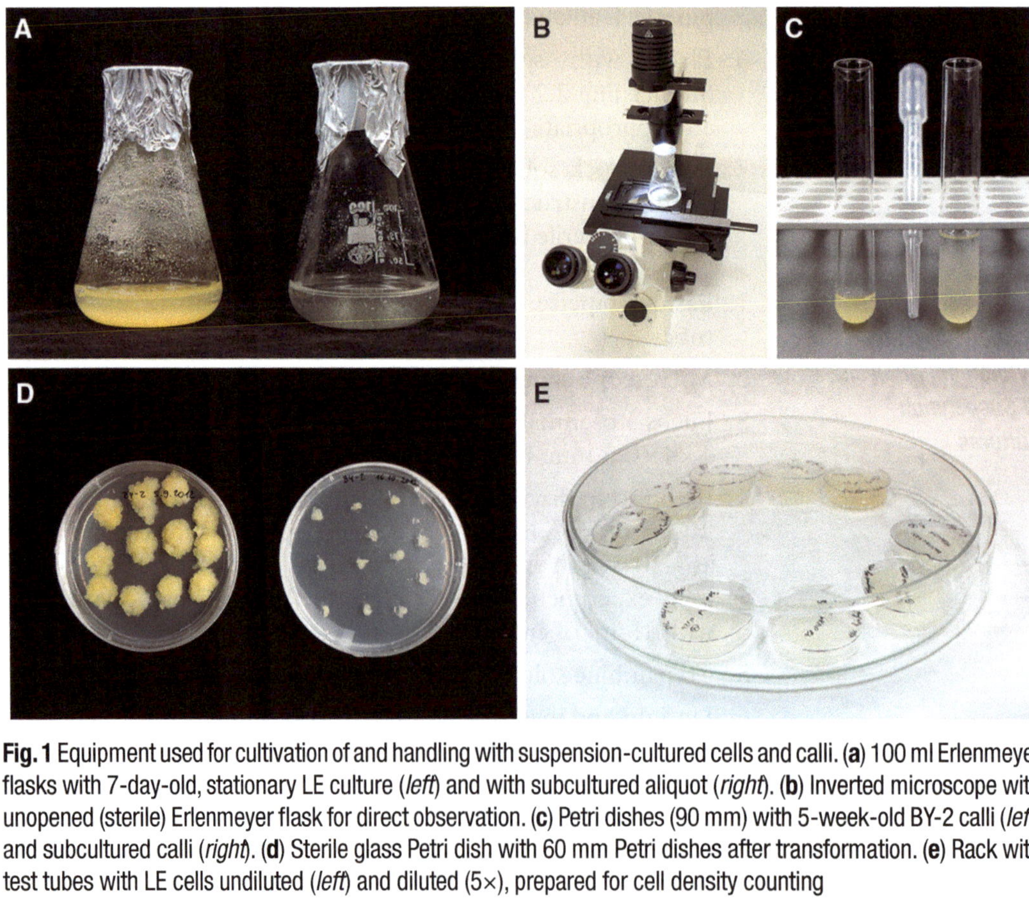

Fig. 1 Equipment used for cultivation of and handling with suspension-cultured cells and calli. (**a**) 100 ml Erlenmeyer flasks with 7-day-old, stationary LE culture (*left*) and with subcultured aliquot (*right*). (**b**) Inverted microscope with unopened (sterile) Erlenmeyer flask for direct observation. (**c**) Petri dishes (90 mm) with 5-week-old BY-2 calli (*left*) and subcultured calli (*right*). (**d**) Sterile glass Petri dish with 60 mm Petri dishes after transformation. (**e**) Rack with test tubes with LE cells undiluted (*left*) and diluted (5×), prepared for cell density counting

3.2 Cultivation of Calli

Perform all steps under sterile conditions in laminar flow-box.

1. Transfer a piece (about 5 mm in diameter, i.e., ca. 50–100 mg fresh weight) of the callus tissue (upper, light-yellow part) onto the solid modified MS (BY-2 and LE lines) or V4 (VBI-0 line) media (*see* **Note 7**). Work with a spatula sterilized by washing in 70 % ethanol followed by flame sterilization.
2. Cultivate the calli at room temperature in darkness.
3. Subculture onto the fresh medium every 4–5 weeks (Fig. 1d).

3.3 Transformation of Cell Cultures

1. Day 0: inoculate 2 ml of BY-2, 4 ml of LE or 16 ml of VBI-0 cell suspension (stationary phase) into 100 ml of modified MS medium in 250 ml Erlenmeyer flask under sterile conditions.
2. One day preceding the transformation procedure, inoculate overnight culture (about 16 h before use) of *A. tumefaciens* strain (i.e., in YEP medium) carrying binary vector with your gene construct.
3. Harvest the cells by filtration on the third day (BY-2, LE) or sixth day (VBI-0) of SBI and resuspend them in the same volume of the fresh modified MS medium.
4. Add acetosyringone stock solution to plant cell culture, to get the final concentration 1 μl/ml.
5. Before proceeding further, it is necessary to prepare plant suspension cells for easier gene transfer by pipetting thoroughly with regular uncut tip (*see* **Note 8**). Use 5 ml or 10 ml pipette with standard tips and aspirate and dispense the full pipette volume about 20 times (BY-2) or 60 times (VBI-0 or LE).
6. Cocultivation: for each sample, mix 4 ml of plant cell suspension with 20–100 μl of *Agrobacterium* overnight culture (*see* **Note 9**) in 90 mm sterile Petri dishes. For each gene construct, prepare at least four parallel plates. Seal the Petri dishes with parafilm and incubate the mixture for 3 days at 27 °C in darkness, without shaking (*see* **Note 10**).
7. Washing: after 3 days, transfer the mixture of the plant cells and bacteria (by pipetting with cut off tips) into a sterile cell filtration device with 20 μm mesh filter, and let the medium flow through under atmospheric pressure. If the suspension cells in all parallel Petri plates are viable, they can be mixed and filtered at once. Add 2.5 ml cefotaxime stock solution into 500 ml of sterile sucrose solution to get 500 μg/ml final concentration and stir. Close the valves on the filtration unit and pour 1/3 of the sucrose/cefotaxime solution to the cells. Let incubate shortly (ca. 10 min) and then let the liquid flow through. Repeat this step twice more (*see* **Note 11**).
8. After the last washing step, close the valves on the filtration device. Add 150 μl of cefotaxime stock solution into 30 ml of

liquid modified MS (BY-2, LE) or V4 (VBI-0) medium to get final concentration 500 μg/ml. Pour this mixture to the cells in the filtration unit and, by opening the valves, let the liquid drain partly. Close the valves to leave approximately 2–3 ml of suspension on the filtration device.

9. Transfer this suspension (by pipetting with cut off tips) onto 60 mm Petri plates with solid modified MS (BY-2, LE) or V4 (VBI-0) medium supplied with appropriate antibiotics (100 μg/ml kanamycin or 20 μg/ml hygromycin). Carefully spread cells over the surface of the agar plate. This is the crucial step of the whole procedure: cells have to be spread in a layer that is dense enough to allow regeneration (about 2 mm high layer of cells equally spread on the whole plate; *see* **Note 12**).
10. Place the Petri dishes into one big sterile Petri dish (Fig. 1e), do not seal with parafilm and incubate in darkness at 27 °C. First regenerating transformed calli (antibiotic resistant) can be observed after 3–4 weeks of incubation in case of BY-2 cells and 6–8 weeks in case of VBI-0 and LE cells.
11. Transfer the regenerated calli onto fresh solid medium supplied with antibiotics and incubate in darkness at 27 °C. Alternatively, prepare cell suspensions by transferring the regenerated small calli (several mm in diameter) into 2–3 ml of liquid medium supplied with selection antibiotics and shake in darkness at 27 °C (tobacco) or 24 °C (Arabidopsis).

3.4 Assessment of the Growth Curve of Suspension-Cultured Cells

Work under sterile conditions in laminar flow-box.

1. Inoculate 8 ml of BY-2 or LE cell suspension into 240 ml of modified MS medium (or 40 ml of VBI-0 cell suspension into 250 ml of V4 medium) in 500 ml Erlenmeyer flask, mix thoroughly by shaking the flask in hand. Remove 1 ml of the freshly inoculated cell suspension with a cut off tip into a test tube (Fig. 1c). Use this sample as the first one for the cell density evaluation (*see* Subheading 3.6) and/or cell viability test (*see* Subheading 3.5).
2. Prepare seven aliquots of BY-2 or LE cell suspensions by pipetting smaller amounts (e.g., 10 ml) consecutively in 100 ml Erlenmeyer flasks. It is important to keep the stock cell culture homogeneous by shaking the 500 ml stock flask well by hand before every transfer. Final volume of cell suspension aliquots in each 100 ml Erlenmeyer flask is 30 ml. Close the flasks with aluminium foil (*see* **Note 13**). For VBI-0 cell culture, prepare two 100 ml aliquots in 250 ml Erlenmeyer flasks. Consecutively remove smaller cell culture aliquots (e.g., 50 ml) with a graduated cylinder to reach the final volume (100 ml). Shake the 500 ml stock flask well by hand before every transfer. Cultivate

cell suspension aliquots under continuous shaking on an orbital incubator (orbital diameter 30 mm) in darkness at 27 °C and 150 rpm (BY-2 and VBI-0) or 25 °C and 130 rpm (LE).

3. Remove 1 ml sample of the cell suspension to a test tube every day for a week (from a new aliquot in case of BY-2 and LE) or every second day for 2 weeks (from the first aliquot the first week and from the second aliquot the second week in case of VBI-0).
4. Use the samples for the cell density determination (*see* Subheading 3.6), alternatively also for the cell viability assessment (*see* Subheading 3.5).

3.5 Assessment of Cell Viability

1. To determine the viability of cells using the trypan (Evans) blue, add 100 μl of trypan blue solution per 1 ml of cell suspension in the test tube. Observe staining via standard light microscope (*see* **Note 14**) and count the number of viable cells as the percentage of the whole amount. Count at least 400 cells per sample in several optical fields.
2. To determine the viability of cells using the FDA assay, mix aliquot of freshly prepared FDA-working solution (1:1 [v/v]) with cell suspension on a microscopic slide and observe via fluorescence microscope (excitation 494 nm; emission 521 nm). Count fluorescing and non-fluorescing cells not later than 30 s after addition of FDA (*see* **Note 15**). Taking image data and postprocessing is advised in this case.

3.6 Cell Density Determination

1. Dilute the cell culture in test tube with modified MS medium (BY-2 and LE) or V4 (VBI-0) (Fig. 2c) to get the final number of cells between 500 and 4,000 cells for LE and 300 and 900 cells for BY-2 or VBI-0 in each counted chamber (*see* **Note 16**).
2. For cell density counting, use Fuchs-Rosenthal hemocytometer. Count viable cells in the whole chamber. For each cell suspension variant, count at least ten repetitions (*see* **Notes 17** and **18**).
3. Evaluate the density of cells according to the formula provided with the hemocytometer or derived from chamber dimensions. The cell number is usually expressed per milliliter of cell suspension. Plot the values in a graph (Fig. 2a).

3.7 Assessment of Mitotic Index of Suspension-Cultured Cells

Inoculate the usual amount of stationary cell suspension into fresh medium (as described in Subheading 3.1)

1. During the SBI remove 1 ml aliquot of the cell culture to a test tube. When necessary, dilute for the appropriate density (*see* Subheading 3.5).
2. Add 10 μl of 10 % solution of Triton X-100 per 1 ml of cell suspension to get 0.1 % final concentration.

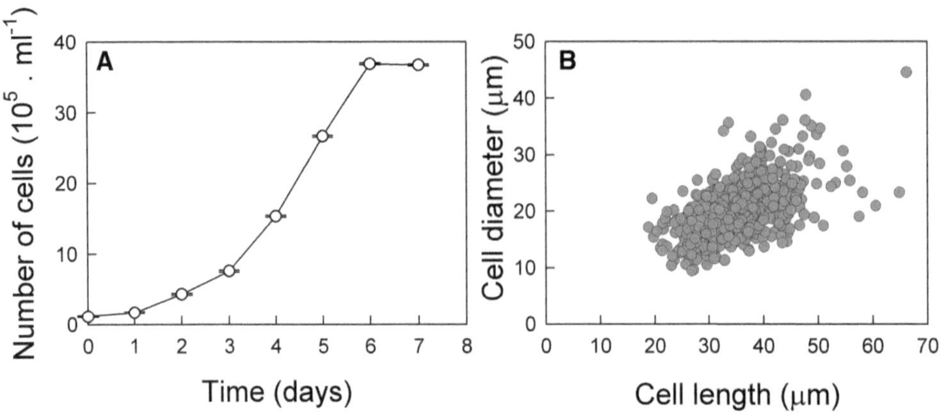

Fig. 2 Growth characteristics of a cell suspension culture. (**a**) Typical growth curve shows the multiplication rate of cell culture during 7-day subculture interval. Error bars = SEs (*n* = 10). Growth phases of cell culture during subculture interval: lag-phase (day 0–1), exponential phase (day 2–5), stationary phase (day 6–7). (**b**) Typical distribution of cell lengths and cell diameters in 2-day-old cell population (*n* = 700)

3. Add 0.2 μl Hoechst dye from stock solution per 1 ml of cell suspension to get 0.2 μg/ml concentration and leave shaking for 10 min.
4. Observe cells under the fluorescence microscope and count the total number of cells and the number of cells in mitotic phase (for mitotic phases, *see* ref. [18]). Count at least 1,000 cells per sample in several optical fields (*see* **Notes 19** and **20**).
5. Calculate mitotic index as the number of cells in mitosis divided by the total number of counted cells, and plot the graph as a function of time (days after subcultivation, e.g., *see* ref. [19]).

3.8 Characterization of Micromorphology of Cell Lines

Commonly determined characteristics are cell size, cell shape, and morphology of cell files or aggregates (*see* **Note 21**).

1. For observation of suspension-cultured cell lines, use upright or inverted light microscope; the application of both fluorescent and Nomarski DIC equipment is highly recommended.
2. For saving and further processing the image data, use adequate camera connected to a computer equipped with image analysis software (e.g., ImageJ and NIS-Elements).
3. Adjust the cell density by dilution of cell suspension so that cells on the microscopic slide do not overlap.
4. The most suitable objectives for observation and capturing of images for further determination of cell length and diameter are 10×, 20× or 40×. Save a set of images of each variant.
5. Determine the length and diameter of the cells using measurement tools in image analysis software. Assess the parameters of at least 400 cells in a set (*see* **Note 22**).

6. Count the number of cells in individual cell files or aggregates, compare the frequency of files/aggregates composed of the similar cell number (e.g., 2–4, 5–8, 9–16) as one of parameters illustrating cell division rate (*see* **Note 21**).
7. Transfer the data to a spreadsheet (e.g., OpenOffice/ LibreOffice Calc and Microsoft Excel). Plot the measured data in a graph (Fig. 2b).

4 Notes

1. Medium can be autoclaved directly in Erlenmeyer flasks or in Duran laboratory glass bottles.
2. Use double aluminium foil of 30 μm thickness.
3. Before adding antibiotics to the autoclaved medium, let the bottle cool so you can hold it in a bare hand. Work aseptically.
4. Use of DL-phosphinothricin (PPT) is not possible in cell suspensions.
5. The aluminium foil and the glass bottle neck are better to be sterilized additionally by gas burner flame during the process.
6. In general, the growth activity of both callus and suspension cultures is determined by two key factors: nutrition (including, sensu lato, also necessary growth regulators) and proper aeration. The use of inocula of subcritical size/fresh weight in case of callus cultures, or subcritical cell density in case of suspension cultures, pronouncedly impairs the viability and multiplication ability of cells as well as their phenotype ("dilution effect"). Therefore, the initial cell density of the fresh BY-2, VBI-0, or LE subculture should not be lower than ca. $1–5 \times 10^5$ cells per ml. On the contrary, too high inoculum density undesirably shortens the exponential phase of the subculture interval owing to the cell competition for oxygen supply. As a rule, the final cell density in the suspension of BY-2, VBI-0, or LE cells reaches max. $10–60 \times 10^5$ cells per ml. Consequently, in case of the use of the abovementioned inoculum density and with respect to equal reproduction ability of the most of cell populations of BY-2, VBI-0, or LE lines, the cultures pass maximally 6–7 subsequent cell divisions during standard subculture interval.
7. There should be about 12 pieces of calli on the 90 mm Petri plate to maintain the optimal hormone and nutrient conditions.
8. By producing small lesions at the surface of the plant cells in this step, we help the successful transformation by Agrobacterium.

9. Approaching the plant cultured cells transformation for the first time, start with at least three Agrobacterium volumes. Suitable volumes are 20, 40, and 60 μl for BY-2 cell culture or 60, 80, and 100 μl for VBI-0 and LE cells. Do not forget to have one plate without Agrobacterium as a control.
10. You may check the condition of cell culture using inverted microscope. If the transformation successfully progresses, attachment of bacteria to the plant cell walls can be seen, and the majority of plant cells remain viable. Medium appears milky because of Agrobacterium growth.
11. Because of the filamentous or cluster-like character of the VBI-0 and LE cell suspensions, it is highly recommended to incubate plant cells for 15 min in the washing sucrose solution to remove Agrobacterium completely.
12. Avoid medium overflow on the plates.
13. Preparation of seven aliquots of BY-2 or LE cells is optimal to avoid suspension density changes due to samples removal.
14. Trypan (Evans) blue "dye exclusion" test is based on the inability of injured or dead cells to exclude the dye actively from their cytoplasm and vacuoles. However, the reliability of the test is not absolute; in a very few cases, also cells with preserved internal structure and even dividing cells can be stained and thus mimic the positive (i.e., nonviable) ones.
15. The fluorescein diacetate (FDA) test [20] is based on the activity of cell esterases which catalyze the release of free fluorescein from FDA and its accumulation in vacuoles of viable cells. Inactive esterases or damaged cell membranes prevent fluorescein accumulation inside cells so that dead cells exhibit no fluorescence.
16. To evaluate cell density in samples containing higher fraction of multicellular spherical aggregates that are hardly observable using standard light microscopy, one can use methods of fluorescent staining of cell nuclei, e.g., using the Hoechst dye. By means of proper software, it is possible to count individual nuclei with reasonable reliability. In such a way, cell number is determined indirectly.
17. Before counting, mix the cell culture with Pasteur pipette to evenly distribute cells in the counted samples.
18. As the most reliable method for determination of actual cell viability in suspension cultures, the combination of standard light microscopy and proper cell staining is recommended. For routine work we prefer to combine Nomarski DIC (differential interference contrast) microscopy either with trypan (Evans) blue or FDA technique. For experienced scientists, who are able to recognize damaged cells (e.g., destruction of the network of cytoplasmic bands), light microscopy alone is sufficient.

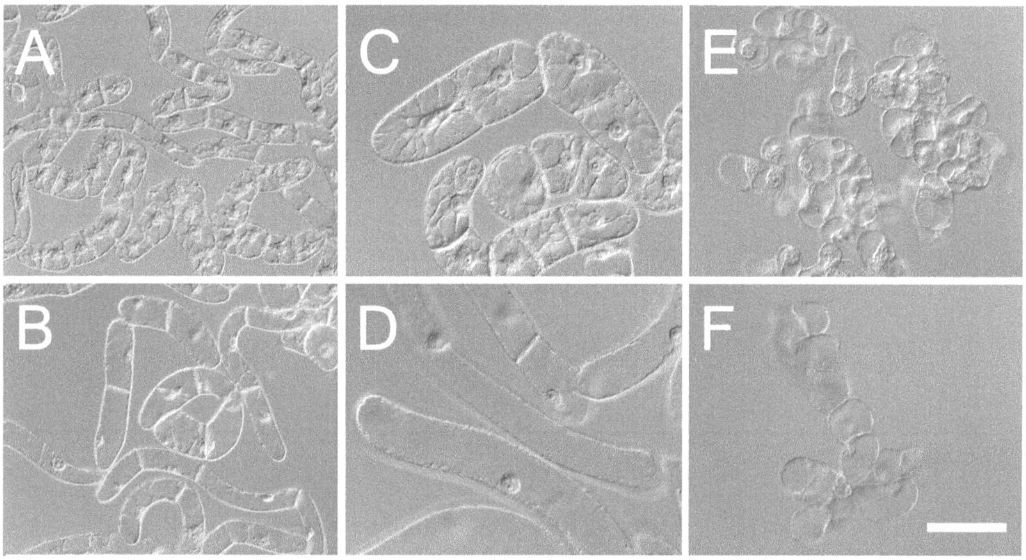

Fig. 3 Phenotype of suspension-cultured cell lines. (**a**, **b**) BY-2. (**c**, **d**) VBI-0. (**e**, **f**) LE. (**a**, **c**, **e**) Exponential phase of the subculture interval. (**b**, **d**, **f**) Stationary phase of the subculture interval. Scale 100 μm

19. Determination of the actual mitotic index is a valuable characteristic, but it cannot substitute determination of cell density dynamics of the culture. One-shot value of the mitotic index illustrates only the actual incidence of cells in the M-phase of the cell cycle that can be affected by various factors, including replication or mitotic blocks.
20. For optimal magnification, use objective 40×.
21. Numerous phenotypic parameters can be determined in high-quality plant cell lines to document their response to the effects of various morphoregulatory factors. Besides the shape and size of individual cells, morphology of cell files can serve as a very sensitive indicator, particularly in case of VBI-0 or BY-2. In both these lines, dynamics of formation and disintegration of the multicellular cell files (filaments) during subculture interval reflects the normality or abnormality of the cell division process (*see* Fig. 3 *for normal appearance of cells at distinct culture stages*). Consequently, even without any direct counting of cell density, one can observe modifications of the standard cell division rate from relative incidence of free cells and cell files composed of 2–4–8 or more cells. Disturbance of cell division polarity in tobacco cell lines is manifested by formation of aberrant cell aggregates, randomly duplicated cell files or irregular, respectively, spherical cell clumps.
22. During subculture intervals, cells of both VBI-0 and BY-2 lines exhibit almost entirely polar cell division and growth (elongation). The diameter of the cells remains almost

unchanged (ca. 30–40 μm and 20–30 μm for VBI-0 and BY-2, respectively), but their length increases 3–5 times. The elongated stationary cells/cell couples forming the inocula start to divide after short lag-phase (1–2 days) in the process of stepwise septation. Consequently, daughter cells become almost isodiametric. Based on the culture conditions, typically the number of cells in files can reach 8–12 but sometimes up to 24. The shortest length of the mitotic cycle oscillates between 12 and 16 h. New cell elongation starts by the end of exponential phase. Later on, the files spontaneously disintegrate again.

Acknowledgments

The authors acknowledge support for their work by the Grant Agency of the Czech Republic (projects P305/11/0797 and P305/11/2476), Ministry of Education, Youth and Sport of the Czech Republic (project MSM00216208858), and Charles University in Prague (project SVV 265203/2012).

References

1. Tulecke W, Nickell LG (1959) Production of large amounts of plant tissue by submerged culture. Science 130:863–864
2. Nickell LG, Tulecke W (1960) Submerged growth of cells of higher plants. Biotechnol Bioeng 2:287–297
3. Street HE, King PJ, Mansfield KJ (1971) Growth control in plant cell suspension cultures. In: Les cultures de tissus de plants. Colloques Internationaux du C.N.R.S. 193. Éditions du Centre National de la Recherche Scientifique, Paris, p 2–40
4. Opatrný Z (1971) Using of tissue cultures in plant genetics. Thesis, Inst. Exp. Bot. ČSAV, Prague (in Czech)
5. Opatrný Z, Opatrná J (1976) The specificity of the effect of 2,4-D and NAA on the growth, micromorphology, and occurrence of starch in long-term *Nicotiana tabacum* L. cell strains. Biol Plantarum 18:359–365
6. Nagata T, Nemoto Y, Hasezawa S (1992) Tobacco BY-2 cell line as the "HeLa" cell in the cell biology of higher plants. Int Rev Cytol 132:1–30
7. Nagata T, Hasezawa S, Inzé D (eds) (2004) Tobacco BY-2 cells. Biotechnology in agriculture and forestry, vol 53. Springer, Berlin, Heidelberg
8. Nagata T, Matsuoka K, Inzé D (eds) (2006) Tobacco BY-2 cells: from cellular dynamics to omics. Biotechnology in agriculture and forestry, vol 58. Springer, Berlin, Heidelberg
9. Petrášek J, Freudereich A, Heuing A et al (1998) Heat-shock protein 90 is associated with microtubules in tobacco cells. Protoplasma 202:161–174
10. Campanoni P, Blasius B, Nick P (2003) Auxin transport synchronizes the pattern of cell division in a tobacco cell line. Plant Physiol 133: 1251–1260
11. Holweg C, Honsel A, Nick P (2003) A myosin inhibitor impairs auxin-induced cell division. Protoplasma 222:193–204
12. Campanoni P, Nick P (2005) Auxin-dependent cell division and elongation. 1-naphtaleneacetic acid and 2,4-dichlorophenoxyacetic acid activate different pathways. Plant Physiol 137: 939–948
13. Paciorek T, Zažímalová E, Ruthardt N et al (2005) Auxin inhibits endocytosis and promotes its own efflux from cells. Nature 435:1251–1256
14. Qiao F, Petrášek J, Nick P (2010) Light can rescue auxin-dependent synchrony of cell division in a tobacco cell line. J Exp Bot 61:503–510
15. Marhavý P, Bielach A, Abas L et al (2011) Cytokinin modulates endocytotic trafficking of PIN1 auxin efflux carrier to control plant organogenesis. Dev Cell 21:796–804

16. Kovařík A, Lim K-Y, Součková-Skalická K et al (2012) A plant culture (BY-2) widely used in molecular and cell studies is genetically unstable and highly heterogeneous. Bot J Linn Soc 170:459–471
17. May MJ, Leaver CJ (1993) Oxidative stimulation of glutathione synthesis in *Arabidopsis thaliana* suspension cultures. Plant Physiol 103:621–627
18. Matsunaga S, Ohmido N, Fukui K (2006) Chromosome dynamics in tobacco BY-2 cultured cell. In: Nagata T, Matsuoka K, Inzé D (eds) Tobacco BY-2 cells: from cellular dynamics to omics. Biotechnology in agriculture and forestry, vol 58. Springer, Berlin, Heidelberg, pp 51–63
19. Frey N, Klotz J, Nick P (2010) A kinesin with calponin-homology domain is involved in premitotic nuclear migration. J Exp Bot 61: 3423–3437
20. Widholm JM (1972) The use of fluorescein diacetate and phenosafranine for determining viability of cultured plant cells. Stain Technol 47:189–194

Chapter 19

Antisense Oligodeoxynucleotide-Mediated Gene Knockdown in Pollen Tubes

Radek Bezvoda, Roman Pleskot, Viktor Žárský, and Martin Potocký

Abstract

Specific gene knockdown mediated by the antisense oligodeoxynucleotides (AODNs) strategy recently emerged as a rapid and effective tool for probing gene role in plant cells, particularly tip-growing pollen tubes. Here, we describe the protocol for the successful employment of AODN technique in growing tobacco pollen tubes, covering AODN design, application, and analysis of the results. We also discuss the advantages and drawbacks of this method.

Key words Antisense oligodeoxynucleotide, AODN, Pollen tube, Gene knockdown, Tip growth

1 Introduction

Angiosperm fertilization requires delivery of sperm cells to embryo sac. This is accomplished by the polarized growth of the male gametophyte, the pollen tube, throughout the pistil tissues into the ovary. Along with root hairs, pollen tube serves as an example and useful model for highly polarized cell expansion with a very high growth rate (reviewed in ref. [1]). This process is regulated by small GTPases, ions, reactive oxygen species, phospholipids, and cytoskeleton remodeling (*see*, e.g., refs. [1–4]). Given the fact that pollen of many species can be easily germinated and cultivated in a liquid media in vitro, it can be utilized as a tool to investigate regulation of polar secretion sustaining polarized cell growth. Arabidopsis, tobacco, and lily pollen are typically used in the research, where Arabidopsis serves as model of tricellular pollen and tobacco and lily as model of bicellular pollen (*see* **Note 1**).

In modern biology, various gene-specific knockdown approaches have become invaluable tools for elucidating the functional role of candidate genes. One particular method uses short (18–22 nucleotides long) antisense oligodeoxynucleotides (AODNs) that are complementary to target gene mRNA and thus suppress gene expression. AODNs are widely used in biomedical

Viktor Žárský and Fatima Cvrčková (eds.), *Plant Cell Morphogenesis: Methods and Protocols*, Methods in Molecular Biology, vol. 1080, DOI 10.1007/978-1-62703-643-6_19,

research and clinical applications [5], whereas this technique is still not commonly used in the plant research. The AODN method is especially useful in species where stable transformation is hard to achieve. Furthermore, AODNs are particularly beneficial for those species with only partially covered genome sequence, because only hundreds of base pairs are usually needed for successful design. On the other hand, the AODN technique has also drawbacks, the most crucial one being only a partial reduction of the target gene expression, where the reduction varies from 40 to 80 % [6, 7].

Single-stranded DNA antisense oligodeoxynucleotides share their underlying basic principle with the widely utilized RNA interference technique: an oligo(deoxy)nucleotide must bind a target sequence through Watson–Crick base pairing ultimately resulting in impaired gene expression. Single-stranded DNA could be very quickly degraded by exonucleases, which are especially abundant in plant cells. For successful gene knockdown, AODNs have to be protected by modifications of its backbone. Phosphorothioate, where one of the nonbridging oxygen is replaced by sulfur, is by far the most frequently used modification. Phosphorothioate AODNs bind to the target sequence and formed hetero-RNA–DNA duplex is recognized and cleaved by RNAse H [8]. Since phosphorodiamidate AODNs are charged molecules, they cannot simply pass through the lipid bilayer by diffusion and one has to use cationic delivery molecules, such as cytofectin. In contrast to phosphorothioate AODNs, phosphorodiamidate morpholino oligomers (morpholinos) are uncharged DNA analogues, although their delivery into intact plant cells is not well studied [9, 10]. Interestingly, morpholinos function through steric hindrance imposed on the ribosome during mRNA translation.

The first report on the usage of AODNs in the plant research is dated to 1992 [11], and the technique was successfully employed in pollen tube biology in 1994 [12]. During the last decade, several reports using phosphorothioate and morpholino AODN strategy in a variety of plant species and cell types have been published [3, 7, 10, 13–19]. Intriguingly, the alternative mechanism of AODN uptake into plant cells through active transport via mono- or disaccharide translocators aided by sucrose itself was proposed [15].

Here we present a detailed description of the gene suppression in tobacco male gametophyte mediated by short AODNs comprising the design, the cultivation procedure, and the analysis. Also the usage of sense ODNs as a negative control is stressed.

2 Materials

2.1 Antisense Oligonucleotides

One of the most crucial part of the successful application of AODN strategy is careful oligo design. The good oligo must be specific, must not form secondary structures, and in order to create functional heteroduplex with mRNA, it must bind into the single-strand-forming region of target sequence fold. To predict secondary structure of designated AODNs, several programs for secondary structure prediction can be used, e.g., Soligo (http://sfold.wadsworth.org/cgi-bin/soligo.pl). As mentioned in the introduction, only partial cDNA may be sufficient for the proper design. In our experience, 18–20-nt-long ODNs are optimal. A selectivity of predicted oligos to target gene has to be verified by BLAST search (http://blast.ncbi.nlm.nih.gov/). For each antisense oligo, corresponding sense ODN is used as a negative control. To achieve the best results, more than one oligo pair should be tested; usually three to four pairs are sufficient. To protect ODNs against nucleases, three bases at each end are synthesized with phosphorothioate backbone. The HPLC purification is usually necessary to reduce nonspecific cytotoxicity.

2.2 Pollen

To obtain pollen, flowers of outdoor- or glasshouse-grown tobacco plants (*Nicotiana tabacum* cv. *Samsun*) are collected in warm and dry weather conditions before opening; anthers are taken out and kept in laboratory conditions on a filter paper for one day to let anthers open and dehydrate (anthers might be surface sterilized and dried in the laminar box to harvest sterile pollen). Dried pollen grains are sifted through to remove anthers. Harvested pollen can be kept frozen at −20 °C without apparent loss of the germination capacity for several years.

2.3 Cultivation Media

To grow tobacco pollen tubes in vitro, several different media could be used [20]. For cultivation times spanning up to 4 h that are required for the AODN experiments, simple liquid growth media are sufficient, such as the following PEG and sucrose media (*see* **Notes 2** and **3**):

1. PEG medium: 20 % w/v polyethyleneglycol 3350, 1.6 mM boric acid.
2. Sucrose medium: 10 % w/v sucrose, 1.6 mM boric acid.

2.4 Reagents for AODN Treatment

1. To obtain working solutions of antisense and sense ODNs, lyophilized oligos from manufacturer are rehydrated with sterile double-distilled water resulting in 1 mM stocks. 20 μL aliquots are kept at −80 °C.
2. Suspension of cationic delivery reagent (cytofectin, Gene Therapy Systems, cat. Nr T610001) is divided into 10 μL aliquots and stored at −80 °C. After thawing, the cytofectin aliquot can be kept at 4 °C for several weeks.

3 Methods

Carry out all procedures at room temperature unless otherwise specified.

3.1 Pollen Tube Cultivation and AODN Treatment

For simple analysis of pollen tube growth parameters, optimal volume per sample is 200 μL of final pollen tube culture.

1. Premix 1.5 μL of cytofectin aliquot (the final concentration is 15 μg/ml) with ODNs to obtain 30–50 μM final ODN concentration in glass or polystyrene test tube and add 20 μl of the cultivation medium. Incubate for 15 min. Do not use polypropylene labware (*see* **Note 4**).
2. Meanwhile, weigh 10 mg of fresh/frozen tobacco pollen per 1 ml of the final pollen tube culture. If freeze-stored pollen is used, keep it at room temperature for 10–15 min for recovery before resuspension.
3. Mix the pollen with the cultivation medium and vortex vigorously to obtain homogenous culture without any clumps of pollen grains.
4. Add appropriate volume of the pollen suspension to ODN-cytofectin premix to get the final volume of 200 μl. Cultivate for 1–4 h on horizontal shaker (~200 rpm) for efficient aeration (*see* **Note 5**).
5. For subsequent analysis, living pollen tubes can be observed immediately or the culture is fixed by addition of formaldehyde solution (3.7 % final concentration). Fixed pollen tube culture can be stored at 4 °C.

3.2 Data Collection and Analysis

To monitor the effects of AODNs on pollen tube growth and morphology, various approaches can be used. A first insight is usually obtained by the light microscope, whereby growth rate, morphological changes of width, or shape could be observed. For the quantification of these parameters, several programs for the image analysis can be used (e.g., ImageJ, http://rsbweb.nih.gov/ij/; NIS Elements, http://www.nis-elements.cz/en/front-page/). Based on the questions studied, an appropriate biochemical analysis (e.g., enzyme essays) of the pollen culture can be performed as well (*see* **Note 5**). To get information about efficiency of the AODN-mediated gene knockdown, quantitative PCR could be utilized.

4 Notes

1. Second pollen mitosis that gives rise to sperm cells takes place either before (in case of tricellular pollen) or after (bicellular pollen) germination.

2. In the PEG medium, only non-metabolized osmotic solute is present and pollen utilizes just internal energy sources. In the case of sucrose medium, sucrose serves as an energy source as well as osmotic agent and abnormal thick callosic depositions are formed in the cell wall of pollen tubes. Many parameters of growth and pollen tube physiology thus differ according to the cultivation medium used.
3. For the short culture times required, the media do not need to be sterilized.
4. Polypropylene test tubes are not recommended as cytofectin binds to polypropylene.
5. Due to the ODN effect on de novo transcription/translation (and possibly also mRNA stability), delayed manifestation of the lack of target protein should be expected.

Acknowledgment

Pollen tube research in the Žárský lab is supported by the Czech Grant Agency grant GACR 13-19073S to M. P.

References

1. Cole RA, Fowler JE (2006) Polarized growth: maintaining focus on the tip. Curr Opin Plant Biol 9:579–588
2. Potocký M, Eliáš M, Profotová B et al (2003) Phosphatidic acid produced by phospholipase D is required for tobacco pollen tube growth. Planta 217:122–130
3. Potocký M, Jones MA, Bezvoda R et al (2007) Reactive oxygen species produced by NADPH oxidase are involved in pollen tube growth. New Phytol 174:742–751
4. Zhang Y, McCormick S (2010) The regulation of vesicle trafficking by small GTPases and phospholipids during pollen tube growth. Sex Plant Reprod 23:87–93
5. Watts JK, Corey DR (2012) Silencing disease genes in the laboratory and the clinic. J Pathol 226:365–379
6. Sun C, Höglund A-S, Olsson H et al (2005) Antisense oligodeoxynucleotide inhibition as a potent strategy in plant biology: identification of SUSIBA2 as a transcriptional activator in plant sugar signalling. Plant J 44:128–138
7. Pleskot R, Potocký M, Pejchar P et al (2010) Mutual regulation of plant phospholipase D and the actin cytoskeleton. Plant J 62:494–507
8. Walder RY, Walder JA (1988) Role of RNase H in hybrid-arrested translation by antisense oligonucleotides. Proc Natl Acad Sci USA 85:5011–5015
9. Moulton HM, Nelson MH, Hatlevig SA et al (2004) Cellular uptake of antisense morpholino oligomers conjugated to arginine-rich peptides. Bioconjug Chem 15:290–299
10. Okuda S, Tsutsui H, Shiina K et al (2009) Defensin-like polypeptide LUREs are pollen tube attractants secreted from synergid cells. Nature 458:357–361
11. Tsutsumi N, Kanayama K, Tano S (1992) Suppression of alpha-amylase gene expression by antisense oligodeoxynucleotide in barley cultured aleurone layers. Jinrui Idengaku Zasshi 67:147–154
12. Estruch JJ, Kadwell S, Merlin E et al (1994) Cloning and characterization of a maize pollen-specific calcium-dependent calmodulin-independent protein kinase. Proc Natl Acad Sci USA 91:8837–8841
13. Moutinho A, Hussey PJ, Trewavas AJ et al (2001) cAMP acts as a second messenger in pollen tube growth and reorientation. Proc Natl Acad Sci USA 98:10481–10486
14. Camacho L, Malhó R (2003) Endo/exocytosis in the pollen tube apex is differentially regulated by Ca2+ and GTPases. J Exp Bot 54:83–92
15. Sun C, Ridderstråle K, Höglund A-S et al (2007) Sweet delivery - sugar translocators as ports of entry for antisense oligodeoxynucleotides in plant cells. Plant J 52:1192–1198

16. de Graaf BHJ, Rudd JJ, Wheeler MJ et al (2006) Self-incompatibility in Papaver targets soluble inorganic pyrophosphatases in pollen. Nature 444:490–493
17. Pleskot R, Pejchar P, Bezvoda R et al (2012) Turnover of phosphatidic acid through distinct signaling pathways affects multiple aspects of pollen tube growth in tobacco. Front Plant Sci 3:54
18. Potocký M, Pejchar P, Gutkowska M et al (2012) NADPH oxidase activity in pollen tubes is affected by calcium ions, signaling phospholipids and Rac/Rop GTPases. J Plant Physiol 169:1654–1663
19. Hafidh S, Breznenová K, Růžička P et al (2012) Comprehensive analysis of tobacco pollen transcriptome unveils common pathways in polar cell expansion and underlying heterochronic shift during spermatogenesis. BMC Plant Biol 12:24
20. Read SM, Clarke AE, Bacic A (1993) Stimulation of growth of cultured Nicotiana tabacum W 38 pollen tubes by poly(ethylene glycol) and Cu(II) salts. Protoplasma 177:1–14

Chapter 20

Lab-on-a-Chip for Studying Growing Pollen Tubes

Carlos G. Agudelo, Muthukumaran Packirisamy, and Anja Geitmann

Abstract

A major limitation in the study of pollen tube growth has been the difficulty in providing an in vitro testing microenvironment that physically resembles the in vivo conditions. Here we describe the development of a lab-on-a-chip (LOC) for the manipulation and experimental testing of individual pollen tubes. The design was specifically tailored to pollen tubes from *Camellia japonica*, but it can be easily adapted for any other species. The platform is fabricated from polydimethylsiloxane (PDMS) using a silicon/SU-8 mold and makes use of microfluidics to distribute pollen grains to serially arranged microchannels. The tubes are guided into these channels where they can be tested individually. The microfluidic platform allows for specific testing of a variety of growth behavioral features as demonstrated with a simple mechanical obstacle test, and it permits the straightforward integration of further single-cell test assays.

Key words Pollen tube, *Camellia japonica*, Cell culture, Lab-on-a-chip, Microfluidics, Microstructures, MEMS, Soft lithography, Tip growth

1 Introduction

In order to reach its target, the ovule, the pollen tube needs to invade the pistillar tissues of the receptive flower and follow guidance cues emitted by the sporophytic tissues and the female gametophyte [1–3]. Studying the roles of chemical, proteic, and mechanical cues that direct pollen tube growth and the mechanism by which the tube turns has become an important aspect of pollen tube research [4–7]. Conventionally, experimentation on pollen tubes is performed on cells germinated in bulk samples and growing in essentially homogeneous and isotropic growth matrices, either a liquid medium or an agarose-stiffened substrate. This in vitro environment is in stark contrast with the in vivo growth conditions which present a microstructured environment consisting of the various cell types and tissues the pollen tube encounters on its path through the pistil [6]. To test the behavior of pollen tubes in structured microenvironments featuring complex geometrical challenges or simple or superimposed chemical gradients, we have

Viktor Žárský and Fatima Cvrčková (eds.), *Plant Cell Morphogenesis: Methods and Protocols*, Methods in Molecular Biology, vol. 1080, DOI 10.1007/978-1-62703-643-6_20, © Springer Science+Business Media New York 2014

developed an experimental platform based on microfluidics and microelectromechanical systems (MEMS) technology, the TipChip.

The TipChip is a lab-on-a-chip device with planar geometry that allows for high-resolution optical microscopy and fluorescence imaging. It consists of a microfluidic network with limited thickness in order to restrain any interactions between two cells or cell and microstructure to a two-dimensional space, to avoid the accumulation of pollen grains into stacks, and to ensure that all growth activity occurs in one focal plane. The design meets several criteria: (1) several cells can be treated and observed simultaneously; (2) positioning of pollen grains occurs through defined fluid flow; (3) the experimental environment is enclosed from all sides, thus preventing evaporation of the growth medium, while allowing continuous flow of medium to supply fresh nutrients and oxygen to pollen tubes; and (4) optical compatibility must allow monitoring of pollen tube growth in bright-field and fluorescence mode.

The fabrication of the design starts with the planning of its layout to ensure the proper, fluid-flow-mediated positioning of the pollen grains at the entrance of the microchannels and the incorporation of the experimental tests within the microchannel. The design pattern is drawn in a CAD software, reproduced on a photomask, and transferred to a silicon/SU-8 mold through photolithography. Next, the microfluidic network is fabricated from polydimethylsiloxane (PDMS) by creating replicas using the silicon/SU-8 mold [8, 9]. The choice of PDMS as material is motivated by its biocompatibility (nontoxicity), optical transparency, relative low cost, and ease of use. Conventional planar microfabrication techniques and soft lithography make redesign loops straightforward since fabrication is systematically performed. Furthermore, the fabrication procedure can be modified easily to include more sophisticated structures, layers, or features.

Using the TipChip in various implementations [10, 11], we obtained successful pollen germination and properly elongating tubes displaying growth morphology and behavior that are indistinguishable from conventional in vitro setups. Pollen tubes grow along the microchannels in the direction enforced by their shape attaining total lengths over 1 mm. Pollen germination and growth rate within the device are consistent with those observed under conventional in vitro conditions confirming that the spatial confinement and associated limitation of the volume of the surrounding growth medium does not interfere negatively with cellular behavior. The interaction of pollen tubes with the microchannel features can elucidate many aspects of pollen tube behavior as demonstrated here through a simple mechanical obstacle test. More importantly, the presented microdevice allows for the design and easy integration of different kinds of microsensors within the microfluidic network to measure various biological parameters at the level of a single pollen tube. This opens multiple new avenues for experimental assays that have not been possible to conduct in conventional bulk experiments.

2 Materials

1. Computer-aided design (CAD) drawing software (*see* **Note 1**).
2. Fluid-flow simulation software (*see* **Note 2**).
3. A class 1000 cleanroom facility (*see* **Note 3**).
4. Silicon wafers (WRS materials) (*see* **Note 4**).
5. Sulfuric acid (H_2SO_4), peroxide (H_2O_2), and a glass container.
6. Hydrofluoric acid (HF), HF antidote (calcium gluconate gel), sodium bicarbonate, and a Teflon container.
7. Negative photoresist SU-8 2035 (MicroChem).
8. SU-8 developer (MicroChem).
9. Isopropyl alcohol (IPA).
10. Deionized water.
11. Hot plate.
12. UV light exposure system.
13. Trichlorosilane.
14. Polydimethylsiloxane (PDMS) (Sylgard® 184 Silicone Elastomer Kit—base and curing agent).
15. Vacuum desiccator.
16. Cutter, revolving punch, syringes.
17. Plasma cleaner.
18. PVC tubes (peristaltic pump tubing).
19. *Camellia japonica* pollen.
20. Growth medium: 1.62 mM H_3BO_3, 2.54 mM $Ca(NO_3)_2{\cdot}4H_2O$, 0.81 mM $MgSO_4{\cdot}7H_2O$, 1 mM KNO_3, 8 % sucrose (w/v), in distilled water.
21. Microscope with image capture.

3 Methods

Carry out all procedures in the cleanroom at room temperature (unless otherwise indicated). Meticulously follow all waste disposal regulations.

3.1 Microfluidic Network Design

1. Design the microfluidic network according to the intended application. Here we develop a microfluidic chip to investigate the instantaneous growth rate of pollen tubes as they encounter a mechanical obstacle consisting of a flat surface oriented at a defined angle relative to the growth direction: 0° (perpendicular to the growth direction), 30°, and 60° (Fig. 1; *see* **Note 5**).
2. Carry out microfluidic simulations to support and validate the platform design. Depending on the simulation result, the

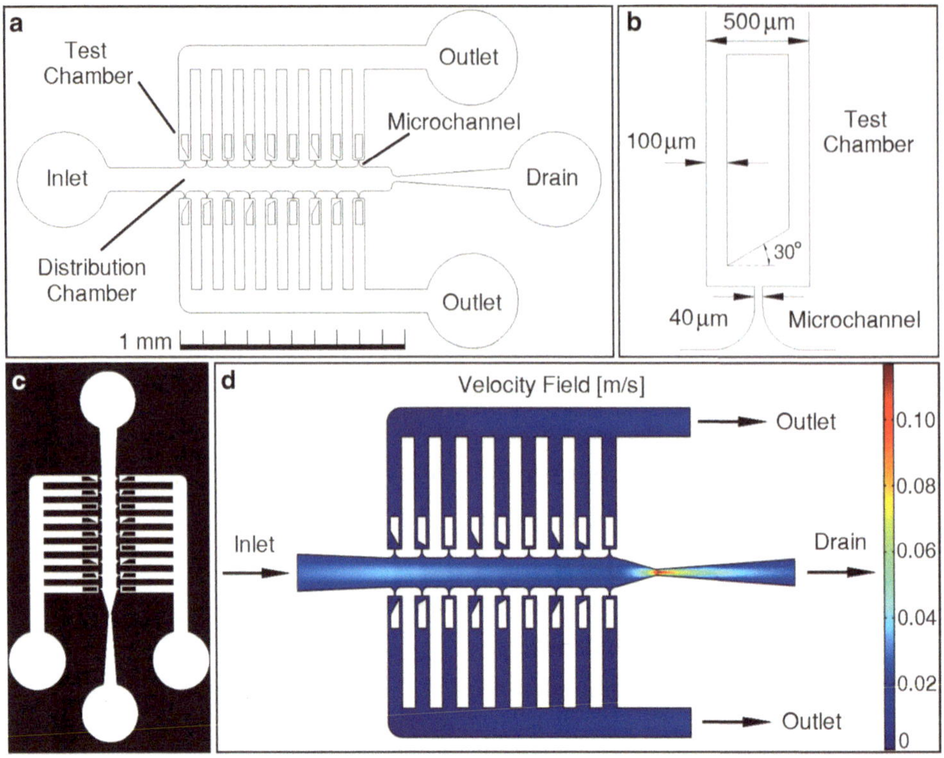

Fig. 1 Overall design of the LOC. (**a**) Schematic. (**b**) Detailed layout of a microchannel and test chamber. (**c**) Photomask. (**d**) Velocity fluid field simulation of the microfluidic platform

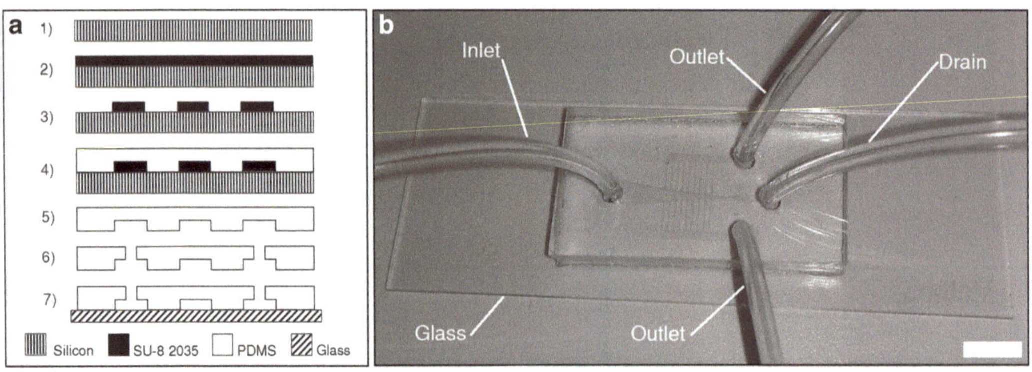

Fig. 2 (**a**) Microfluidic platform fabrication: (*1*) Silicon wafer cleaning, (*2*) SU-8 2035 photoresist spin-coating, (*3*) Photolithographic patterning and photoresist development, (*4*) PDMS pouring and curing, (*5*) PDMS layer detachment, (*6*) Microfluidic access drilling, (*7*) PDMS-glass bonding. (**b**) Fabricated LOC. Scale bar = 1 mm. Reproduced from [11] with permission from IOP Publishing Limited

overall design of the microfluidic network might need to be revisited. Figure 2 shows the microfluidics simulation for the current platform which ensures a uniform distribution of pollen grains over the microchannel entrances (*see* **Note 6**).

3. Draw the microfluidic design in Fig. 1 in the CAD software (*see* **Note 7**).
4. Reproduce the design on a photomask (*see* **Note 8**).

3.2 Silicon/SU-8 Mold

1. The fabrication process of the silicon/SU-8 mold is illustrated in Fig. 2a. Start by cleaning a silicon wafer with a piranha bath. The solution is dangerously aggressive and corrosive; use protective gear. Carefully mix three parts of sulfuric acid (H_2SO_4) and one part of peroxide (H_2O_2) in a glass container. A total volume of 200 ml is enough for cleaning one or two wafers. Use tweezers to place the silicon wafer slowly inside the solution. The exothermic reaction of the solution is good for cleaning only for about one hour. Afterwards, move the wafer to deionized water and air-dry with filtered, pressurized air or ideally with a N_2 gun (*see* **Note 9**).
2. Perform an HF cleaning. Hydrofluoric acid is a lethal solution; handle with extreme care. Use protective gear: butyl rubber gloves, face shield, safety glasses, leather closed shoes, lab coat, and chemical apron. Perform the cleaning inside a fume hood. Have a safety shower and HF antidote (calcium gluconate gel) nearby in case of skin contact. Mix 10 ml of HF with 200 ml of deionized water in a Teflon container. Place the silicon wafer slowly inside the solution with tweezers and leave for 3 min. Next move the wafer to deionized water and air-dry. Neutralize the HF with copious amounts of diluted sodium bicarbonate. HF should be disposed as a corrosive hazardous waste (*see* **Note 10**).
3. Spin-coat 4 ml of SU-8 on the 10 cm silicon wafer at 1,500 rpm for 30 s. Soft-bake for 5 min at 65 °C and then 10 min at 95 °C on a hot plate to harden the photoresist (by evaporating the photoresist solvent) and to increase adhesion to the substrate. Next, leave the wafer to cool down at room temperature (*see* **Note 11**).
4. Expose the negative photoresist SU-8 to UV light using the photomask (*see* **Note 12**).
5. Perform a postexposure bake (PEB) directly after exposure to enhance the chemical linking induced by the UV light. Bake for 5 min at 65 °C and then 10 min at 95 °C on a hot plate.
6. Develop the photoresist layer to obtain the final SU-8 pattern. Pour enough developer in a glass container to fully cover the silicon/SU-8 mold. Immerse the silicon/SU-8 mold in the SU-8 developer to dissolve the areas not exposed to UV light. Agitate gently. The development time depends directly on the thickness of the SU-8 layer. For an 80 μm thick SU-8 layer, the development time is about 8 min. Next, rinse with IPA and again with fresh developer. Air-dry. Monitor the state of

development by microscope and continue development if necessary (*see* **Note 13**).

7. Hard-bake at 150 °C for 10 min to solidify the remaining photoresist and reduce mechanical stresses in the structure (*see* **Note 14**).
8. Verify the mold under the microscope (*see* **Note 15**).
9. Silanize the silicon/SU-8 mold. The mold is exposed to trichlorosilane (or simply silane) vapors to prevent the PDMS replica from sticking to the mold. Use protective gear and handle with care. Silane is highly flammable (flash point of 87 °C), highly corrosive, and reacts violently with water. Using a plastic dropper (or syringe), place four drops of silane in a glass dish close to the silicon wafer. Close the glass dish and place it on a hot plate at 70 °C to evaporate the silane. Leave for at least 4 h. Cool at room temperature before opening the glass dish to allow the vapors to settle (*see* **Note 16**).

3.3 PDMS Microfluidic Chip

1. Thoroughly mix ten parts of PDMS polymer base with one part of PDMS curing agent in a disposable container (*see* **Note 17**).
2. Pour the PDMS mix onto the silicon/SU-8 mold. Place the mold/PDMS ensemble in a vacuum desiccator for 15 min to degas the PDMS and next cure in an oven at 80 °C for 2 h.
3. Carefully excise each PDMS replica from the mold (*see* **Note 18**).
4. Punch inlet and outlet ports of the PDMS replica. Rinse with IPA and air-dry to clean any dirt particle (*see* **Note 19**).
5. Bond the PDMS replica to a glass slide to seal the microfluidic chip (*see* **Note 20**).
6. Insert inlet and outlet PVC tubes from the top of the structure to obtain the microfluidic platform shown in Fig. 2b (*see* **Note 21**).

3.4 Microfluidic Platform Testing

1. Collect, dehydrate, and store *Camellia japonica* pollen on silica gel at −20 °C for later use (*see* **Note 22**).
2. Prior to experimentation, thaw and rehydrate a few milligrams of the pollen in humid atmosphere for 1 h (*see* **Note 23**).
3. Prepare liquid growth medium (*see* **Note 24**).
4. Place the microfluidic platform under the microscope (or any other imaging setup).
5. Immerse the pollen grains in 1 ml liquid growth medium. Agitate gently to mix the suspension; pollen grains are very sensitive to excess mechanical stress.
6. Using a syringe, carefully inject the pollen suspension through the PVC tube into the microfluidic device. Monitor the injection through the microscope. Figure 3a shows a typical pollen grain distribution.

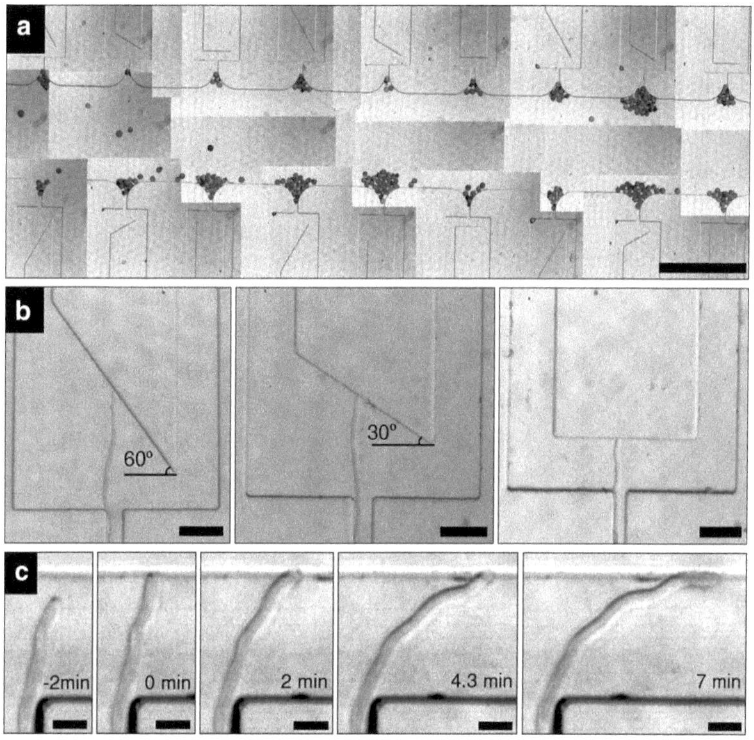

Fig. 3 Mechanical obstacle test. (**a**) Initial pollen grain distribution (image is stitched from several high-magnification micrographs. Scale bar = 1 mm. (**b**) Pollen tubes colliding with flat wall oriented at 60°, 30°, and 0° relative to the growth direction. Scale bars = 100 μm. (**c**) Time-lapse sequence of a type 0° collision. Scale bars = 25 μm. Reproduced from [11] with permission from IOP Publishing Limited

7. Leave the pollen tubes to grow. *Camellia* pollen tubes usually germinate after 30 min imbibition in growth medium, elongate at average growth rates of 12 μm/min, and easily attain more than 1 mm in total length after 2 h. Figure 3b, c shows pollen tubes encountering mechanical obstacles at defined angles [11].

4 Notes

1. The current design is drawn in AutoCAD, but any CAD software can be used as long as the output format is compatible with the photomask creation process.
2. The current design uses COMSOL multiphysics, a finite element analysis (FEA) solver of partial differential equations for various coupled phenomena.
3. Although a class 1000 cleanroom or better (maximum of 1,000 particles of size of 0.5 μm or larger in a cubic foot of air)

is ideal, a class 10000 cleanroom is sufficient to properly fabricate a microdevice with feature sizes in the order of a few micrometers. A less clean environment would often result in fabrication defects.

4. The current design uses 100 mm diameter standard type silicon wafers in particular. For the current application the dopant and orientation are not an issue. The size of the silicon wafer is highly dependent on the UV exposure system at hand. Silicon is preferred because of the good adhesion between silicon and the SU-8 photoresist, but the process can be performed on other substrates as well.
5. The current design of the microfluidic network is based on an inlet, a linear distribution chamber, two series of symmetrical microchannels and test chambers (top and bottom), two outlets, and a drain outlet at the end of the distribution chamber. Only straight-shaped microchannels are used in order to impose an initial direction of growth on the pollen tubes and to obtain a homogeneous fluid flow among the microchannels [11]. In the setup shown here, the sizes of structural features are chosen to fit the dimensions of *Camellia japonica* pollen grains and tubes. However, the design can be easily adapted to serve different applications. The test chamber, for example, can be modified to allow for the integration of microelectrodes. Examples for designs can be downloaded from the Optical-Bio Microsystems Laboratory website (http://users.encs.concordia.ca/~mpackir/index.html) and from the Geitmann Lab website (*see* Publications at http://www.geitmannlab.org).
6. In order to predict the fluid-flow behavior within the microfluidic network, and particularly the movement of pollen grains along the streamlines, a 2D Finite Element Method (FEM) fluid analysis implementing the incompressible Navier–Stokes and continuity equations was carried out using COMSOL. A velocity of 0.02 m/s was selected as the boundary condition at the inlet to reflect typical medium injection by syringe, and the outlets were set to atmospheric pressure. Since the liquid medium used here consists mostly of water [7], the density (ρ) and the dynamic viscosity (μ) are set to 10^3 kg/m^3 and 10^{-3} Pa s, respectively.
7. Be careful to use simple curves. Lines and arcs alone can be used to draw most designs. Keep the amount of vertices to a minimum and do not overlap any. Make sure the design is composed of closed curves; by definition, no single curve should be open. Once the skeleton of the drawing is done, shade every region as necessary. In the current design we use a negative photoresist (SU-8) to fabricate the mold; hence, those regions that are not to be permanent in the mold must be dark (*see* Fig. 1c).

8. Low-resolution (low cost) photomasks can be easily obtained by using high-resolution digital printing on a transparent film at 3,600 dpi. However, this method results in rough walls for feature sizes in the range of 10–50 μm and completely misses features below 10 μm. For smoother, well-defined features down to 1 μm, a more precise (and expensive) alternative is Direct Writing Laser Lithography on a glass/chrome mask.
9. The piranha bath is advised but optional depending on the cleanliness state of the silicon wafer and the cleanroom. However, we found that it is a must if the silicon wafer is being recycled.
10. The HF cleaning is used to eliminate any native oxide layer on the silicon wafer. We have noticed that this native oxide layer often prevents the SU-8 from adhering firmly to the wafer during the developing process. It is advised to keep the silicon wafers always inside a hermetic box in the cleanroom to avoid contact with the ambient air as much as possible. HF cleaning should be carried out if and only if adhesion problems between silicon and SU-8 arise since HF needs to be handled with extreme care due to its potentially lethal effects. Although not advised, the Teflon container can be replaced with a regular plastic container if necessary.
11. We found that it is best to dispense the SU-8 directly from the bottle to avoid the formation of air bubbles in the SU-8 layer. Be particularly careful at the end of the dispensing when pulling out the bottle since a narrow stream of photoresist might easily create bubbles in the already dispensed SU-8. Stop the dispensing with tissue if necessary. The spin speed is set to obtain a thickness of 80 μm since *Camellia* pollen grains vary from 50 to 60 μm in diameter. However, the thickness can be changed to accommodate different-sized specimens. A hot plate is preferred over a convective oven to ensure uniform temperature across the wafer during soft-bake. Thicker layers require longer soft-bake times (see manufacturer's SU-8 datasheet for a complete table of suggested soft-bake times). After proper soft-bake the SU-8 layer must be smooth; extend the duration of the soft-bake should wrinkles, bumps, or bubbles appear on the surface (an expired resist can also produce similar issues).
12. UV light exposure is critical in the mold fabrication. Make sure the SU-8 layer is completely flat and hardened before exposure. Any air bubble or dirt in the SU-8 layer will result in a loss of features in the area. Although exact exposure energies can be found in the SU-8 datasheet, an exposure matrix must be carried out in order to determine the optimal exposure parameters for the current setup of UV exposure system, substrate, and SU-8 thickness. Enough exposure should produce a latent image on the SU-8 layer within the first minute of postexposure bake. It is useful to realize that since SU-8 is a

negative photoresist, the areas exposed to UV light are meant to remain on the silicon wafer, whereas the areas covered by the photomask will be dissolved during development. Therefore, it is generally preferred to overexpose (ensure the exposed SU-8 will remain at the expense of enlarging the features due to light scattering) than to underexpose (lose parts of the design).

13. If there are adhesion problems between the silicon wafer and the SU-8 layer, the photoresist will peel off from the wafer during development. Should this detachment occur, an HF cleaning may be required.
14. Since the photoresist is a thermal resin, the hard-bake temperature should be carried out at a temperature slightly higher than the expected microdevice operating temperature in order to ensure the mechanical properties of the photoresist. The visual effect of the hard-bake is to "smooth" the SU-8 layer.
15. Direct visual inspection of the mold is required to determine what step of the process went wrong or can be improved. Verify the thickness of the mold to adjust the SU-8 spin-coating speed accordingly. Verify the dimensions of the mold to adjust the exposure time and possibly the PEB temperature. Pay particular attention to the smoothness of the SU-8 walls and any lost feature since it may indicate a photomask with a resolution that is too low. Any undeveloped SU-8 can be removed by extending the development time. Any cracks in the mold bulk can be reduced by increasing the hard-bake temperature or duration.
16. We found that a minimum of 65 °C is needed for proper silanization. The level of silanization can be varied by dispensing more or less drops or by using shorter or longer times. Adjust if necessary. We found that glass can also be silanized in the same way.
17. The total amount of poured PDMS depends on the size of the container of the mold. In order to not waste PDMS, place the silicon/SU-8 mold in a glass container of approximately the same size. Aluminum foil can also be used as a container with folded "walls." The PDMS on top of the mold will determine the thickness of the PDMS layer. This thickness of the PDMS layer is not critical, usually 2–3 mm, as long as the inlet and outlet drilling can be done properly. However, excessive thickness may compromise the optical properties when using high-resolution microscopy. Since the PDMS volume for a 100 mm diameter mold is in the range of a few tens of milliliters (usually between 40 and 60 ml), it is recommended to use syringes to measure the PDMS base and curing agent quantities. Use separate syringes for base and curing agent to avoid cross-contamination.

18. Handle with care. Use a clean, sharp cutter to dice the cured PDMS replicas before extracting them. The cutter should touch the silicon wafer as the dicing takes place. If silanization was properly performed, the PDMS clearly detaches from the mold as the PDMS is being cut. Removing the replicas from the mold should be done effortlessly. If PDMS is stuck to the mold, the most likely reason is a problem with the silanization. Unfortunately, the mold is almost inevitably lost if the PDMS is stuck since it is next to impossible to remove the PDMS without destroying the SU-8 layer. However, should this happen, a complete wafer cleanup can be performed (piranha bath and HF cleaning) and the silicon wafer can be recovered to restart mold fabrication.
19. Punching of inlets and outlets can be easily done with a revolving punch. The size of the round holes must match the PVC tubes used.
20. Oxygen plasma bonding is recommended for fast and reliable results. A matrix test must be carried out to determine the optimal parameters for bonding glass and PDMS. In our setup (Harrick Plasma PDC-001), the bonding time for a glass/PDMS interface is 40 s with a high voltage application. Another (low-cost) alternative is to spin-coat a thin layer of PDMS on the glass slide (2,000 rpm for 30 s), semi-cure the PDMS layer, make the bond, and then completely cure. The duration of the semi-curing depends on many factors; however, a good starting point is 3 min at 90 °C. We found that it is preferable to leave it longer since if the bonding fails, another thin layer of PDMS can be added on top, whereas if less time is used then the features on the PDMS replica will be filled by the PDMS gel.
21. Since the fluid pressures inside the microfluidic platform are relatively low, the PVC tubes do not need to be glued to the PDMS replica. Friction is sufficient to keep the tubes in place (given the PDMS is at least 1 mm thick). Inlets and outlets on the side of the chip are discouraged since this requires more complex connections.
22. Although fresh *Camellia japonica* pollen is ideal, this may be difficult to obtain as this species flowers only once a year for a few weeks.
23. An easy way to hydrate pollen grains is to wet a piece of paper towel with hot water and place both pollen and tissue in an enclosed glass container (Petri dish). Very importantly, avoid any direct contact between pollen and water to prevent the pollen grains from absorbing liquid water at this point.
24. Liquid growth medium has already been optimized for *Camellia japonica* pollen [7, 12]. Usually, we prepare 10 ml of medium and use 1 ml plastic capsules for testing.

References

1. Malhó R (2006) The pollen tube: a cellular and molecular perspective. Springer, Berlin
2. Palanivelu R, Preuss D (2000) Pollen tube targeting and axon guidance: parallels in tip growth mechanisms. Trend Cell Biol 10: 517–524
3. Geitmann A, Palanivelu R (2007) Fertilization requires communication: signal generation and perception during pollen tube guidance. Floricult Ornamental Biotechnol 1:77–89
4. Mascarenhas J, Machlis L (1964) Chemotropic response of the pollen of *Antirrhinum majus* to calcium. Plant Physiol 39:70–77
5. Jaffe L, Nuccitelli R (1977) Electrical controls of development. Annu Rev Biophys Bioeng 6:445–476
6. Chebli Y, Geitmann A (2007) Mechanical principles governing pollen tube growth. Funct Plant Sci Biotechnol 1:232–245
7. Bou DF, Geitmann A (2011) Actin is involved in pollen tube tropism through redefining the spatial targeting of secretory vesicles. Traffic 12:1537–1551
8. Vanapalli S, Duits M, Mugele F (2009) Microfluidics as a functional tool for cell mechanics. Biomicrofluidics 3:012006
9. Ziaie B, Baldi A, Lei M et al (2004) Hard and soft micromachining for BioMEMS. Adv Drug Deliv Rev 26:145–172
10. Agudelo C, Sanati A, Ghanbari M, Naghavi M, Packirisamy M, Geitmann A (2012) TipChip—a modular, MEMS (microelectromechanical systems)-based platform for experimentation and phenotyping of tip growing cells. Plant J 73:1057–1068
11. Agudelo C, Sanati A, Ghanbari M, Packirisamy M, Geitmann A (2012) A microfluidic platform for the investigation of elongation growth in pollen tubes. J Micromech Microeng 22:115009. doi:10.1088/0960-1317/22/11/115009
12. Brewbaker J, Kwack B (1963) The essential role of calcium ion in pollen germination and pollen tube growth. Am J Bot 50:859–865

Chapter 21

Laser Microdissection of Plant Cells

Yvonne Ludwig and Frank Hochholdinger

Abstract

Different plant cell types express unique transcriptomes, proteomes, and metabolomes. Therefore, the isolation of specific cell types prior to molecular analyses is important to understand the specification, differentiation, and function of these cells. Isolation of specific plant cell types from composite organs can be achieved by laser microdissection (LMD). A wide variety of methods to fix and embed tissues prior to LMD and downstream molecular analyses have been developed for different plant species and tissues. The present review summarizes and highlights the most recently applied LMD approaches in plant science.

Key words Laser microdissection, Plant, Cell type, Sectioning, Fixation, Embedding

1 Introduction

All cells of a plant contain the same nuclear genome. However, each cell type expresses a unique subset of these genes at a specific developmental stage or environmental condition. The set of active genes reflects the status of a cell type with respect to specification, differentiation, and functionality. Therefore, the subset of all biomolecules such as transcripts, proteins, and metabolites expressed by a specific cell widely varies between different cell types of a plant. Due to these cell-type-specific expression variations, monitoring gene activity in composite plant tissues or organs masks individual expression profiles of specific cells because such surveys integrate expression over all cell types [1]. This is in particular a problem when a specific cell type represents only a small fraction of all cells of an organ. For instance, pericycle cells, which divide and give rise to lateral roots, are estimated to constitute <1 % of all root cells in maize [2]. Therefore, monitoring gene expression profiles in such specific cells is important to better understand their specification, differentiation, and function [3]. To study specific cell types from composite tissues in more detail requires separating them from other cells to reduce contamination to a minimum.

Viktor Žárský and Fatima Cvrčková (eds.), *Plant Cell Morphogenesis: Methods and Protocols*, Methods in Molecular Biology, vol. 1080, DOI 10.1007/978-1-62703-643-6_21, © Springer Science+Business Media New York 2014

The isolation of specific cells is straightforward for only a few very exposed plant cell types such as root hairs [4] and leaf trichomes [5]. However, most cell types are interconnected with other cells deep inside a complex tissue. Specific cell types expressing fluorescent marker genes can be separated by fluorescent-activated cell sorting (FACS) after protoplasting [6]. If no such marker lines are available, cells that can be microscopically distinguished from other cell types can be mechanically separated from their cellular context by laser microdissection (LMD). For laser microdissection in general, two types of techniques have been developed to separate specific cell types from the surrounding tissue. In the first "touch-free" approach, cells of interest are cut out and transported away from their tissue context either by gravity or by catapulting them away e.g., Veritas Laser Microdissection LCC1704 (Arcturus), PALM MicroBeam (Zeiss), or AS LDM (Leica). In an alternative approach, cells of interest are melted to a plastic membrane or are brought in contact with a sticky surface and are subsequently removed mechanically from the remaining tissue e.g., PixCell II LCM (Arcturus).

The plant kingdom consists of a wide variety of diverse plant species with a considerable number of functionally diverse cell types (*see* **Note 1**). Laser microdissection remains challenging because there is no general protocol available that is suitable for all plant species and cell types. Hence, optimal protocols have to be established for each application. The quality of tissue sections and the integrity of biomolecules such as RNA or proteins in these samples is the major prerequisite for successful LMD analyses. The initial decision on the fixation and embedding method is already decisive for the success of the experiment. In general, two types of fixation are applied: precipitative and cross-linking fixation. Typically, precipitative fixation (Farmer's fixative, *see* **Note 2**) yields higher RNA concentrations (*see* **Note 3**) than cross-linking fixation (formaldehyde-acetic acid-ethanol; [7, 8]). In most experiments, either frozen plant tissues (embedded in OCT (optimal cutting temperature compound) or CMC (carboxymethylcellulose)) or paraffin-embedded plant tissues are sectioned. The morphology of sections is often better preserved in paraffin-embedded tissues [9]. However, in many instances frozen tissues provide higher yield and quality of RNA or other biomolecules under analysis (*see* **Note 4**). Hence, in each experiment a balance between conservation of morphology and preservation of biomolecules needs to be achieved.

Another crucial step that determines the quality and integrity of nucleic acids is the mounting of paraffin sections to the slide. RNA quality can be affected during the stretching of paraffin ribbons on water and the drying process afterwards ([10]; *see* **Note 5**).

Thus far, only a small number of LMD studies combined with proteomics experiments have been performed. Farmer's fixative and embedding of the samples in OCT to perform cryo-sectioning was used [3, 11]. Alternatively, no fixation or embedding chemicals were employed to avoid polymer peaks during the mass spectrometric identification of peptides [12].

Table 1 summarizes the major characteristics of recently performed plant LMD studies including the analyzed plant species, cell type, the used LMD system, fixation, tissue-embedding approach, RNA extraction, amplification step, and downstream analysis.

2 Materials

2.1 Paraffin Embedding

1. Farmer's fixative: 3:1 ethanol: acetic acid (*see* **Note 6**).
2. Ethanol series: 70, 80, 90, and 100 % (v/v).
3. Ethanol/xylene or ethanol/isopropanol (dehydration solution): 75 %:25 %; 50 %:50 %; 25 %:75 %; 100 % (v/v).
4. Melted paraffin chips (*see* **Note 7**).
5. Membrane slides (*see* **Note 8**) or plain slides.
6. 100 % Xylene.

2.2 Microwave Technique

1. Farmer's fixative: 3:1 ethanol: acetic acid (*see* **Note 6**).
2. Ethanol: 50, 70, and 100 % (v/v).
3. Ethanol: isopropanol: 50:50.
4. 100 % Isopropanol.
5. Melted paraffin chips (*see* **Note 7**).

2.3 OCT Embedding

1. Farmer's fixative: 3:1 ethanol: acetic acid. (*see* **Note 6**).
2. 10 % Sucrose solution: 10 % sucrose (w/v) in phosphate-buffered saline (PBS; *see* **Note 9**).
3. 15 % Sucrose solution: 15 % sucrose (w/v) in PBS (*see* **Note 9**).
4. OCT (optimal cutting temperature compound) or CMC (carboxymethylcellulose).
5. Membrane slide (*see* **Note 8**) or plain slides.
6. 70 % Ethanol: store at –20 °C.
7. Ethanol: 95, 100 % (v/v).
8. 100 % Xylene.

3 Methods

3.1 Fixation and Embedding of Plant Tissues for Paraffin Sectioning

1. Most frequently acetone [13, 14] or Farmer's fixative (3:1 ethanol:acetic acid) [8, 11, 15, 16] are used for fixation as they provide the best results for a large number of different tissues. Plant tissue samples are typically fixed at 4 °C between 1 and 24 h (*see* **Note 10**).

Table 1
Overview of LMD application in plants

Species	Tissue	LMD system	Fixation	Embedding	RNA extraction	Amplification	Downstream analysis	References
Arabidopsis thaliana	Embryo	PixCell II (Arcturus)	None	OCT	Absolutely RNA Nanoprep Kit (Stratagene)	MessageAmp Kit (Ambion)	Microarray	[25]
Brassica napus	Seed	PALM MicroBeam (Zeiss)	None	OCT	Total RNA (columns)/ mRNA (beads)	According to ref. [26]	Lipids, starch, metabolites, transcriptomics	[23]
Citrus clementina	Leaf	AS LMD (Leica)	None	OCT	RNeasy Micro Kit (Qiagen)	TargetAmp Kit (Epicentre Biotech.)	Microarray	[16]
Hordeum vulgare	Grain	PALM MicroBeam	None	None	NA	NA	Proteomic	[12]
Medicago truncatula	Root cortex	PALM MicroBeam	None	TFM	RNeasy Plant Extraction Kit (Qiagen)	WT-Ovation Kit (NuGEN)	Microarray	[27]
Oryza sativa	Shoot apex	Veritas LMD LCC1704 (Arcturus)	3:1 EtOH: AA	Paraffin—microwave	PicoPure RNA isolation kit (Arcturus)	Quick Amp Labelling Kit (Agilent)	Microarray	[22]
	Crown roots	Veritas LMD LCC1704	99.5 % EtOH or acetone	Paraffin—microwave	PicoPure RNA isolation kit	Quick Amp Labelling Kit	Microarray	[13]
	Anthers	Veritas LMD LCC1704	3:1 EtOH: AA	Paraffin—microwave	PicoPure RNA isolation kit	None	Microarray	[15]
	Aleurone, endosperm	AS LMD	Acetone, 3:1 EtOH: AA	CMC, paraffin	PicoPure RNA isolation kit	Based on ref. [7, 8]	qRT-PCR	[17]

Zea mays	Shoot	PALM MicroBeam	Acetone	Paraffin	PicoPure RNA isolation kit	RiboAmpTM HS kit (Arcturus)	Microarray	[14]
	Primary root pericycle	PixCell II	3:1 EtOH: AA	OCT	NA	NA	Proteomics	[11]
	Coleoptile	PixCell II	3:1 EtOH: AA	OCT	Absolutely RNA Microprep Kit	According to ref. [28]	Microarray	[8]
	Primary roots	PixCell II	3:1 EtOH: AA	OCT	RNaqueous-micro kit (Ambion)	BD SMART PCR cDNA synthesis kit (BD Bioscience)	Microarray/ Proteomics	[3]
	Shoot apex	PALM MicroBeam	Acetone	Paraffin	PicoPure RNA isolation kit	According to ref. [8]	qRT-PCR/ Microarray	[19]
	Primary root pericycle	PixCell II	3:1 EtOH: AA	OCT	RNaqueous-micro kit	According to ref. [8]	Microarray	[2]

EtOH:AA ethanol:acetic acid, *CMC* carboxymethylcellulose, *qRT-PCR* quantitative real-time PCR, *TFM* tissue freezing medium, *OCT* optimal cutting temperature compound, *NA* not applicable

2. The plant tissue is dehydrated by an ethanol series (70, 80, 90, and 100 %), followed by an ethanol:xylene or ethanol:isopropanol treatment (3–4 h (v/v) 75 %:25 %; 50 %:50 %; 25 %:75 %; 100 %) at room temperature [7, 17, 18].
3. The samples are slowly embedded directly into paraffin by a gradient, starting with a 3:1 ratio of dehydration solution and paraffin for 3–4 h at 58 °C.
4. The dehydration solution:paraffin ratio is changed to 1:1 and the samples are incubated twice in this solution prior to replacement by 100 % paraffin [7, 14, 19, 20]. Each step of this sequence has an incubation time of 3–4 h at 58 °C.
5. The paraffin-embedded samples can be stored at 4 °C prior to downstream analyses.

4 Fixation and Embedding for Paraffin Sectioning via Microwave Oven Technique

1. An alternative way of embedding plant samples in paraffin is the microwave oven technique. In this approach the samples are fixed by Farmer's solution in a specific microwave oven (e.g., PELCO BioWave), for 15 min at 37 °C.
2. Farmer's solution is replaced by an ethanol series (50, 70, and 100 %) at 67 °C for 75 s per step.
3. Ethanol is replaced by a 50:50 ethanol:isopropanol mixture for 90 s.
4. Samples are incubated in pure isopropanol at 77 °C for 90 s.
5. Isopropanol is replaced by a 1:1 isopropanol:paraffin mixture at 77 °C for 10 min followed by at least four exchanges of paraffin at 67 °C for 30 min each step [21].
6. Several modifications of this basic protocol have been applied to different plant tissues [13, 15, 22].

5 Fixation and Embedding for Cryo-sectioning

1. Plant tissue for cryo-sectioning is fixed in Farmer's solution (see above for paraffin sections) [2, 3, 8, 11]. Alternatively, no fixation is applied [23–25, 27].
2. Samples are incubated in 10 % sucrose solution (10 % sucrose (w/v) in PBS) (*see* **Note 11**) after fixation at 4 °C for 1 h (*see* **Note 12**).
3. The 10 % sucrose solution is replaced by 15 % sucrose (w/v) in PBS at 4 °C between 1 and 24 h.

4. The plant tissue is embedded in "optimal cutting temperature" compound (OCT) or carboxymethylcellulose (CMC; [17]) and frozen in liquid nitrogen.
5. Specimens are stored at −80 °C until sectioning.

5.1 Sectioning of Paraffin: Embedded Samples

1. Sectioning of the paraffin-embedded specimens is performed with a microtome. Typically 10–15 μm sections are generated.
2. The sections are transferred to specific membrane or plain microscopic slides and are dried at 42 °C O/N (*see* **Note 5**).
3. Sections are stored at 4 °C under dehydrating conditions [7, 14, 19].
4. Prior to laser microdissection the specimens have to be deparaffinized two times with xylene for 5 min [13, 18].
5. Then samples are air-dried.

5.2 Sectioning of OCT-Embedded Samples

1. OCT-embedded tissues are sectioned with a cryo-microtome and mounted to membrane slides or on plain microscopic slides.
2. To remove the embedding medium, the sections on the slides are treated with 70 % ethanol at −20 °C for 1 min.
3. Sections are dehydrated with ethanol (95, 100 %) at room temperature for 1 min and subsequently with xylene for 2 min [2, 3, 8].
4. A modified protocol to remove OCT is described in ref. [24].

6 Notes

1. The yield of RNA in plant cells and tissues can vary between two- and threefold which requires optimized LMD protocols adapted to each tissue and cell type under analysis [18].
2. Two precipitative fixation methods (Farmer's fixative and methacarn) combined with classical paraffin embedding and microwave paraffin embedding were tested for *Arabidopsis thaliana* leaves. After fixation, cells were sampled and the isolated RNA was tested by RT-PCR. The results did not show any significant differences between the applied approaches [18]. However, the amplification of transcripts did not provide any information on the quality of the underlying RNA [10].
3. Since LMD yields typically only minute amounts of biomolecules, it is recommended that their quality is assessed for instance by Bioanalyzer analyses prior to subsequent downstream analyses [10, 16, 17].

4. Quality differences between paraffin- and CMC-embedded samples were shown in Bioanalyzer surveys. In these experiments conventional paraffin-embedded samples displayed a higher degree of RNA degradation compared to the CMC-embedded specimens [17].
5. In a comparative study, paraffin ribbons were exposed to water for different time periods with and without RNase inhibitor and drying temperatures of 42 and 4 °C. It was demonstrated that avoiding the 42 °C drying step resulted in decent RNA quality and the RNase inhibitor was most efficient in the first hour. Even better results were obtained after removing the water and drying the specimens at 4 °C for at least 1 h [16]. A different way to avoid the water and drying step is using the paraffin-tape transfer system [10]. The system is based on adhesive tapes to capture sections instead of using a brush for transferring them to the slides, which avoids rolling or breaking.
6. Prepare fresh before use.
7. Place paraffin chips at 58 °C a day before use.
8. A wide variety of membrane slides are provided by different suppliers. Typically they are coated with PEN: polyethylene naphthalate (Zeiss, Leica, Arcturus); PET: polyethylene terephthalate (Zeiss, Leica); PPS: polyphenylene sulfide (Leica); POL: polyester (Leica); FLUO: fluocarbon (Leica).
9. 1× PBS (1,000 ml): 8 g NaCl, 0.2 g KCl, 1.44 g Na_2HPO_4, 0.24 g KH_2PO_4; adjust to pH 7.4.
10. A partial vacuum (200–400 mbar for 15–20 min) facilitates the fixative to penetrate the tissue [2, 8, 15, 16].
11. Plant cells in contrast to mammalian cells have big vacuoles for water storage. This is problematic for cryo-sectioning, because during the fast freezing process ice crystals can form and destroy the cellular structure. This can be avoided by treating the plant tissue in sucrose solution.
12. Vacuum infiltration (200–400 mbar on ice for 15 min) facilitates the penetration of the solution into the tissue.

Acknowledgment

Root research in the laboratory of FH is supported by the DFG (Deutsche Forschungsgemeinschaft) and the 7th Framework Programme of the European Union.

References

1. Schnable PS, Hochholdinger F, Nakazono M (2004) Global expression profiling applied to plant development. Curr Opin Plant Biol 7:50–56
2. Woll K, Borsuk LA, Stransky H et al (2005) Isolation, characterization, and pericycle-specific transcriptome analyses of the novel maize lateral and seminal root initiation mutant *rum1*. Plant Physiol 139:1255–1267
3. Dembinsky D, Woll K, Saleem M et al (2007) Transcriptomic and proteomic analyses of pericycle cells of the maize primary root. Plant Physiol 145:575–588
4. Nestler J, Schütz W, Hochholdinger F (2011) Conserved and unique features of the maize (*Zea mays L.*) root hair proteome. J Proteome Res 10:2525–2537
5. McDowell ET, Kapteyn J, Schmidt A et al (2011) Comparative functional genomic analysis of Solanum glandular trichome types. Plant Physiol 155:524–539
6. Birnbaum K, Shasha DE, Wang JY et al (2003) A gene expression map of the Arabidopsis root. Science 302:1956–1960
7. Kerk NM, Ceserani T, Tausta SL (2003) Laser capture microdissection of cells from plant tissues. Plant Physiol 132:27–35
8. Nakazono M, Qiu F, Borsuk LA et al (2003) Laser-capture microdissection, a tool for the global analysis of gene expression in specific plant cell types: identification of genes expressed differentially in epidermal cells or vascular tissues of maize. Plant Cell 15:583–596
9. Barcala M, Fenoll C, Escobar C (2012) Laser microdissection of cells and isolation of high-quality RNA after cryosectioning. Methods Mol Biol 883:87–95
10. Cai S, Lashbrook CC (2006) Laser capture microdissection of plant cells from tape-transferred paraffin sections promotes recovery of structurally intact RNA for global gene profiling. Plant J 48:628–637
11. Liu Y, von Behrens I, Muthreich N et al (2010) Regulation of the pericycle proteome in maize (*Zea mays L.*) primary roots by RUM1 which is required for lateral root initiation. Eur J Cell Biol 89:236–241
12. Kaspar S, Weier D, Weschke W et al (2010) Protein analysis of laser capture micro-dissected tissues revealed cell-type-specific biological functions in developing barley grains. Anal Bioanal Chem 398:2883–2893
13. Takehisa H, Sato Y, Igarashi M et al (2012) Genome-wide transcriptome dissection of the rice root system: implications for developmental and physiological functions. Plant J 69:126–140
14. Brooks L III, Strable J, Zhang X et al (2009) Microdissection of shoot meristem functional domains. PLoS Genet 5:e1000476
15. Suwabe K, Suzuki G, Takahashi H et al (2008) Separated transcriptomes of male gametophyte and tapetum in rice: validity of a laser microdissection (LM) microarray. Plant Cell Physiol 49:1407–1416
16. Takahashi H, Kamakura H, Sato Y et al (2010) A method for obtaining high quality RNA from paraffin sections of plant tissues by laser microdissection. J Plant Res 123:807–813
17. Ishimaru T, Nakazono M, Masumura T et al (2007) A method for obtaining high integrity RNA from developing aleurone cells and starchy endosperm in rice (*Oryza sativa L.*) by laser microdissection. Plant Sci 173:321–326
18. Inada N, Wildermuth MC (2005) Novel tissue preparation method and cell-specific marker for laser microdissection of Arabidopsis mature leaf. Planta 221:9–16
19. Zhang X, Madi S, Borsuk L et al (2007) Laser microdissection of narrow sheath mutant maize uncovers novel gene expression in the shoot apical meristem. PLoS Genet 3:e101
20. Emrich SJ, Barbazuk WB, Li L et al (2007) Gene discovery and annotation using LCM-454 transcriptome sequencing. Genome Res 17:69–73
21. Schichnes D, Nemson J, Sohlberg L et al (1998) Microwave protocols for paraffin microtechnique and in situ localization in plants. Microsc Microanal 4:491–496
22. Kobayashi K, Yasuno N, Sato Y et al (2012) Inflorescence meristem identity in rice is specified by overlapping functions of three AP1/FUL-like MADS box genes and PAP2, a SEPALLATA MADS box gene. Plant Cell 24:1848–1859
23. Schiebold S, Tschiersch H, Borisjuk L et al (2011) A novel procedure for the quantitative analysis of metabolites, storage products and transcripts of laser microdissected seed tissues of *Brassica napus*. Plant Methods 7:19
24. Agusti J, Merelo P, Cercos M et al (2009) Comparative transcriptional survey between laser-microdissected cells from laminar abscission zone and petiolar cortical tissue during ethylene-promoted abscission in citrus leaves. BMC Plant Biol 9:127

25. Spencer MWB, Casson SA, Lindsey K (2007) Transcriptional profiling of the Arabidopsis embryo. Plant Physiol 143:924–940
26. Eberwine J, Yeh H, Miyashiro K et al (1992) Analysis of gene expression in single live neurons. Proc Natl Acad Sci USA 89: 3010–3014
27. Gaude N, Bortfeld S, Duensing N et al (2012) Arbuscule-containing and non-colonized cortical cells of mycorrhizal roots undergo extensive and specific reprogramming during arbuscular mycorrhizal development. Plant J 69: 510–528
28. Luo L, Salunga RC, Guo H et al (1999) Gene expression profiles of laser-captured adjacent neuronal subtypes. Nat Med 5: 117–122

Chapter 22

Optical Trapping in Plant Cells

Tijs Ketelaar, Norbert de Ruijter, and Stefan Niehren

Abstract

Optical tweezers allow noninvasive manipulation of subcellular compartments to study their physical interactions and attachments. By measuring (delay of) displacements, (semi-)quantitative force measurements within a living cell can be performed. In this chapter, we provide practical tips for setting up such experiments paying special attention to the technical considerations for integrating optical tweezers into a confocal microscope. Next, we describe some working protocols to trap intracellular structures in plant cells.

Key words Optical tweezers, Optical trap, Confocal microscope, Noninvasive manipulation, Plant cell, Cytoskeleton, Endomembrane system

1 Introduction

Optical tweezers use gradient light forces for attracting (sub-) micrometer-sized particles in a highly focused laser beam. Differences in refractive indices cause a total force pointing towards the center of the diffraction-limited spot. If the diffraction index of the particle is higher than the index of the surrounding medium, the force is attractive and creates an optical trap. Generally, infrared laser radiation is used for optical trapping since the absorbance of light with wavelengths in the infrared range is extremely low in most biological materials. For more information about optical trapping, *see* ref. [1].

Optical trapping is often used in in vitro experiments in which conditions can be controlled. Nevertheless, optical trapping of subcellular compartments with a high refractive index is also possible in intact, live cells. In plant cells, these types of experiments have greatly improved our understanding of physical aspects of intracellular organization [2, 3].

In in vitro studies, optical trapping is often combined with wide-field fluorescence microscopy to obtain positional information of fluorescently tagged molecules or beads. Plant cells are often large, embedded in tissues, and cell walls or other cell constituents can give autofluorescence. Thus, imaging of fluorescence

Viktor Žárský and Fatima Cvrčková (eds.), *Plant Cell Morphogenesis: Methods and Protocols*, Methods in Molecular Biology, vol. 1080, DOI 10.1007/978-1-62703-643-6_22, © Springer Science+Business Media New York 2014

in plant cells highly benefits from the reduced out-of-focus blur obtained with confocal microscopy. Although the integration of optical tweezers setup into a confocal microscope is nontrivial, it is essential to perform simultaneous optical trapping experiments and imaging of fluorescently tagged structures in plant cells. In this chapter, we describe our system, consisting of a Zeiss LSM510 META confocal microscope, mounted on a Zeiss Axiovert 200 M inverted microscope stand with an optical tweezers setup (Molecular Machines and Industries, Glattbrugg, Switzerland).

The combination of optical trapping and confocal imaging has only become available in the last decade but has so far failed to produce a large amount of published data of intracellular manipulation of plant cells [4, 5], mainly due to the limited availability of suitable microscope systems and the challenges to prepare plant samples that are suitable for combined optical trapping and confocal microscopy. Even so, this type of experimentation opens fantastic opportunities to gain insight in subcellular force generation. In our work with the optical tweezers/confocal microscope system, we have explored physical aspects of actin organization in Tobacco Bright Yellow-2 (BY-2) cells [5], the connection between Golgi bodies and the ER, and ER organization [4]. This chapter describes the technical specifications of our integrated confocal microscope and optical tweezers and provides hints to prepare plant tissues for these experiments and a description of the subsequent steps we follow during a typical experiment.

2 Materials

2.1 Description of Our System

A Molecular Machines & Industries (MMI, Glattbrugg, Switzerland) CellManipulator optical trap, consisting of an infrared Nd-YAG solid-state laser (1,064 nm, 3,000 mW CW) and x-y galvo scanner, is connected to the backport of an Axiovert 200 M inverted microscope (Zeiss, Jena, Germany). At the same backport, a Uniblitz shutter (Vincent Associates, Rochester, USA) is used to control 100 W HBO illumination. An MMI expander is used to fill the back focal plane of a 100×/N.A. 1.45 α-Plan Fluar or a 63×/N.A. 1.4 Plan-Apochromat objective. At the Axiovert 200 M base port a scan box of a Zeiss LSM510 META confocal is connected. The confocal is equipped with a 25 mW 405 nM laser, a 30 mW Ar laser (458, 477, 488, 514 nm), a 1 mW green HeNe laser (543 nm), and a 5 mM red HeNe laser (633 nm), which allows simultaneous trapping and confocal imaging of various fluorescent probes. A schematic overview of our system is given in Fig. 1.

During combined confocal imaging and optical trapping, 1,064 nm laser light from the optical tweezers is reflected into the optical path for confocal imaging. This is achieved by placing a beam splitter in the Axiovert 200 M reflector module that combines maximal transmittance for excitation and emission at

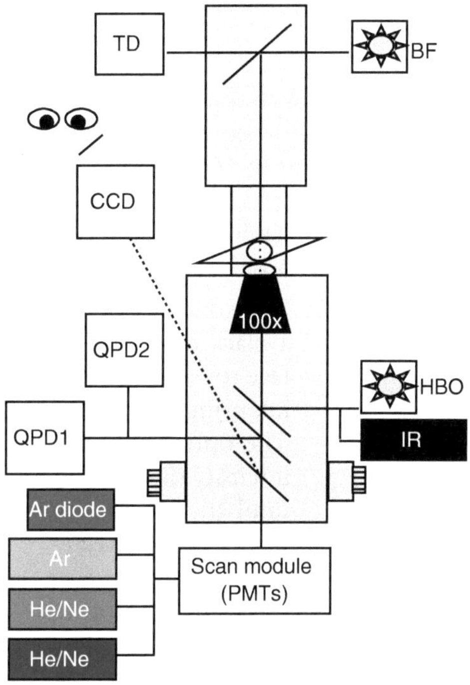

Fig. 1 Schematic overview of the described confocal microscope (Zeiss LSM510 META) with integrated optical tweezers (MMI CellManipulator). The *grey structure* represents a Zeiss Axiovert 200 M inverted microscope. The *colored boxes* represent lasers, as described in the text, and the *boxes* marked with BF and HBO represent, respectively, the bright-field halogen (BF) and wide-field fluorescence (HBO) illumination sources. CCD, charge-coupled device camera for collection of wide-field images during tweezers experiments; TD, transillumination detector of the LSM510 system; QPD1 and QPD2, quadrant photo detectors; PMTs, photomultiplier tubes. The dichroic mirror that integrates confocal with trapping lasers (see description system) is positioned at the HBO/IR optical path

400–800 nm from/to the LSM510 scan module (positioned at the base port) with maximal reflection for optical trapping at 1,064 nm (positioned at the back port).

The hardware is operated by two separate computers, one running the Zeiss LSM510 operating software and the other computer running the CellTools (MMI) software that controls the optical tweezers. A switch box links both systems to a Märzhauzer XY SCAN IM 120–100 stage (Märzhauzer stage, Wetzlar, Germany) to control *x,y* positioning with step sizes down to 75 nm. After calibration (*see* Subheading 3) the CellTools software can position up to ten quasi-simultaneous, timeshared traps to any location in the field of view taking advantage of the ultrafast galvos that position the laser.

The system is equipped with two quadrant photo detectors (QPDs; Spectral Applied Research, Ontario, Canada) and an additional 13× magnification at the left-side port to allow accurate bright-field (DIC) position detection for force calibration.

3 Methods

3.1 Calibration of the Optical Tweezers

Prior to trapping experiments, the focal point of the trapping laser is adjusted in the *z* direction to the visual focus of the microscope using the CellTools software (MMI), to allow imaging and trapping at identical *z*-planes. Afterwards, the *x,y* position of the trapping laser is calibrated such that the position on the screen and the real position correspond. We use the following approach:

1. Paint one face of a large (50 × 24 mm or similar) coverslip with a black marker and mount it dry on a slide with the painted face towards the slide and focus with bright-field optics on the ink using the objective that requires calibration (*see* **Note 1**). The optical tweezers laser is switched on and the focus and intensity of the laser are adjusted using the CellTools software until the ink absorbs sufficient energy to locally decompose the ink (*see* **Note2**).
2. Once the laser position has been detected, iterate adjustment of the laser power and *z* position at different *x,y* positions by moving the stage, until the laser produces a small and focused point using minimal laser power. Keep this *z* position and continue to adjust the *x* and *y* positions using the CellTools software by first repositioning the trapping laser to the center position of the field of view and subsequently calibrating the *x,y* direction and amplitude (*see* **Note 3**).
3. When the trapping laser has been focused and its position has been calibrated for the objective lenses that will be used, switch to your biological sample or perform additional testing/calibration (*see* **Note 4**).

3.2 Laser Alignment

It is possible to align a low-intensity laser beam in the visible light range with the trapping laser to detect its position while imaging. After successfully trapping a structure, the visible laser can be switched off during confocal imaging. Our system is not equipped with such a laser. However, a CCD camera can be used to detect the infrared trapping laser that gives a green reflection on glass surfaces such as the coverslip. To position the trapping laser, we collect a bright-field or a wide-field fluorescence image using the CellTools software prior to a trapping experiment. By displaying this image on the monitor of the computer that runs CellTools, we use it as a positional reference for confocal imaging. Then we switch to confocal mode (imaging settings should be adjusted before the experiment) and acquire confocal time series during a trapping experiment.

3.3 Trapping in Plant Samples

High-numerical-aperture objectives maximize the focusing of the trapping laser. We have obtained good results with a Zeiss 100×/N.A. 1.45 α-Plan Fluar and a Zeiss 63×/N.A. 1.4 Plan-Apochromat objective. At lower magnifications or numerical

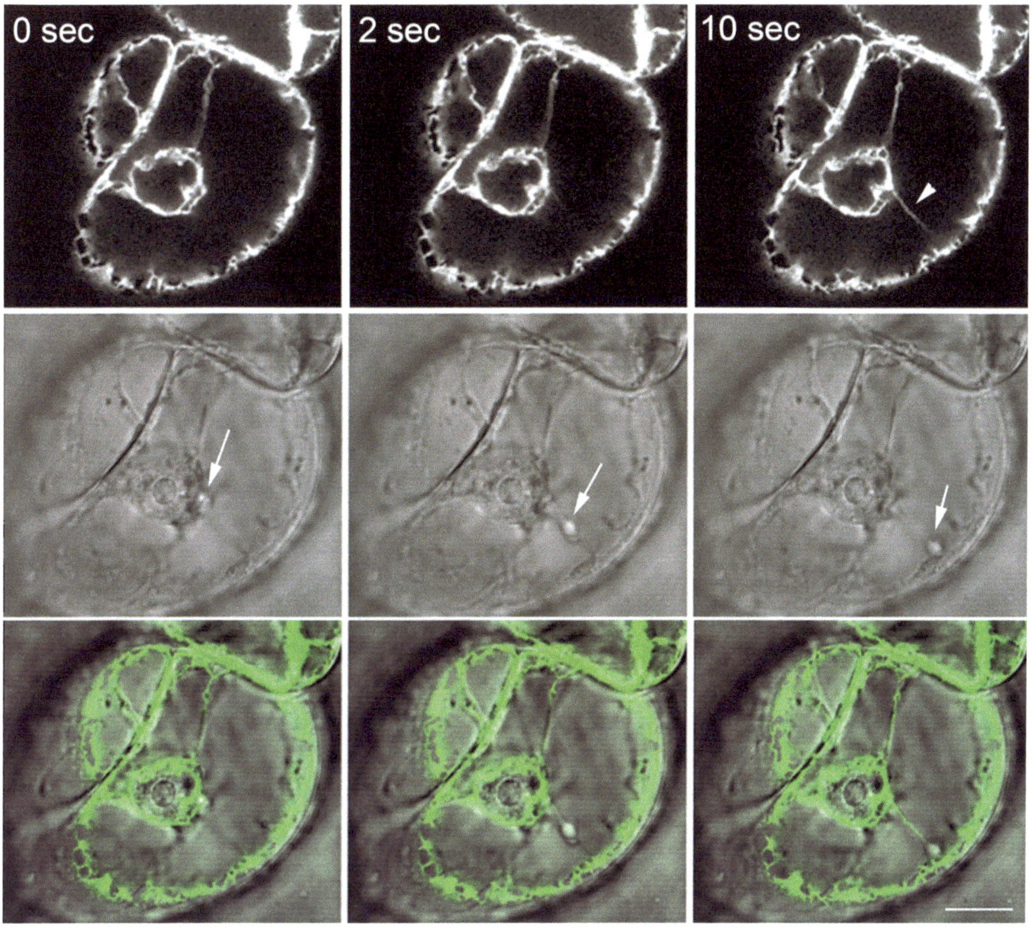

Fig. 2 Moving a trapped, unidentified organelle into the vacuolar lumen by relocating the position of the optical tweezers causes the formation of a cytoplasmic strand that acquires ER. The *arrows* indicate the position of the tweezers in the Nomarski images, and the *arrowhead* points to the ER that has appeared in the tweezers-formed strand after 10 s. The *first row* of images shows the confocal fluorescence images, the *second row* the Nomarski images, and the *third row* a merge of the upper two rows in which the fluorescence image is displayed in *green*. Bar: 10 μm

apertures, we were not able to trap structures within plant cells. During the trapping experiments, the IR-laser power was varied between 25 and 100 %, which corresponds to a laser intensity in the sample that lies in the 25–130 mW range. Several considerations for sample preparation are given below. As an example of a typical imaging sequence, we have included three images taken from a time series in which an unidentified organelle is trapped in a Tobacco BY-2 cell expressing the live cell ER marker HDEL-GFP. When the organelle is moved into the vacuolar lumen, a tweezers-formed cytoplasmic strand is produced that is surrounded by a tonoplast membrane. This particular experiment shows that an ER strand rapidly appears in the tweezers-formed strand (Fig. 2).

3.4 Preparation of Slides

Only use coverslips with a thickness of 1.0 (equivalent to 0.13–0.16 mm), prepare slides with plant material just before trapping, and mount the sample as close and flat to the coverslip as possible, since the strength of the tweezers drops rapidly with distance from the coverslip.

We use custom-made slides covered with gas-permeable Biofoil ([6]; for Biofoil ordering details, see the root hair growth chapter in this volume) and use VALAP (1:1:1 vaseline:lanolin:paraffin) to prevent slides from drying out by sticking a glass Pasteur pipette approximately 1 cm into solidified VALAP at room temperature. When the tip of the Pasteur pipette containing the VALAP is briefly heated in a flame, it melts. By tracing the outline of the coverslip with the tip of the pipette with the molten VALAP, the slide can be sealed.

We noticed that not only transparent refractive bodies but also light-absorbing structures such as chloroplasts or colored lipid droplets can be trapped. Trapping of these structures leads to rapid local heating due to light absorption. In cells that show signs of vitality loss or degradation, for example, by increased contrast and cytoplasmic clumping, it is more difficult or not possible at all to trap or move structures.

Interactions of organelles with the actin cytoskeleton, such as cytoplasmic streaming, produce forces in the same range as the optical tweezers and can interfere with trapping in some cell types. These interactions can be inhibited by actin depolymerization when it does not interfere with the research questions (*see*, e.g., ref. [4]).

Objects that are irregularly shaped are difficult to trap because they are often repulsed from the focal point of the trapping laser. This occurs, for example, with polystyrene beads that have been stored under non-sterile conditions and have acquired an irregular coating of bacteria. Elongate micron-sized objects can be trapped but tilt with their long axis in the *z*-axis of the focal point of the optical tweezers laser. Such reorientations in the trap focus should be considered in experiments.

Since the amount of trapping force correlates with the difference in refractive index of the trapped structure and the surrounding medium, the visibility of such a structure when using Nomarski optics often is an indication whether an object can be trapped or not, although we have successfully trapped objects that could not be detected at all using Nomarski imaging. An excessively high intensity of the trapping laser can cause accumulation of multiple refractile bodies in the focal point of the trap. In plant cells this is manifested by an accumulation of organelles around the position of the trap. To avoid this, the minimal power that is required for trapping the desired organelle should be determined.

4 Notes

1. It is important that the coverslip surface is flat. This can be achieved by fixing the coverslip with super glue or vacuum grease on a metal frame.
2. Safety considerations: a 3,000 mW infrared laser beam is both invisible and very dangerous; especially exposure to the eyes should be avoided at all times. Although most optical trapping systems require the use of infrared-blocking safety goggles, we have equipped our system with safety detectors that automatically switch off the infrared laser when the appropriate filters are not in place, when the trapping laser is not properly connected to the back port, or when the arm of the microscope is tilted backwards.
3. If desired, the output intensity of the trapping laser can be measured by placing a photon flux sensor, adjusted to sensitivity at 1,064 nm, in a flat position in the focal plane of an objective lens. The output intensity should be linear in a large output range of the laser (25–100 % output).
4. If the trap does not appear to trap anything in plant samples, a slide with fluorescent beads can be prepared to further test the tweezers. Freshly prepared fluorescent, polystyrene beads of 0.5–5 μm should be easy to trap. Freshly prepare a dilution of 1 drop polystyrene beads (Polysciences, carboxylate fluorescent microspheres) in 5 ml water. Prepare a slide with diluted beads. A fraction of the beads will attach to the charged glass surface and cannot be trapped.

References

1. Neumann KC, Block SM (2004) Optical trapping. Rev Sci Instrum 75:2787–2810
2. Grabski S, Xie XG, Holland JF et al (1994) Lipids trigger changes in the elasticity of the cytoskeleton in plant cells: a cell optical displacement assay for live cell measurements. J Cell Biol 126:713–726
3. Grabski S, Arnoys E, Busch B et al (1998) Regulation of actin tension in plant cells by kinases and phosphatases. Plant Physiol 116:279–290
4. Sparkes IA, Ketelaar T, de Ruijter NC et al (2010) Grab a Golgi: laser trapping of Golgi bodies reveals in vivo interactions with the endoplasmic reticulum. Traffic 10:567–571
5. van der Honing HS, de Ruijter NC, Emons AM et al (2010) Actin and myosin regulate cytoplasm stiffness in plant cells: a study using optical tweezers. New Phytol 185:90–102
6. Vos JW, Dogterom M, Emons AMC (2004) Microtubules become more dynamic but not shorter during preprophase band formation: a possible 'search-and-capture' mechanism for microtubule translocation. Cell Motil Cytoskeleton 57:246–258

Chapter 23

Heterologous Expression in Budding Yeast as a Tool for Studying the Plant Cell Morphogenesis Machinery

Fatima Cvrčková and Michal Hála

Abstract

The budding yeast (*Saccharomyces cerevisiae*) can serve as a unique experimental system for functional studies of heterologous genes, allowing not only complementation of readily available yeast mutations but also generation of overexpression phenotypes and in some cases also rescue of such phenotypes. Here we summarize the main considerations that have to be taken into account when using the yeast expression system for investigating the function of plant genes participating in cell morphogenesis; outline the strategies of experiment planning, yeast strain selection (or construction), and expression vector choice; and provide detailed protocols for yeast transformation, transformant selection, and phenotype evaluation.

Key words *Saccharomyces cerevisiae*, Heterologous gene expression, Inducible expression system, Complementation, Phenotype rescue, cDNA library screening

1 Introduction

After decades as a versatile model system of its own merit, the budding yeast *Saccharomyces cerevisiae* has become firmly established also in laboratories focusing on other eukaryotes, though mainly only as a "living test-tube" utilized for the detection of protein-protein interactions in various modifications of the yeast two-hybrid system, or Y2H [1]. While the Y2H itself is out of scope of this chapter, it helped to introduce routine work with yeast into many laboratories, making the use of additional yeast-based experimental strategies easier.

Evolutionarily conserved components of the cellular machineries—such as the cytoskeleton or the membrane-trafficking system—provide the most promising targets for yeast-based functional experiments. Several experimental strategies can utilize the knowledge gained in yeast to study plant cell morphogenesis; specific examples are provided in the references cited below.

Viktor Žárský and Fatima Cvrčková (eds.), *Plant Cell Morphogenesis: Methods and Protocols*, Methods in Molecular Biology, vol. 1080, DOI 10.1007/978-1-62703-643-6_23,

If yeast mutants defective in the process of interest, and preferentially exhibiting a conditional—usually temperature-sensitive, but sometimes also pharmacologically enhanced—growth defect, are available, such as the secretory pathway *sec* mutants [2], expression of a functionally equivalent plant gene may result in complementation of the mutant defect [3]. While even close sequence similarity between genes from different organisms does not guarantee successful complementation, failure to complement may be used as a starting point for identification of functionally important parts of the protein of interest by mutagenesis and/or construction of chimeric proteins and their subsequent testing for complementation in yeast [4].

While dead yeast is, on the first glance, not an encouraging starting point, even lethal loss-of-function mutations can be complemented, since a lethal allele can be maintained in heterozygous diploids. Introduction of a plasmid carrying a wild-type allele of the gene in question into such a heterozygote will allow subsequent isolation of haploids carrying the loss-of-function mutation and kept alive by the plasmid, which then cannot be lost. Complementation of the mutation by a second plasmid carrying the heterologous gene would enable plasmid loss, which can be easily followed if the first plasmid carries a marker affecting colony color [5]. Since use of this technique includes isolation of haploids by tetrad analysis, it goes beyond the scope of methods usually available in a plant laboratory, and we recommend collaboration with a specialized yeast laboratory.

In the absence of mutants suitable for complementation, novel phenotypes may be elicited by heterologous expression of proteins interfering with the intrinsic cellular machinery. If such phenotypes result in a growth defect, they can be rescued by co-expression of a second heterologous protein interacting with the first one [6].

Complementation, phenotype induction, and rescue of induced phenotypes can be used either for testing already available candidate plant clones preselected, e.g., on the basis of homology with their yeast counterparts (i.e., in test mode), or for screening cDNA libraries to isolate novel cDNAs (screen mode).

Especially in screening experiments, additional verification of the identified clones by plasmid rescue from yeast and repeated transformation of the initial yeast strain is crucial. An outline of a general experimental strategy is shown in Fig. 1.

2 Materials and Equipment

Besides standard equipment for molecular biology (pipettes, incubators and shakers for microbial cultures, an autoclave, a laminar flow cabinet, a microcentrifuge, a vortex, a low-velocity centrifuge for larger volumes, a spectrophotometer, a refrigerator, freezers,

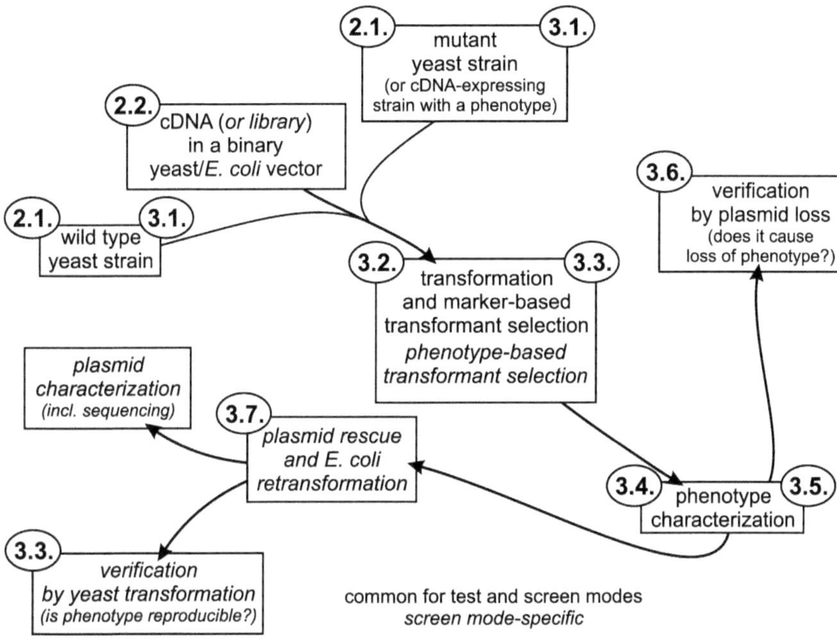

Fig. 1 Outline of common experimental strategies of yeast-based functional studies. Steps shown in *plain text* are common to test- and screen-mode experiments; those shown in *italics* are obligatory for screen experiments but commonly omitted in test ones. *Numbers* denote relevant sections of the present chapter

equipment for DNA cloning), a stand-mounted digital camera, a microscope with a 40× to 60× dry objective (preferentially equipped with phase contrast), and a hemocytometer (counting chamber) will be required. A replica-plating tool with sterile velvet squares is recommended but not essential (replica plating can be replaced by more laborious transfer of individual colonies using sterile toothpicks).

Use an incubator set at 28–30 °C for routine agar cultures of yeast. For some temperature-sensitive strains, the permissive temperature may have to be set lower (25 °C) to ensure optimal growth. For eliciting temperature-sensitive phenotypes, follow strain documentation; 37 °C is often used as restrictive temperature. Use the same temperatures for liquid cultures (*see* **Note 1**).

Prepare all solutions using deionized or distilled water and analytical or "molecular biology grade" chemicals, sterilize them by autoclaving unless stated otherwise, and handle aseptically.

2.1 Yeast Strains

Genetically defined yeast strains with a variety of mutations, including those resulting from systematic projects of deleting every open reading frame of the yeast genome, can be obtained from several public repositories whose catalogs can be accessed through the "Methods and Reagents" section of The *Saccharomyces* Genome Database (SGD, http://www.yeastgenome.org), which also holds a comprehensive collection of experimental data and literature [7].

Viable recessive mutations, in particular those causing temperature-sensitive growth defects, are obvious candidates for complementation studies. We shall use complementation of a temperature-sensitive mutation as an example throughout, albeit also mutations causing, e.g., increased sensitivity to various chemicals may be complemented as well (adjust media composition appropriately for such cases).

Use preferentially haploid strains (often referred to as mating type *MATa* or *MAT*α) and obtain also isogenic (parental) strains for control purposes together with the mutants. Pay attention to the metabolic requirements of your yeast strains. Most laboratory strains are polyauxotrophic. Make sure that the strains are compatible with the selection marker(s) of the vectors you will be using and with your inducible gene expression system, if you intend to use one (*see* **Note 2**).

2.2 Expression Vectors and Libraries

For heterologous expression, the gene of interest must be cloned into a suitable binary (shuttle) *Escherichia coli*-yeast vector that allows DNA amplification in bacteria, as well as maintenance, selection, and expression in *S. cerevisiae* (*see* **Note 3**). Many such vectors can be obtained from public repositories such as the ATCC (http://www.atcc.org), where a large collection of yeast expression vectors [8] is available (order No. 87669). Addgene (http://www.addgene.org) provides another very large collection of plasmids [9] designed for use with the Gateway cloning system (Invitrogen).

Complementation of a host yeast strains' auxotrophic mutation is usually used to select for yeast transformants. Among the commonly used markers, the *URA3* gene complementing the *ura3* uracil auxotrophy provides an extra advantage of possible selection not only for plasmid acquisition (leading to uracil-independent growth) but also for plasmid loss, resulting in resistance towards 5-fluoroorotic acid, or 5-FOA [10].

Both constitutive and inducible promoters can be used for heterologous protein expression in yeast. Among the later, the most common is the galactose-inducible "two-headed" promoter of the *GAL1/GAL10* genes and its derivatives. If preparing constructs based on existing cDNAs, avoid introducing extraneous sequences into the 5′UTR. If possible, locate the start of transcription at the beginning of the original plant 5′UTR, or, alternatively, maintain as much of the yeast-derived vector sequence upstream of the ATG as possible.

For screening experiments, a good plant cDNA library in a binary vector fulfilling the above requirements is essential. Besides constructing the library de novo (which is out of scope of this chapter), some libraries, such as the "Davis library" of *Arabidopsis thaliana* seedling aboveground organ cDNAs in the Lambda YES vector featuring the *GAL1* promoter and the *URA3* markers [11], may be obtained on request by laboratories that produced them. The "Lacroute library" of young *A. thaliana* seedling-derived cDNAs under the control of the constitutive *PGK1* promoter in the pFL61 vector including the *URA3* selection marker [12] can be acquired from the ATCC (No. 77500).

2.3 Yeast Cultivation, Storage, and Manipulation

1. YEPD medium: 10 g/l yeast extract, 20 g/l peptone, 20 g/l glucose. For strains with an adenine requirement, add 40 mg/l adenine to improve growth (*see* **Note 4**). Autoclave the glucose solution separately (*see* **Note 5**).
2. YEPD plates: 10 g/l yeast extract, 20 g/l peptone, 20 g/l agar, 20 g/l glucose. For strains with an adenine requirement, add 40 mg/l adenine to improve growth (*see* **Note 4**). Autoclave the glucose solution separately and pour plates (*see* **Note 5**).
3. YEPGal plates: 10 g/l yeast extract, 20 g/l peptone, 20 g/l agar, 20 g/l galactose. Autoclave the galactose solution separately and pour plates (*see* **Note 5**).
4. –URA plates: 7.3 g/l yeast nitrogen base without amino acids, 20 g/l agar, 10 g/l casamino acids, 50 mg/l tyrosine, 50 mg/l adenine, 20 g/l glucose (replace with galactose if required for galactose-inducible expression under selective conditions). Autoclave the sugar solution separately and pour plates (*see* **Note 5**).
5. +URA plates: prepare like –URA plates but add uracil to final concentration 50 mg/l from an autoclaved or filter-sterilized 100× concentrated stock prior to pouring plates.
6. 10× Dropout solutions: 20 mg/100 ml L-adenine hemisulfate salt, 20 mg/100 ml L-arginine HCl, 20 mg/100 ml L-histidine HCl monohydrate, 30 mg/100 ml L-isoleucine, 100 mg/100 ml L-leucine, 30 mg/100 ml L-lysine HCl, 20 mg/100 ml L-methionine, 50 mg/100 ml L-phenylalanine, 200 mg/100 ml L-threonine, 20 mg/100 ml L-tryptophan, 30 mg/100 ml L-tyrosine, 20 mg/100 ml L-uracil, 150 mg/100 ml L-valine. Leave out the appropriate amino acid(s) or nucleotides to obtain selective media (e.g., leaving out tryptophan produces a –TRP dropout for use in media selecting for clones carrying the *TRP1* marker in a *trp1* background); a complete solution without leaving out any components should be used for controls.
7. SD dropout medium: 6.7 g yeast nitrogen base without amino acids, 100 ml 10× dropout solution (see above), 850 ml water. Adjust the pH to 5.8 if necessary and autoclave. Allow to cool to 55 °C and then add 50 ml of filter-sterilized 40 % glucose, or other appropriate carbon source to a final concentration of 2 % (*see* **Note 6**).

8. SD dropout plates: 6.7 g yeast nitrogen base without amino acids, 20 g agar, 100 ml 10× dropout solution (see above), 850 ml water. Adjust the pH to 5.8 if necessary and autoclave. Allow to cool to 55 °C and then add 50 ml of filter-sterilized 40 % glucose (or other appropriate carbon source to final concentration 2 %, *see* **Note 6**) and pour plates.
9. 5-FOA plates: solution A—14 g/l yeast nitrogen base without amino acids, 100 mg/l uracil, 110 mg/l adenine, 40 g/l glucose, 20 ml/l solution B, sterilize by filtration; solution B—400 mg/100 ml tryptophan, 600 mg/100 ml leucine, 100 mg/100 ml histidine, 100 mg/100 ml methionine, sterilize by filtration (*see* **Note 7**); solution C—40 g/l agar, autoclaved. For twenty 35 mm Petri dishes, which is a reasonable amount, dissolve 70 mg of 5-FOA (e.g., Sigma F5013) in 35 ml of solution A. It is quite difficult to dissolve; heating the solution to approx. 40 degrees is recommendable. Mix with 35 ml of melted agar (solution C) and pour approx. 3 ml per 35 mm dish (use larger dishes if treating many strains at the same time).
10. 15 % glycerol, autoclaved.
11. Sterile 1.5 ml screw-top Eppendorf vials or cryogenic storage tubes.
12. Sterile wooden toothpicks (*see* **Note 8**), glass rods for plating, and a bacteriological wire loop.
13. Sterile water.
14. Sterile plastic Petri dishes, pipette tips, Eppendorf vials, glass bacteriological tubes with stoppers, and Erlenmeyer flasks.

2.4 Yeast Transformation

1. Sterile water.
2. 2×YEPD medium: 20 g/l yeast extract, 40 g/l peptone, 40 g/l glucose, 80 mg/l adenine. Autoclave the glucose solution separately (*see* **Note 5**).
3. Single-stranded carrier DNA: can be obtained from different sources as ready-to-use solution with concentration ranging from 2 mg/ml to 10 mg/ml (e.g., Sigma D7656). Store aliquots in −20 °C. Immediately before transformation or while cells are growing, thaw 1 ml carrier DNA, incubate for 5 min at 99 °C (or boiling water bath), then place immediately on ice. Repeat boiling and cooling twice for high-efficiency transformation (for rapid transformation, once is enough).
4. 10× TE: 100 mM Tris, 10 mM EDTA, pH 7.5. Use 0.5–1 M Tris stock solution with pH adjusted to pH 7.5 by HCl to prepare this buffer.
5. LiAc/TE mix: 1.1 ml/10 ml 1 M lithium acetate solution, 1.1 ml/10 ml 10× TE, 7.8 ml/10 ml water, sterilize by filtration.

6. PEG/LiAc mix: 1.5 ml/15 ml 1 M lithium acetate solution, 1.5 ml/15 ml 10× TE, 12 ml/15 ml 50 % PEG4000, sterilize by filtration.
7. Dimethyl sulfoxide (DMSO). No need to sterilize.
8. 0.9 % NaCl, autoclaved.
9. Sterile pipette tips, Eppendorf vials, centrifuge tubes, Erlenmeyer flasks, and materials and media for pre- and post-transformation yeast cultivation (*see* Subheading 2.3).

2.5 Plasmid Rescue

1. GTE buffer: 50 mM glucose, 25 mM Tris-Cl, 10 mM EDTA, pH 8.0.
2. Acid-washed glass beads approx. 500–600 μm in diameter (e.g., Sigma G-8772).
3. NaOH/SDS: 200 mM NaOH, 1 % SDS. Do not sterilize.
4. KAc/HAc: 3 M potassium acetate, 2 M acetic acid.
5. Isopropanol. No need to sterilize.
6. 75 % Ethanol. No need to sterilize.
7. 96 % Ethanol. No need to sterilize.
8. Sterile water.
9. Sterile pipette tips, Eppendorf vials, centrifuge tubes, Erlenmeyer flasks, and materials and media for yeast cultivation (*see* Subheading 2.3).
10. Competent *E. coli* cells and materials for *E. coli* transformation (*see* **Note 9**).

3 Methods

Follow local biohazard rules and regulations when working with transgenic yeasts. Laboratory strains of *S. cerevisiae* are usually considered low risk, but some paperwork may be needed if introducing yeast into the lab the first time.

3.1 Yeast Strain Handling and Maintenance

Handle yeast using standard microbiology techniques, similar to those well established for *E. coli*.

1. For long-term storage, grow a fresh agar culture from a reliable source (*see* **Note 10**). Use YEPD plates for plasmid-free strains and selective media for strains containing plasmids. Scrape off a couple of colonies using the blunt end of a toothpick (resulting in a mix of several single-colony isolates), resuspend in 1 ml of 15 % glycerol in a screw-top vial, and store in −80 °C. Such stocks last for several years, but renew them from a fresh verified culture once you notice falling viability.

2. To revive a strain from storage, scratch the surface of the still frozen stock (kept in a freezer block or on ice) with the blunt end of a sterile toothpick and streak on a fresh YEPD plate to obtain single colonies (use a repeatedly fire-sterilized wire loop or additional two new toothpicks to ensure colony separation). Alternatively, keep the stock in liquid nitrogen and scratch its surface with a yellow pipette tip; while this is more hassle, it is less destructive for the stock.
3. Use a refrigerator for short-term storage of yeast cultures on plates (up to 2 weeks).

3.2 High-Efficiency Yeast Transformation for Library Screening

This protocol is based on the standard library transformation protocol from Dualsystems Biotech (http://www.dualsystems.com); other transformation protocols for two hybrid library screening may work as well.

1. Inoculate yeast from the master plate into 10 ml YEPD and grow for 8 h at 28 °C with shaking (25 °C should be used for temperature-sensitive strains in this and all following steps).
2. Inoculate 100 ml YEPD with the entire 10 ml culture from **step 1** and grow overnight at 28 °C with shaking.
3. Take a 1 ml aliquot of the culture, centrifuge at 2,500 × *g* for 5 min, and resuspend the pellet in 1 ml of water. Measure OD_{550} against a water blank.
4. Calculate the amount of culture needed for 30 OD units. Aliquot this amount of the overnight culture into 50 ml Falcon tubes and spin down at 700 × *g* for 5 min.
5. Resuspend the pellet in 200 ml 2×YEPD (pre-warmed to 28 °C) in a 1 l Erlenmeyer flask and remove a 1 ml aliquot. Centrifuge this 1 ml aliquot at 2,500 × *g* for 5 min, discard the supernatant, and resuspend the pellet in water. Measure OD_{550} against a water blank; the value should be around 0.15 (30 OD units in 200 ml correspond to OD = 0.15).
6. Grow the cells at 28 °C with vigorous shaking to $OD_{550} = 0.6$ (two cell divisions; typically this takes 3–5 h but may vary depending on the yeast strain used). Boil the single-stranded carrier DNA in the meantime.
7. After reaching the required OD, divide the 200 ml culture into four 50 ml Falcon tubes and centrifuge at 700 × *g* for 5 min.
8. Resuspend each pellet in 30 ml of sterile water by vortexing. Centrifuge at 700 × *g* for 5 min.
9. Remove the supernatant, resuspend each pellet in 1 ml LiAc/TE mix, and transfer to an Eppendorf tube. Centrifuge at 700 × *g* for 5 min.

10. Remove the supernatant and resuspend each pellet in 600 μl of LiAc/TE mix.
11. Set up four 50 ml Falcon tubes, each of them containing, in this order: cDNA library 7 μg, single-stranded carrier DNA solution 200 μg (in a volume to up to 100 μl), cell suspension from the previous step 600 μl, PEG/LiAc mix 2.5 ml.
12. Vortex for 1 min to thoroughly mix all components and incubate at 28 °C for 45 min, mixing briefly every 15 min.
13. In the meantime, rinse one of the Falcon tubes containing the rest of cells from **step 10** with 100 μl of sterile water and plate this suspension on a clearly labeled selection plate (same as in **step 18**) as DNA free control.
14. Add 160 μl DMSO to each tube and mix immediately by shaking. Incubate at 42 °C for 20 min.
15. Pellet cells at 700 × *g* for 5 min. Resuspend each pellet in 3 ml of 2×YEPD and pool all cell suspensions. Let the cells recover at 28 °C for 90 min with shaking.
16. Pellet the cells at 700 × *g* for 5 min and resuspend the pellet in 4.5 ml of 0.9 % NaCl.
17. Plate onto the appropriate selection plates (depends on the plasmid used in the library; if using an inducible expression system, make sure that the plates contain the right inducer, e.g., galactose). Use 300 μl of resuspended cells per 15 cm plate or 200 μl per 9 cm plate. Take aside a 45 μl aliquot (1/100 of total volume), dilute it into 450 μl of 0.9 % NaCl (resulting in 10^{-2} dilution), repeat 4–5 times to obtain serial dilutions with the factor of 10, and plate three to four highest dilutions on a separate clearly labeled control plate for estimating transformation efficiency. This protocol can give about 10^6–10^7 transformants in total, depending on the strain and library.
18. If complementing a temperature-sensitive defect, keep all plates for a couple of hours (at most overnight) at the permissive temperature and transfer the library screen to the restrictive temperature next morning. Colonies may appear gradually in the course of a week. Leave both controls at the permissive temperature; the DNA-free control should produce no colonies, while the efficiency control is expected to show abundant growth. Count colonies on the later and calculate transformation efficiency.
19. If the growth defect of the mutant to be complemented is not temperature sensitive but, e.g., elicited by a pharmacological treatment, include the appropriate chemicals into the selective media.

3.3 Rapid Yeast Transformation

This procedure is faster and simpler than Subheading 3.2 but gives significantly lower yields—use only for test-mode experiments if enough plasmid DNA is available. The following recipe is for one transformation; scale up as necessary and divide samples from **step 4** onwards.

1. Grow cells in 3 ml YEPD overnight at 28 °C with shaking.
2. Centrifuge at 2,500 × *g* for 7 min (*see* **Note 11**) and remove supernatant.
3. Wash with 3 ml of sterile water and spin down at 2,500 × *g* for 7 min; remove supernatant.
4. If doing multiple transformations from a scaled-up culture, divide the suspension into aliquots. To each aliquot, add 100 μg denatured single-stranded carrier DNA and 1–2 μg plasmid DNA in minimal volume. It is recommendable to include also mock transformation without plasmid DNA as a control for absence of contamination.
5. Vortex the mixture briefly and add 500 μl PEG/LiAc mix. Incubate at room temperature with gentle mixing (300 rpm on a vortex shaker) for 15 min.
6. Incubate at 42 °C for 10 min.
7. Add 96 % ethanol to a final concentration of 10 % and continue incubation for another 5 min.
8. Pellet cells at 2,500 × *g*, 5 min. Wash cells with sterile water and centrifuge again as before.
9. Resuspend cells in 200 μl of sterile water and plate onto appropriate selective agar medium. Select for the presence of plasmid only, i.e., incubate the plates at a permissive temperature afterwards if working with temperature-sensitive strains.

3.4 Evaluating Effects of Heterologous Protein Expression: Experiment Design

When studying the effects of heterologous gene expression in yeast, always include appropriate controls.

1. As a minimum, compare two identically treated cultures, ideally with several biological replicates such as independent transformed clones: (a) a transformant carrying and expressing your gene of interest (i.e., a complemented mutant or a wild-type strain with an overexpression phenotype) and (b) an empty vector transformant, carrying the plasmid used for overexpression but without your gene of interest. In case of complementing a mutation causing a conditional growth defect, add (c) also an isogenic wild-type strain, preferentially also transformed with the empty vector (*see* **Note 12**).
2. For additional verification, you may also include (d) derivatives of the original transformant devoid of the plasmid (*see* Subheading 3.6) together with (e) the original

pre-transformation strain. For rechecking a plasmid isolated from a library screen, compare the minimum set of strains (a) and (b) with (f) a retransformed host strain harboring the plasmid isolated from the primary transformant (*see* Subheading 3.7).

3. For conditional defects (such as temperature-sensitive growth) grow all cultures in parallel at permissive and restrictive conditions. Similarly, when using an inducible expression system, compare inducing and non-inducing conditions.
4. Evaluate first the effects of heterologous protein expression on gross growth of yeast cultures on agar media (*see* Subheading 3.5). Depending on the result, progress towards other methods such as more detailed quantitative growth evaluation or microscopic studies. Protocols for visualization of various cell components in the budding yeast are available, e.g., in ref. [13].

3.5 Drop Test for Evaluating Effects of Heterologous Protein Expression on Yeast Growth

While streaking cultures onto appropriate plates (as for colony isolation) may in some cases provide initial insight into the effects on growth, more accurate information is obtained using a drop test:

1. Grow fresh plate cultures of the strains to be tested on media selective for plasmid retention (–URA or SD dropout) but otherwise as gentle as possible (i.e., permissive temperature for temperature-sensitive strains, non-inducing conditions for inducible gene expression).
2. If (and only if) your strains grow very slowly and yield little biomass under selective conditions, transfer a fairly massive inoculum from the selective plate onto YEPD agar using the blunt end of a toothpick and grow for a day or two.
3. For each strain, harvest a pinhead-size amount of biomass using the blunt end of a toothpick and resuspend in 1 ml of sterile water in an Eppendorf tube. Vortex well. Take a sample of this suspension and measure its optical density (OD_{550}) after 20× dilution. Adjust all samples to the same OD using sterile water.
4. Prepare a dilution series in sterile water with a 30× factor (i.e., 30×, 900×, and 27,000× dilution) for each strain.
5. Spot 10 μl droplets in an array of strains vs. concentrations onto nonselective plates (*see* **Note 13**), incubate at appropriate conditions until colonies develop, and document photographically. The number and kind of plates depends on experimental setup—e.g., for complementation of a temperature-sensitive defect by a galactose-regulated construct, plant two YEPD and two YEPGal plates and incubate one plate of each kind at the permissive and the other at the restrictive temperature.

3.6 Verification of Complementants by Plasmid Loss

To make sure that the observed phenotype(s) are caused by the plasmid carrying the heterologous gene rather than a modifier mutation (*see* **Note 10**), verify that your transformant reverts to the phenotype of the original host strain after evicting the plasmid.

1. Grow a fresh YEPD plate culture of the transformed yeast strain at permissive temperature until nice colonies develop (approx. 2–3 days). Alternatively grow a 2 ml culture to stationary phase (about 2 days with shaking) in liquid YEPD.
2. If the *URA3* marker was used for plasmid selection, use the blunt end of a sterile toothpick to transfer a pinhead-sized amount of biomass from the YEPD plate from **step 1** to a 5-FOA plate and spread it to the size of a thumbnail. Alternatively, centrifuge 1 ml of the liquid culture from **step 1** in an Eppendorf tube, resuspend in 1 ml of sterile water, and plate 0.1 ml on a 5-FOA plate. Incubate at a permissive temperature. Colonies formed by plasmid-free cells will develop in a couple of days. Pick them with the sharp end of a toothpick and re-streak on a YEPD plate to obtain more material. Verify that they are *ura-* by parallel streaking to +URA and −URA plates.
3. If you used any other marker, scratch the biomass off the YEPD plate from **step 1** or wash it off with sterile water to obtain a cell suspension. Dilute this suspension, or the liquid culture from **step 1**, into 1 ml of water to a slightly cloudy appearance and determine cell count using a hemocytometer (the counting step may be skipped once you acquire experience). Brief sonication may help to break cell clumps.
4. Prepare serial dilutions in sterile water to achieve 500–1,000 and 100–250 of cells per 0.2 ml. Spread 0.1 and 0.2 ml aliquots of each dilution on YEPD plates and incubate at permissive temperature till colonies develop.
5. Select plates with well-developed, mutually separated colonies and replica-plate those in parallel on appropriate complete SD and "minus" dropout plates. Alternatively, transfer about 100 well-separated colonies in parallel to complete SD and "minus" plates using sterile toothpicks (*see* **Note 14**). Pick colonies that only grow on the complete medium and re-streak them on a YEPD plate to obtain more material.
6. Compare the relevant growth-related and morphological phenotype(s) of the resulting plasmid-free segregants with the original transformant and the non-transformed host strain.

3.7 Plasmid Rescue from Yeast

1. Grow cells in 3 ml of selective media (e.g., SD-Ura for an *URA3*-containing plasmid) overnight at 28 °C with shaking.
2. Pellet cells at maximum speed for 2 min (*see* **Note 11**). Discard supernatant.
3. Add 180 μl of GTE buffer, resuspend, and add 100 μl of acid-washed glass beads. Vortex at full speed for 5 min.

4. Add 360 μl of the NaOH/SDS solution, mix by inverting several times, and incubate at room temperature for 5 min.
5. Add 270 μl of the KAc/HAc solution, mix by inverting several times, and centrifuge at full speed for 10 min.
6. Transfer supernatant carefully to a fresh tube, add 560 μl of isopropanol, and mix. Centrifuge at full speed for 5 min and remove supernatant.
7. Wash the pellet with 75 % ethanol and centrifuge as above. Remove supernatant carefully.
8. Dissolve the pellet in 20 μl of sterile water and use an aliquot for transformation of competent *E. coli* cells (*see* **Note 9**).

4 Notes

1. A dedicated incubator is better than one shared with other cultures, although 37 °C cultures can be grown together with *E. coli* and low temperature ones with *Agrobacterium*. However, beware that yeast cultures produce CO_2 that might affect other contents of the incubator and that they also attract free-living fruit flies that may become a nuisance. Depending on incubator type, you will probably have to seal the plates with Parafilm to prevent drying out during prolonged cultivation.
2. Commonly used selection markers are auxotrophies (e.g., the *trp1*, *leu2*, *his3*, *ade2*, *ura3* mutations conferring dependence on exogenous tryptophan, leucine, histidine, adenine, or uracil, respectively).The standard yeast nomenclature labels recessive mutations in lowercase italics and dominant alleles in uppercase italics; distinct genes of the same or related function are numbered (i.e., *ADE2* and *ADE3* are two different genes), and distinct alleles of the same locus are denoted by numbers after a hyphen (i.e., *leu2-3,112* is a recessive allele of the *LEU2* gene carrying two different point mutations). Beware of multiple mutations in the same pathway (i.e., a double *ade2 ade3* mutant would have to be complemented by two genes to regain adenine independence) and of undocumented mutations introduced by crossing in the strains' history. If using a galactose-inducible expression system, make sure your strain does not carry *gal* mutations preventing utilization of galactose.
3. DNA constructs may be maintained in yeasts either integrated into the yeast chromosome *via* homologous recombination (integrative vectors) or as episomes of a varying copy number (replicative plasmid vectors). Avoid integrative vectors if you plan to rescue the construct back from yeasts. Replicative vectors usually contain the replication origin of the yeast endogenous 2 μm plasmid, or *2 μ ORI* (leading to high-copy amplification,

which makes plasmid rescue easier), or a chromosome-derived autonomously replicating sequence (*ARS*) that is copied exactly once per cell cycle, resulting in a single copy plasmid easily lost unless it also contains a centromere (*CEN*).

4. Strains carrying the *ade1* or *ade2* mutations, and no additional mutations in the adenine biosynthesis pathway, accumulate a red metabolic intermediate on media with a limited amount of adenine (provided by the yeast extract without an extra supplement). While this may serve as a visible marker for strain verification, red biomass color is a sign of futile attempts to synthesize adenine and a cause of concern for the culture's well-being. Adding extra adenine to the media is important especially when transforming adenine auxotrophs and aiming for high efficiency (i.e., in library screens).
5. Mix all components of the medium except the sugar in half of the volume of water in a sufficiently large autoclavable bottle with an autoclavable stirring bar until any clumps disappear (agar will not dissolve at room temperature). Dissolve the sugar in another bottle in the remaining half of the water. Autoclave both components (leaving the stirring bar in) and mix them together after the content has cooled below 60 °C. For agar-containing media, pour plates (approx. 40 from 1 l if using 9 cm Petri dishes) immediately after mixing. Keep liquid media in reasonably sized aliquots to avoid prolonged storage of opened bottles.
6. A 40 % solution of sufficiently pure (p.a.) glucose may be autoclaved without adverse effects. Check with a small amount of your sugar stock—if the solution does not turn yellow during autoclaving, you can use this method of sterilization.
7. If your strain has an additional auxotrophic requirement, add the appropriate amino acid at 100 mg/100 ml of solution B.
8. Flat toothpicks are better than round ones. Sterilize them vertically in a small glass beaker covered by an aluminum foil. Random orientation ensures that you can choose between the flat end for transferring larger amount of yeasts or spreading them over a large surface and the pointed end for picking small colonies.
9. Use the bacterial strain and transformation protocol you would routinely use for cloning. In our hands, 0.5–1 μl work well for electroporation in 50 μl aliquots of *E. coli* cells.
10. Yeast cultures can undergo epigenetic or genetic changes that may lead to phenotypic alterations upon prolonged cultivation (this is sometimes referred to as "picking up modifiers"). They also tend to lose plasmids on nonselective media and lose viability on selective ones upon storage. Retest the strains' phenotype (including microscopic appearance) periodically

and use only fresh plate cultures (preferentially less than a week old) to start experiments. Avoid using cultures exhibiting obvious heterogeneity in colony appearance. In liquid cultures, beware of unusual smell or too fast growth, as both may be signs of bacterial contamination. Normal doubling time of a haploid, polyauxotrophic laboratory yeast strain in liquid YEPD at 28 °C is around 2 h.

11. It is possible to use 1.5 ml Eppendorf tubes at this and following steps, adding an extra round of centrifugation: spin down half of the sample, remove supernatant, add a second half of the sample to the pellet, and spin again.
12. Empty vector transformation introduces the vector's selective marker, eliminating thus effects of differing auxotrophic requirements of the original (non-transformed) strain and the transformant carrying the gene to be expressed.
13. Consider using larger, square Petri dishes when evaluating many strains. In this case, you also can use mutichannel pipettes and sterile microtiter plate for serial dilutions and plating. Make sure the layout of drops is identical for all dishes.
14. Use a fresh toothpick for each colony. To maintain the same layout on both plates, draw two copies of a template on two disposable Petri dish lids using a permanent marker and lay these templates underneath the dishes when transferring the colonies.

Acknowledgments

This work has been supported by the MSM0021620858 project. We thank Marta Čadyová for expert technical assistance and members of the A. Ragnini laboratory (University of Vienna, Austria) for the rapid yeast transformation protocol.

References

1. Fields S (2009) Interactive learning: lessons from two hybrids over two decades. Proteomics 9:5209–5213
2. Novick P, Field C, Schekman R (1980) Identification of 23 complementation groups required for post-translational events in the yeast secretory pathway. Cell 21: 205–215
3. Žárský V, Cvrčková F, Bischoff F et al (1997) At-GDI1 from *Arabidopsis thaliana* encodes a rab-specific GDI that complements the *sec19* mutation of *Saccharomyces cerevisiae*. FEBS Lett 403:303–308
4. Hála M, Eliáš M, Žárský V (2005) A specific feature of the angiosperm Rab escort protein (REP) and evolution of the REP/GDI superfamily. J Mol Biol 348:1299–1313
5. Kranz JE, Holm C (1990) Cloning by function: an alternative approach for identifying yeast homologs of genes from other organisms. Proc Natl Acad Sci USA 87: 6629–6633
6. Ueda T, Matsuda N, Anai T et al (1996) An *Arabidopsis* gene isolated by a novel method for detecting genetic interaction in yeast encodes the GDP dissociation inhibitor of Ara4 GTPase. Plant Cell 8:2079–2091
7. Cherry JM, Hong EL, Amundsen C et al (2012) *Saccharomyces* genome database: the genomics resource of budding yeast. Nucleic Acids Res 40:D700–D705

8. Mumberg D, Muller R, Funk M (1995) Yeast vectors for the controlled expression of heterologous proteins in different genetic backgrounds. Gene 156:119–122
9. Albert S, Gitter AD, Lindquist S (2007) A suite of gateway cloning vectors for high-throughput genetic analysis in *Saccharomyces cerevisiae*. Yeast 24:913–919
10. Boeke JD, Trueheart J, Natsoulis G et al (1987) 5-Fluoroorotic acid as a selective agent in yeast molecular genetics. Methods Enzymol 154:164–175
11. Elledge SJ, Muligan JT, Ramer SW et al (1991) Lambda YES: a multifunctional cDNA expression vector for the isolation of genes by complementation of yeast and *Escherichia coli* mutations. Proc Natl Acad Sci USA 88: 1731–1735
12. Minet M, Dufour ME, Lacroute F (1992) Complementation of *Saccharomyces cerevisiae* auxotrophic mutants by *Arabidopsis thaliana* cDNAs. Plant J 2:417–422
13. Hašek J (2006) Yeast fluorescence microscopy. Methods Mol Biol 313:85–96

INDEX

A

Acrylic resin. *See* also LR White and Lowicryl 5, 44, 45, 47, 149, 186
Actin. *See* also Microfilaments 87, 88, 92–93, 95, 260, 264

Antibody
monoclonal (MAB) 31, 41–43, 45, 47–49, 51, 185, 191
Arabidopsis thaliana 101, 106–108, 112–118, 121, 172, 184, 195–197, 217, 222, 231, 252, 255, 270, 271

B

Biofoil 196–198, 264
Biolistics. *See* Particle bombardement

C

Calibration 68, 69, 71–73, 76, 137, 142, 204, 262
Callose 26–28, 31–32, 35
Camellia japonica 237–248
Cell
counting 223–227, 278
culture 149–151, 172, 176, 185, 215–226
density 220, 222–227
growth 87, 100, 195–198, 201–212, 231
line
BY-2 .... 148–151, 154, 173, 174, 216–218, 220–223, 225–228, 260, 263
VBI-0 .... 216–218, 220–223, 225–228
phenotyping 219
polarity 227
shape 69, 73–75, 121–133, 224–225
suspension 215, 217, 220–226, 275, 278
viability 142, 222, 223, 226
wall 25–38, 41–52, 63, 73, 87, 88, 107, 114, 116, 123, 130, 135, 136, 140, 143, 144, 154, 172
Cellulose 34, 41, 51, 87
Coated pit 148, 174
Coating, slides 8, 14, 20, 44, 47, 142
Complementation 78, 202, 203, 205–206, 209, 211, 268, 270, 275–279
Counting frame 68, 73, 75
Critical point drying 175, 177–178
Cryotome 6, 16–18, 35
Cytoskeleton 87–95, 171–180, 264

D

Development 28, 29, 53, 58, 68, 101, 108, 116, 201

E

Elasticity 135, 136, 143, 144
Embedding
paraffin 3–6, 11–15, 36, 250, 251, 254–256
resin 5, 19, 41–52, 153, 154, 159, 183–192
Enzyme activity detection (enzyme histochemistry) 5, 29
Epidermis/epidermal 28, 69, 73, 74, 77–85, 106–110, 118, 133, 137, 141, 143, 155, 184, 195
Epoxy gel or resin 100–109, 189
Experiment design 70, 276–277

F

Fixation 1–3, 6–7, 9–11, 16, 19, 20, 26–28, 36, 37, 43–46, 50, 52, 147, 148, 154, 160, 172, 175–177, 179, 184, 186, 190, 191, 250–255
Fluorescent protein 42, 48, 54, 77–85, 87–95, 116, 184, 202, 250
Forward genetics 53–65
Freeze substitution (FS) 147–149, 151–155
Freezing 5, 6, 16, 22, 147–155, 160, 186, 256

G

Gene
expression
heterologous 88, 267–281
inducible 270, 271, 275, 277, 279
pattern of 61, 63, 111, 129
transient 77–85, 201–203, 205
knock-down (*see* also RNA interference) 231–235
Gold 44, 49, 151, 153, 161, 162, 164, 168, 172, 180, 183–192
GUS staining 54–58, 63, 112

H

Hemicellulose 26–28, 41, 43
Histochemistry 5, 25–38

Viktor Žárský and Fatima Cvrčková (eds.), *Plant Cell Morphogenesis: Methods and Protocols*, Methods in Molecular Biology, vol. 1080, DOI 10.1007/978-1-62703-643-6, © Springer Science+Business Media New York 2014

I

Image analysis.................... 56, 60–61, 65, 67–76, 79, 82, 85, 89, 94, 121–133, 219, 224, 234
Imaging
live 112, 117, 195–198
time lapse.............................95, 111–118, 140, 143, 243
Immunolabeling (immunodetection, immunolocalization, immunostaining)............................ 5, 27, 31–32, 35, 41–52, 112, 151, 153, 161, 172, 183, 192, 210, 211

L

Lab-on-a-chip (LOC)...237–247
Lignin...26, 28–29, 36–37, 52
Lipids 20, 28–29, 36, 252, 264

M

Mechanical properties. *See* also Stiffness 112, 135–145, 245
Membrane.......................... 38, 81, 114, 116, 118, 123, 133, 148, 160, 171–180, 226, 263, 267
Meristem.................................. 99, 100, 103, 106, 108, 109, 111–118, 121–133, 142, 198
Microdissection, laser 249–256
Microelectromechanical systems (MEMS) 238
Microfilaments. *See* also Actin..................... 91–95, 148, 155
Microfluidics 237–246
Micro-indentation...135–145
Microscopy
automated (.slide) 53–65
cellular force (CFM)...135–145
confocal laser scanning (CLSM) 87–95, 100, 101, 107, 111–118, 184
electron
field emission scanning (FESEM)................171–180
scanning (SEM)................................. 100–102, 105, 106, 109, 110, 171, 172, 178, 180
transmission (TEM).......................... 148, 151, 159, 160, 162, 168, 169
fluorescence 42, 74, 88, 186, 223, 224, 259
immunofluorescence (*see* also immunolabeling)............45
total internal reflection (TIRFM)...............................88
variable angle epifluorescence (VAEM).................87–95
Microtome (ultramicrotome).........................3–5, 10, 19, 46, 47, 149, 153, 161, 186, 255
Microtubules 87, 88, 92–95, 112, 148, 155, 174, 198

N

Nuclear envelope 173
Nucleus......................... 26, 83, 107, 171–180, 207, 219, 226

O

Oligodeoxynucleotide
antisense (AODN) 231–235
Onion 77–85, 137, 139, 148–150, 153–155
Optical trap 259–265
Optical tweezers 259–264
Organelle markers 78, 82, 83, 184

P

Particle bombardement (biolistics) 78–79, 210
Pectin..26–28
Physcomitrella patens (moss) 201–212
Pollen tube... 231–235, 237–247
Protonema 202

R

Razor
blades.................3, 9–11, 47, 78, 102, 105, 106, 114, 150
straight...3, 4, 11
RNA interference (RNAi)....................... 201–207, 209–211
Root hair 20, 87, 94, 195–198, 231, 250, 264

S

Saccharomyces cerevisiae. See also Yeast.............. 267, 270, 273
Sampling 2, 9, 67–71, 73–75, 95
Screening
cDNA library...270
mutants..54
Sections
free hand...3, 4, 10, 37, 72
frozen...5–6, 184
paraffin 5, 8, 11–15, 35–37, 250, 251, 254, 255
resin-embedded........................19, 41–52, 159, 183–192
serial 3, 159–169
ultra-thin 44, 46, 47, 153, 155, 183, 186, 188
Segmentation.......................................68, 73, 116, 126–130
Sequential replicas 99–109
Software
Etomo... 160, 162–169
Fiji 90, 123, 203
ImageJ 69, 73, 75, 76, 82, 85, 90–94, 123, 202–204, 206–208, 210–212, 219, 224, 234
IMOD..160, 168
MorphoGraphX 112, 121–133
Stereology...68, 69, 71
Stiffness 100, 136–138, 140, 142–144
Subculture interval (SBI)...........216, 220, 221, 223–225, 227
Suberin 28, 29

T

3D reconstruction........ 99, 100, 105, 107, 112, 123, 161–167
Tip growth 87, 195–198, 201–212

Tobacco. *See* also Nicotiana tabacum........................45, 142, 154, 172–174, 215–228, 231–234, 260, 263
Tomography..148, 155, 159–169
Transformation
biolistic (*see* also particle bombardment)
transient..201, 202, 210
Turgor......................................135, 136, 138, 140, 143, 144

W

Wet fracturing...172, 173, 177
Whole-mount..1–3, 7, 9–10, 51

Y

Yeast..219, 267–281

MIX
Papier aus verantwortungsvollen Quellen
Paper from responsible sources
FSC® C105338

If you have any concerns about our products,
you can contact us on
ProductSafety@springernature.com

In case Publisher is established outside the EU,
the EU authorized representative is:
Springer Nature Customer Service Center GmbH
Europaplatz 3, 69115 Heidelberg, Germany

Printed by Libri Plureos GmbH
in Hamburg, Germany